AF595343

Remote Sensing for Land Administration 2.0

Remote Sensing for Land Administration 2.0

Editors

Mila Koeva
Rohan Bennett
Claudio Persello

MDPI • Basel • Beijing • Wuhan • Barcelona • Belgrade • Manchester • Tokyo • Cluj • Tianjin

Editors
Mila Koeva
University of Twente
The Netherlands

Rohan Bennett
Swinburne University of Technology
Australia

Claudio Persello
University of Twente
The Netherlands

Editorial Office
MDPI
St. Alban-Anlage 66
4052 Basel, Switzerland

This is a reprint of articles from the Special Issue published online in the open access journal *Remote Sensing* (ISSN 2072-4292) (available at: https://www.mdpi.com/journal/remotesensing/special_issues/Land_Administration_2).

For citation purposes, cite each article independently as indicated on the article page online and as indicated below:

LastName, A.A.; LastName, B.B.; LastName, C.C. Article Title. *Journal Name* **Year**, *Volume Number*, Page Range.

ISBN 978-3-0365-5407-5 (Hbk)
ISBN 978-3-0365-5408-2 (PDF)

Cover image courtesy of Mila Koeva

Contents

About the Editors

Mila Koeva

Mila Koeva is an Associate Professor at University Twente, International Institute of Geo-Information Science and Earth Observation ITC, The Netherlands. She holds Ph.D. and M.Sc. degrees from the University of Architecture, Civil engineering, and Geodesy in Sofia. Her career includes ten years of work at the photogrammetry department of GIS Sofia (www.gis-sofia.bg), producing cadastral and topographic maps, and three years of work at the private organization Mapex (www.mapex.bg), leading three EU projects related to geodetic, cadastral, and photogrammetric activities. Her main areas of expertise include Digital Twins/3D modelling, image data acquisition, and processing techniques (satellite, aerial, and UAVs), and automatic feature extraction for cadastral mapping and urban planning, among others. More specifically, her research focuses on the implementation of innovative geospatial and machine learning methods based on remotely sensed data in support of 3D urban modelling and cadastral applications.

Rohan Bennett

Rohan Bennett is an Associate Professor of Information Systems at the Swinburne University of Technology, Melbourne, Australia. He specializes in spatial information systems and land information management. He holds or has held posts with Swinburne School of Business, Law, and Entrepreneurship (Australia); Kadaster International (The Netherlands); the University of Twente (The Netherlands); and the University of Melbourne (Australia). He holds degrees in Science and Engineering, and a Ph.D. from the University of Melbourne. He is currently involved with projects on the digital transformation of land systems, including the application of blockchain, UAVs, and auto feature extraction in land rights management.

Claudio Persello

Claudio Persello is an Associate Professor at the Faculty of Geo-Information Science and Earth Observation (ITC), University of Twente, The Netherlands. He received his Laurea (B.Sc.) and Laurea Specialistica (M.Sc.) degrees in telecommunications engineering and his Ph.D. degree in communication and information technologies from the University of Trento, Trento, Italy, in 2003, 2005, and 2010, respectively. Before joining ITC in 2014, he was a Marie Curie research fellow, conducting research at the Max Planck Institute for Intelligent Systems and the Remote Sensing Laboratory of the University of Trento. His main research interests include machine learning and deep learning for information extraction from remotely sensed images and geospatial data. His research activity includes the investigation and development of dedicated deep learning techniques for various remote sensing sensor data and multiple applications. He is particularly interested in combining deep learning and Earth observations to address and monitor progress towards the sustainable development goals.

MDPI

Editorial

Remote Sensing for Land Administration 2.0

Mila Koeva [1,*], Rohan Bennett [2,3] and Claudio Persello [1]

[1] ITC Faculty, University of Twente, 7514 AE Enschede, The Netherlands
[2] Swinburne School of Business, Law and Entrepreneurship, Swinburne University of Technology, Hawthorn, VIC 3122, Australia
[3] Kadaster International, The Netherlands Cadastre, Land Registry and Mapping Agency, 7300 GH Apeldoorn, The Netherlands
* Correspondence: m.n.koeva@utwente.nl

Abstract: Contemporary land administration (LA) systems incorporate the concepts of cadastre and land registration. Conceptually, LA is part of a global land management paradigm incorporating LA functions such as land value, land tenure, land development, and land use. The implementation of land-related policies integrated with well-maintained spatial information reflects the aim set by the United Nations to deliver tenure security for all (Sustainable Development Goal target 1.4, amongst many others). Innovative methods for data acquisition, processing, and maintaining spatial information are needed in response to the global challenges of urbanization and complex urban infrastructure. Current technological developments in remote sensing and geo-spatial information science provide enormous opportunities in this respect. Over the past decade, the increasing usage of unmanned aerial vehicles (UAVs), satellite and airborne-based acquisitions, as well as active remote sensing sensors such as LiDAR, resulted in high spatial, spectral, radiometric, and temporal resolution data. Moreover, significant progress has also been achieved in automatic image orientation, surface reconstruction, scene analysis, change detection, classification, and automatic feature extraction with the help of artificial intelligence, spatial statistics, and machine learning. These technology developments, applied to LA, are now being actively demonstrated, piloted, and scaled. This Special Issue hosts papers focusing on the usage and integration of emerging remote sensing techniques and their potential contribution to the LA domain.

Keywords: UAV; LiDAR; automated feature extraction; cadaster; land registration; land use planning

Citation: Koeva, M.; Bennett, R.; Persello, C. Remote Sensing for Land Administration 2.0. *Remote Sens.* **2022**, *14*, 4359. https://doi.org/10.3390/rs14174359

Received: 24 July 2022
Accepted: 23 August 2022
Published: 2 September 2022

1. Introduction

The United Nations sustainable development goal (SDG), particularly indicator 1.4.2, promotes tenure security for all [1]. Despite this, the majority of the world lacks formal land and property registration. Land administration (LA) incorporates the concepts of cadastre and land registration. This includes processes such as recording, securing, storing, and disseminating data related to land tenure, value, use, and development [2]. Delivering tenure security is essential for LA systems that aim to reduce poverty and ensure food security by promoting land investments [3]. Conventionally, cadastral data has been collected by high precision ground-based methods, which are often time-consuming, demand high levels of expertise, and therefore unaffordable for many countries. With the technological development, and the aim to speed up the land recordation process, many countries have adopted the so-called fit-for-purpose LA (FFP-LA) approach [4]. It promotes multisensory data collection and an interdisciplinary approach providing outputs that serve the purpose of the concrete country case [5].

The FFP-LA approach is being actively supported by international organizations such as the International Federation of Surveyors (FIG) and Global Land Tool Network (GLTN). The crucial concept of FFP-LA also includes the use of remote sensing data acquisition techniques such as satellite, aerial, LiDAR, RADAR, or UAV data. In addition, related cutting-edge innovative analytical technologies such as machine learning [6] for data processing and automatic feature extraction based on imagery data are used to accelerate the cadastral mapping process [7]. Geo-cloud services are also playing big part in data storage, management, dissemination, and e-service delivery [8].

The abovementioned approaches are the focus of the current Special Issue as a continuation of the "Remote Sensing for Land Administration" first edition [9]. Starting with a broad review paper, focusing on the developments of photogrammetric methods and remote sensing applied to the LA domain, the Special Issue provides an overview of diverse experiments, demonstrations and implementations in a range of case contexts. This includes papers related to (i) design and testing of the flight configurations of UAVs; (ii) quantitative and qualitative methods for assessment and comparison of different remotely sensed techniques for valuation and taxation; (iii) innovative machine learning methods development and integration for cadastral data extraction, including also a social-economic parameters; and (iv) usage of LiDAR data and analysis techniques for building extraction and agricultural land delineation.

The Special Issue consists of nine (9) individual works; developed by multi-disciplined research by researchers from Europe, Asia, Oceania, and Africa. The works use a variety of qualitative and quantitative research methods applied to diverse social and remotely sensed data types, in support of lend tenure recordation, land valuation, and land use planning (Table 1). The next section outlines each work and synthesizes the overarching contribution of the Special Issue.

Table 1. Remote sensing applications for LA presented in this issue.

Source	Title	Geographic Focus	Applications	Techniques	Data
Rohan Mark Bennett et al.	Review of Remote Sensing for Land Administration: Origins, Debates, and Selected Cases	Global and historic	Land tenure, photogrammetry and remote sensing	Review paper	LiDAR, Aerial, UAV, topographic and cadastral maps among many other
Claudia Stöcker et al.	High-Quality UAV-Based Orthophotos for Cadastral Mapping: Guidance for Optimal Flight Configurations	Europe and Africa	Land Tenure from UAV images	Comparative analysis	UAV imagery
Koeva et al.	Remote Sensing for Property Valuation: A Data Source Comparison in Support of Fair Land Taxation in Rwanda	Rwanda	Land valuation and taxation	GNSS survey, semi structured interviews, focus group discussion	Satellite digital aerial and UAV imagery, cadastral data
Cheonjae Lee et al. Damian Wierzbicki et al.	Testing and Validating the Suitability of Geospatially Informed Proxies on Land Tenure in North Korea for Korean (Re-)Unification	North Korea	Land Tenure; cadastral mapping	Geospatially informed analysis, questionnaire	Google Earth images
Bujar Fetai et al.	Deep Learning for Detection of Visible Land Boundaries from UAV Imagery	Slovenia	Multi-purpose cadastre; Map creation and updating	U-Net ENVINet5	UAV imagery, cadastral boundaries

Table 1. *Cont.*

Source	Title	Geographic Focus	Applications	Techniques	Data
Shih-Hong Chio et al	Application of a Hand-Held LiDAR Scanner for the Urban Cadastral Detail Survey in Digitized Cadastral Area of Taiwan	Taiwan	Land tenure; cadastral mapping	Pointcloud filtering, RANSAC	Handheld LiDAR, cadastral data
Natalia Borowiec et al.	Using LiDAR System as a Data Source for Agricultural Land Boundaries	Poland	Land Tenure; cadastral mapping of agricultural lands	Edge detectors, Hough-Transform	LiDAR, cadastral data
Damian Wierzbicki, et al.	Polish Cadastre Modernization with Remotely Extracted Buildings from High-Resolution Aerial Orthoimagery and Airborne LiDAR	Poland	Cadastral map creation, verification and updating	Fully convolutional network U-Shape Network (U-Net)	LiDAR, cadastral data high resolution aerial orthoimagery
Dušan Jovanović et al	Building Change Detection Method to Support Register of Identified Changes on Buildings	Serbia	Cadastral updating, building change detection	Pixel-based and object-based analysis	VHR imagery with RGB and NIR bands

2. Overview of the Contributions

- Paper 1

Rohan Mark Bennett et al. present a systematic review paper that enriched a complete synthesis of the developments of photogrammetric methods and remote sensing applied to the domain of LA. It incorporates developments from early phototopography and aerial surveys through to analytical photogrammetric methods, the emergence of satellite remote sensing, digital cameras, and latterly lidar surveys, UAVs, and feature extraction. The synthesis illustrates the debates over the benefits of the techniques. Various comparative analyses on criteria relating to time, cost, coverage, and quality are presented. Apart from providing this more holistic view and a timely reminder of previous works, this paper brings contemporary practical value in further demonstrating to LA practitioners that remote sensing for data capture, and subsequent map production, are an entirely legitimate, if not essential, part of the domain.

- Paper 2

Claudia Stöcker et al. provide a detailed investigation of different flight configurations to guide efficient and reliable UAV data acquisition in support of cadastral map creation and updating. Imagery from six study areas across Europe and Africa provide the basis for an integrated quality assessment, including three main aspects: (1) the impact of land cover on the number of tie-points as an indication of how well bundle block adjustment can be performed, (2) the impact of the number of ground control points (GCPs) on the final geometric accuracy, and (3) the impact of different flight plans on the extractability of cadastral features. The results suggest that scene context, flight configuration, and GCP setup significantly impact the final data quality and subsequent automatic delineation of visual cadastral boundaries. This study reveals large discrepancies in the accuracy and the completeness of automatically detected cadastral features for orthophotos generated from different flight plans. With its unique combination of methods and integration of various study sites, the results and recommendations presented in this paper can help land professionals and bottom-up initiatives alike to optimize existing and future UAV data collection workflows.

- Paper 3

Koeva et al. in their study, assess different remote sensing data in support of developing a new approach for property valuation for taxation in Rwanda. Three different remote sensing technologies, (i) aerial images acquired with a digital camera, (ii) WorldView2 satellite images, and (iii) unmanned aerial vehicle (UAV) images obtained with a DJI Phantom 2 Vision Plus quadcopter, are compared and analyzed in terms of their fitness to fulfill the requirements for valuation for taxation purposes. Quantitative and qualitative methods are applied for the comparative analysis. Primary data is collected via semi-structured interviews and focus group discussions. The results show that UAVs have the highest potential for collecting data to support property valuation for taxation. The main reasons are the prime need for accurate-enough and up-to-date information.

- Paper 4

Cheonjae Lee et al. investigate in their research the role of remote sensing data in detecting, estimating, and monitoring socio-economic status (SES) such as quality of life dimensions and sustainable development prospects. In the context of Korea, the main challenge is the lack of complete and adequate information when it comes to clarifying unknown land tenure relations and land governance arrangements. Deriving informative land tenure relations from geospatial data in line with socio-economic land attributes is currently the most innovative approach. Therefore, the authors provide empirical evidence of whether the proposed proxies are scientifically valid, policy-relevant, and socially robust. They revealed differences in the distributions of agreements relating to land ownership and land transfer rights identification among scientists, bureaucrats, and stakeholders. Moreover, the authors measured intrinsic, contextual, representational, and accessibility attributes of information quality regarding the associations between earth observation (EO) data and land tenure relations in North Korea from several different viewpoints.

- Paper 5

Bujar Fetai et al. with their research, aim to accelerate cadastral mapping through innovative and automated approaches for the creation and updating of cadastral maps. Using deep learning, they explored algorithms to automatically detect visible land boundaries from unmanned aerial vehicle (UAV) imagery. In addition, the authors evaluated the advantages and disadvantages of programming-based deep learning compared to commercial software-based deep learning. They used the convolutional neural network U-Net, implemented in Keras, written in Python using the TensorFlow library. For commercial software-based deep learning, they used ENVINet5. The results showed that both models achieved an overall accuracy of over 95%. The high accuracy is due to the problem of unbalanced classes, which is usually present in boundary detection tasks. U-Net provided a recall of 0.35 and a precision of 0.68 when the threshold was set to 0.5. A threshold can be viewed as a tool for filtering predicted boundary maps and balancing recall and precision. The authors concluded that programming-based deep learning provides a more flexible yet complex approach to boundary mapping than software-based, which is rigid and does not require programming.

- Paper 6

Shih-Hong Chio and Kai-Wen Hou present a study that investigates the feasibility of a handheld LiDAR scanner to collect 3D point clouds in an efficient way for a detail survey in urban environments with narrow and winding streets. After point cloud filtering and the ranging systematic error correction that was determined by a plane-based calibration method, the collected point clouds are transformed to the local cadastral coordinate system using control points. Using the detail points surveyed by a total station to verify the detail line data digitized from the corrected handheld LiDAR point cloud, 97% error of the digitized detail data was less than 15 cm The results demonstrated the feasibility of using a handheld LiDAR scanner to perform an urban cadastral detail survey in digitized graphic areas.

- Paper 7

Natalia Borowiec and Urszula Marmol explore LiDAR sensor data to identify agricultural land boundaries. Their study focuses on accurately determining the edges of parcels using only the point cloud, which is an original approach because the point cloud is a scattered set, which may complicate finding those points that define the course of a straight line defining the parcel boundary. To detect automatically the edges of parcels, the author's first step is to do classification then use edge detectors to define land use boundaries. The obtained boundaries are compared with the boundaries from the Polish land registry database. The proposed algorithm allowed the detection of inconsistencies in farmers' declarations. These mainly concerned areas of field roads that farmers misclassified as subsidized land when in fact, they should be excluded from subsidies.

- Paper 8

Damian Wierzbicki et al. dived into automatic building extraction from remote sensing data for cadastre verification, modernization and updating. They explored deep learning algorithms, particularly fully convolutional network U-Shape Network (U-Net), for high resolution aerial orthoimagery segmentation and dense LiDAR data to extract building outlines automatically. They reached 89.5% overall accuracy and an 80.7% completeness compared to the reference data, which made it very promising for cadastre modernization in Poland. In addition to the numerical achievements, the authors discuss the possibilities and limitations of the automated approaches that could help local authorities decide on the use of remote sensing data in LA.

- Paper 9

Dušan Jovanović et al. based on satellite imagery and existing cadastral data proposed a method based on a comparison of object-based and pixel-based image analysis approaches to automatically detect newly built, changed, or demolished buildings and import these data into extended cadastral records. Using only VHR images containing only RGB and NIR bands, the results showed object identification accuracy ranging from 84% to 88%, with kappa statistics from 89% to 96%. The accuracy of obtained results is satisfactory for the purpose of developing a register of changes on buildings to keep cadastral records up to date and to support activities related to the legalization of illegal buildings, etc.

Author Contributions: Conceptualization and writing—original draft preparation, M.K. and R.B.; writing—review and editing, M.K., R.B. and C.P. All authors have read and agreed to the published version of the manuscript.

Funding: This research received no external funding.

Conflicts of Interest: The authors declare no conflict of interest.

References

1. Nations, U. Home | Sustainable Development. Available online: https://sdgs.un.org/ (accessed on 25 May 2021).
2. UNECE. *Land Administration Guidelines;* With Special Reference to Countries in Transition; United Nations Publication: New York, NY, USA; Geneva, Switzerland, 1996.
3. Conforti, P. *Looking Ahead in World Food and Agriculture: Perspectives to 2050;* Food and Agriculture Organization of the United Nations (FAO): Rome, Italy, 2011.
4. Enemark, S.; Bell, K.; Lemmen, C.; McLaren, R. *Fit-for-Purpose Land Administration;* International Federation of Surveyors: Copenhagen, Denmark, 2014.
5. UN-GGIM. *Framework for Effective Land Administration;* A Reference for Developing, Reforming, Renewing, Strengthening or Modernizing Land Administration; United Nations Committee of Experts on Global Geospatial Information Management: New York, NY, USA, 2019.
6. Persello, C.; Wegner, J.D.; Hänsch, R.; Tuia, D.; Ghamisi, P.; Koeva, M.; Camps-Valls, G. Deep Learning and Earth Observation to Support the Sustainable Development Goals. *IEEE Trans. Geosci. Remote Sens.* **2021**, *10*, 172–200. [CrossRef]

7. Xia, X.; Persello, C.; Koeva, M. Deep Fully Convolutional Networks for Cadastral Boundary Detection from UAV Images. *Remote Sens.* **2019**, *11*, 1725. [CrossRef]
8. Koeva, M.; Humayun, M.I.; Timm, C.; Stöcker, C.; Crommelinck, S.; Chipofya, M.; Bennett, R.; Zevenbergen, J. Geospatial Tool and Geocloud Platform Innovations: A Fit-for-Purpose Land Administration Assessment. *Land* **2021**, *10*, 557. [CrossRef]
9. Bennett, R.; Oosterom, P.; van Lemmen, C.; Koeva, M. Remote sensing for land administration. *Remote Sens.* **2020**, *12*, 2497. [CrossRef]

MDPI

Review

Review of Remote Sensing for Land Administration: Origins, Debates, and Selected Cases

Rohan Mark Bennett [1,2,*], Mila Koeva [3] and Kwabena Asiama [4]

1 Swinburne School of Business, Law, and Entrepreneurship, Swinburne University of Technology, Hawthorn, VIC 3122, Australia
2 Kadaster, The Netherlands Cadastre, Land Registry and Mapping Agency, 7311 KZ Apeldoorn, The Netherlands
3 Faculty of Geo-Information Science and Earth Observation (ITC), University of Twente, 7514 AE Enschede, The Netherlands; m.n.koeva@utwente.nl
4 Geodetic Institute (GIH), Leibniz University Hannover, 30167 Hannover, Germany; asiama@gih.uni-hannover.de
* Correspondence: rohanbennett@swin.edu.au

Abstract: Conventionally, land administration—incorporating cadastres and land registration—uses ground-based survey methods. This approach can be traced over millennia. The application of photogrammetry and remote sensing is understood to be far more contemporary, only commencing deeper into the 20th century. This paper seeks to counter this view, contending that these methods are far from recent additions to land administration: successful application dates back much earlier, often complementing ground-based methods. Using now more accessible historical works, made available through archive digitisation, this paper presents an enriched and more complete synthesis of the developments of photogrammetric methods and remote sensing applied to the domain of land administration. Developments from early phototopography and aerial surveys, through to analytical photogrammetric methods, the emergence of satellite remote sensing, digital cameras, and latterly lidar surveys, UAVs, and feature extraction are covered. The synthesis illustrates how debates over the benefits of the technique are hardly new. Neither are well-meaning, although oft-flawed, comparative analyses on criteria relating to time, cost, coverage, and quality. Apart from providing this more holistic view and a timely reminder of previous work, this paper brings contemporary practical value in further demonstrating to land administration practitioners that remote sensing for data capture, and subsequent map production, are an entirely legitimate, if not essential, part of the domain. Contemporary arguments that the tools and approaches do not bring adequate accuracy for land administration purposes are easily countered by the weight of evidence. Indeed, these arguments may be considered to undermine the pragmatism inherent to the surveying discipline, traditionally an essential characteristic of the profession. That said, it is left to land administration practitioners to determine the relevance of these methods for any specific country context.

Keywords: photogrammetry; aerial imagery; UAV; HRSI; lidar; artificial intelligence

Citation: Bennett, R.M.; Koeva, M.; Asiama, K. Review of Remote Sensing for Land Administration: Origins, Debates, and Selected Cases. *Remote Sens.* **2021**, *13*, 4198. https://doi.org/10.3390/rs13214198

Academic Editor: Parth Sarathi Roy

Received: 22 September 2021
Accepted: 14 October 2021
Published: 20 October 2021

1. Introduction

In the context of this work, 'land administration' incorporates the concepts of cadastre and land registration and is understood as the process of recording, securing, and disseminating information about land tenure, value, use, and development, within a jurisdiction [1]. Its core purposes are to support land rights securitisation, land market governance, credit access, fair land taxation, and responsible spatial planning, amongst other societal concerns [2]. 'Photogrammetry' incorporates methods and tools for extracting multi-dimensional geospatial information from images needed for mapping activities [3] (for further origins and etymology, see Polidori L. 'Words as tracers in the history of science and technology: the case of photogrammetry and remote sensing'. *Geo-spatial Information*

Science. 2021, 24, 167–177). 'Remote sensing' is the process of scanning or monitoring the physical characteristics of a terrestrial surface, measuring the emitted radiation at a distance [4]. Both photogrammetry and remote sensing have grown out of photographic mapping and aerial survey traditions.

The driver for the work is to further consolidate arguments for the use of photogrammetric and remote sensing methods in the domain of land administration [5], particularly when used in a complementary fashion with ground-based surveying methods. Whilst photogrammetric and remote sensing methods are used within the field in some contexts, arguably, they are heavily underutilised, especially given the amount of imagery data and collection ability now available at a relatively low cost, with vast spatial coverage, and good temporal qualities [6]. Compared to other related fields, such as construction and agriculture, rates of the application of image-based mapping, at scale, remain low in the land administration domain. The argument can be made for both developed and developing contexts, where field-based data collection techniques prevail in many circumstances [1]. Overall, it is argued that the strong bias towards the use of ground methods alone is driven by the existing land administration practitioner community for reasons of financial expedience and industry inertia [6].

The justification for the work is that such a review has never been undertaken, at least in the contemporary era. The opportunity to undertake this work is now available due to the increased availability of archival journals and records, thanks to digitisation, scanning, and online availability. This enables a more complete understanding of the historical developments within the domain to be presented to a new audience, thereby informing future developments, and creating a better appreciation of the close relationship between the fields of land administration photogrammetry and remote sensing, which have often operated disparately. Accordingly, the structure of this paper is as follows. First, an outline of the approach and methods used for the review is provided. Second, the presentation of the review results, using a chronological approach, commencing from the 1700s and swiftly moving into the 1900s, using a combination of theme and decade, is provided. Third, a summary of the synthesis of developments is delivered in a concise fashion. Finally, conclusions relevant for contemporary discourse on the use of photogrammetry and remote sensing in the land administration field are articulated.

2. Materials and Methods

To enable the achievement of the objective to provide a comprehensive review of photogrammetry and remote sensing applied in land administration, a research synthesis methodology was applied [7]. Couched somewhere within—or between—the positivist, constructivist, and pragmatic research paradigms [8], this approach seeks critical analysis of a scoped body of literature, synthesizing the results, to deliver a previously unrecognised model or description. The approach is used widely in the domain of land administration [9], amongst others, particularly since the 2000s, due to the greater availability of historical sources, and an increasing amount of empirical literature more generally [10].

For this review, an unlimited starting date, and up to August 2021 for the conclusion date, were selected. This rather expansive epoch enabled the most comprehensive coverage of documents, and was still considered to be achievable in terms of available time and resources. For practical purposes, the initial search and selection of documents was conducted by decade, commencing with pre-1900s, and subsequently 1900–1909, 1910–1919, and so on, up until 2021.

Using [10] as a model, the repositories examined included those exploited in other research syntheses from the land administration domain, including Google Scholar, Scopus, Science Direct, and the OICRF website (International Office of Cadastre and Land Records (See: https://www.oicrf.org/search, accessed between July and August 2021 website, a searchable index and repository maintained by the Dutch Cadastre, Land Registry and Mapping Agency (Kadaster). As in [10], non-Scholar Google searches were completed alongside the academic database searches, so that relevant grey literature, from industry and governments, could also be captured. In general, the grey literature was given less weighting. An important limitation of this approach is that non-English language documents received less attention, primarily documents written in French and German, which were certainly prominent languages in terms of developments in the late 1800s, the 1900s, and between the world wars. It is left for other scholars to fill these gaps, yet it is expected that a similar trajectory in technological developments, albeit based on different country experiences, will be observed.

Specific search terms and search string combinations included 'land administration', 'land registry', 'land registration', 'cadastre', 'cadastral boundaries', 'cadastral surveying', 'land surveying', 'land parcel', 'property', 'monuments', 'photography', 'balloon survey', 'remote sensing', 'photogrammetry', 'photogrammetric methods', 'aerial photography', 'aerial survey', 'high-resolution satellite imagery' (and variations, e.g., VHRSI), and 'indirect methods'—and later 'UAVs', 'RPAS', 'lidar', 'SAR-radar', 'oblique imagery', 'feature extraction' and 'pictometry'. During this process, it was determined that different terms increased and decreased in popularity over time. This fact was considered when conducting the searches. The approach produced thousands of returned results; however, snowballing [11] and expert knowledge was used to determine the final constellation of approximately 300 relevant articles. The authors took the opportunity to present this bibliography in the references section. Whilst making the paper more cumbersome, it was felt that this complete provision of sources increases the utility of the paper for readers and invites the reader to undertake their own explorations.

The review, critique, and synthesis were initially undertaken and reported in chronological order, and the results are presented in Section 3 to Section 8. The further synthesis of salient ideas and development of an overarching synthesis model was then undertaken and is presented in Section 9.

3. First Forays (1700s to 1909)

Within the scope of a single journal paper, attempting a complete analysis of all converging developments in land administration and photographic methods prior to the 1900s is at best overly ambitious and perhaps naïve. That said, to not attempt to include coverage would constitute a disservice to the pioneering work. Here, a humble attempt is made to provide a potted overview of key developments and examples.

3.1. 1700s

Whilst the contemporary view that photographic approaches, applied to land administration, only developed significant impetus in the 20th century, certainly the potential for the developing science of photography, applied to land surveying and mapping, was well recognised in the 19th century. The European Age of the Enlightenment of the 17th and 18th centuries spurred the development and application of many of the applied sciences, including those tools and techniques relating to geometry and land surveying, particularly as the era of colonisation advanced, and there was an increasing need to map new territories. Here, the works and treatises of Martindale [12], Love [13], Breaks [14], and Laybourne [15], amongst others, are remarkable, ushering in the emerging era of more wide-scale accurate plot measurement via the use of theodolites, chain, and other plain surveying methods.

3.2. 1800s

Similar works, out of the United Kingdom, North America, and other colonies, followed into the early part and middle part of the 19th century: Ainslie [16] provides an example out of the Scottish Enlightenment and the subsequent industrialisation period.

Likewise, from the same motivations came the first texts on developments and practical guides on photography and the Daguerreotype [17,18]. However, it was not until the middle and later part of the 19th century that applications of photography in the domains of land surveying and mapping were first documented. Tissandier [19], writing in 1877, hypothesises (alongside explanations of the science, tools, and applications of photographs) that future applications of photography will prominently include 'land surveying'. He goes on to explain how it was already considered *"possible to combine surveying with photography"* by placing *"a camera on a land surveyor's stand, fixing it upon an axis so it can be turned around in any direction ... "* enabling the creation of complete panorama of the landscape. Surely, herein lies a very early envisioning for 'Cyclorama' and Google's 'Street View', that would appear well over a century later. Meanwhile, Reed [20], writing in 1889, provides a historical account of developments across that century, from the use of perspective drawings for topographic mapping, to the development of telemetrography in the 1850s, through to the development of photographic surveying in the 1860s in France. An outline of various methods, including via plane, via cylindric, via radial, and balloon photography is provided (although the latter was more readily used for survey reconnaissance purposes). Thomas [21] details similar examples in geographic and engineering surveys out of America. Devillle [22] provides similar techniques as applied in Canada (for a comprehensive visual overview, see: https://www.isprs.org/society/history/100Jahre.pdf, accessed on 18 October 2021). The work of the Uniting Kingdom's Ordnance Survey is also noteworthy in this period. Although not using photogrammetric methods in the field, as early as the 1840s, it was using photographic methods, specifically photozincography, to produce and reproduce topographic maps of various scales [23–25] (note, in the 1970s, Mumford would provide a full synthesis of these developments. See *Mumford I. Lithography, photography and photozincography in English map production before 1870. The Cartographic Journal. 1972 Jun 1;9(1):30-6)*. The same techniques were later applied to map production in India [26], apparently with great success. The approach was used in Sweden, and presumably elsewhere across Europe [27].

By the end of the century, the techniques of aerial photographic methods were recognized as legitimate [28] for surveying, and were being taken advantage of, particularly in more rugged and inaccessible landscapes. Flemer [28], writing in 'Science', explains the 'phototopography' method being used in Alaska. Whilst it would only be in the following century that these innovations truly impacted land administration functions, thanks to the work of these early photographic pioneers, the vision, tools, and methods were now in place.

3.3. 1900s

Moving into the first decade of the 1900s, there appear only limited relevant works (recognizing, however, that works in French are not included in this review, for which there appears to be numerous works on 'cadastre' during this period), although several are truly worth noting. Writing in 1908 [29], the remarkable Vivian Thompson, apparently later killed during World War 1 [30], picks up from the work of Deville [22] to provide a full detailed account of the tools and techniques involved in Stereo-Photo Surveying (Figure 1). Perhaps of most novelty is the discussion on the relative merits of the method versus plane-table surveying, in terms of accuracy, cost, and time—a discourse quickly settled upon by others [31], and that was to be oft returned to throughout the century, as more photogrammetric advances emerged. Thompson explains how: *"The objective of photographic surveying is to map the detail of a triangulated area at minimum expenditure of time and labour in the field, and at a total cost so far below that involved in plane-tabling as to warrant the sacrifice of that that high degree of accuracy attenable in good plane-tabling"*. He also clarifies: *" ... it might appear that photographic surveying is necessarily less accurate than plane-tabling. This is not the case; but, to attain the same degree of accuracy in detail" ... "the plotting would be so tedious" and "less economical"*. He suggests that photographic surveying has not proved more popular due to it being wrongly applied: the economic benefits only increase as the scale of the map decreases, and the ruggedness of the landscape increases. He summarises that small-scale contour maps (2 inches = 1 mile) are the most economical, taking one-tenth or one-fifth of the time as compared to ground-based techniques.

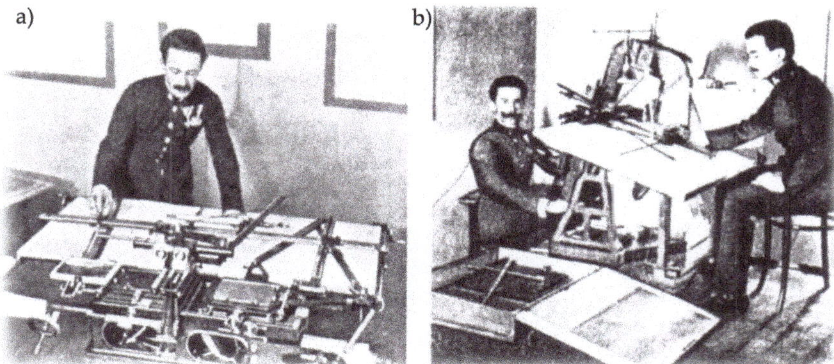

Figure 1. The first mechanical stereographs, developed in parallel by (**a**) Van Orel of Austria and (**b**) Vivian Thomson of England (Adapted from https://www.asprs.org/wp-content/uploads/pers/1985journal/jul/1985_jul_919-933.pdf, accessed on 18 October 2021), had a significant impact on cadastral mapping in later decades.

That said, notwithstanding the prioritisation of topographic mapping over cadastral mapping in the British Isles, the use of photographic methods in cadastral work is not countenanced at this point: frontier and railway reconnaissance work were the prime applications. Indeed, Johnston [32,33] confirms the emerging distinction between geographic, topographic, and cadastral mapping, in terms of the practitioners and techniques employed. He also suggests that within a generation, most parts of the world will have been accurately mapped topographically—as in France, Germany, and the United Kingdom—and that mapping of the landscape is likely to become a regular or repeating activity, rather than a singular occurrence. Importantly, he recognises that the great expenses in resources and time used to map those jurisdictions may be avoided using emerging techniques, and presumably, photographic methods are front of mind here.

4. New Era Begins (1910 to 1929)

4.1. 1910s

Major developments in aerial surveying and photogrammetric mapping techniques occurred in the 1910s, largely driven by the Great War, or World War I, 1914–1918 [34] (Figures 2 and 3). The strategic importance of these developments with regard to the conflict—including aerial photography, sound-ranging, and flash-spotting—meant that they remained largely unpublished until after the war [35]. Thereafter, the different developments from the German, British and French perspectives were eventually shared [36,37]. The close combat nature of trench warfare demanded large-scale and highly accurate topographic maps, but also made conventional mapping techniques impossible. Remote techniques, such as resection and aerial survey, were therefore developed out of necessity.

The use of these new and enhanced techniques was then considered for non-war applications. It was duly recognised that large-scale topographic maps could be readily produced, even for cadastral applications [38,39], although the role of the surveyor for more detailed work remained recognised: "*One may safely sum up the situation by saying that the aeroplane is already a valuable instrument for both exploration and accurate survey in flat country, and that it should not be long before its application will be universal, and one may venture to predict that in survey, as in other matters, the Great War will mark the beginning of a new era.*" The analysis here included cost breakdowns and comparisons between ground methods and aerial survey [39], summing at between 5 and 15 pounds sterling, per square mile, at 1/2500 scale, for aerial survey, and anywhere from 10 to 1100 pounds sterling per square mile at 1/2500 for a ground survey. In terms of costs, the new methods are argued to abolish significant costs around traverses, bookings, calculations, and plotting, and the associated fieldwork expenses. Instead, these dense survey networks could be replaced with trig stations every 1–2 miles. With regard to speed, examples from Italian work in Damascus are pointed to: *"three Italian engineers took two years to produce a 1/4000 map of Damascus with its winding streets; an aeroplane produced a picture in a few hours, a rough-scale map or mosaic in a day, and an accurately finished map on 1/2500 could be completed with triangulation within a month of starting"*.

Figure 2. By the 1920s, new-era cadastral mapping was exploiting World War I aerial photography advances to produce higher-accuracy parcel maps (adapted from [36]). (**a**) Triangulation with a Lucas signalizing lamp; (**b**) printing office of a field survey battalion; (**c**) map distribution by car; (**d**) first establishment of flash-spotting post in a trench after an advance.

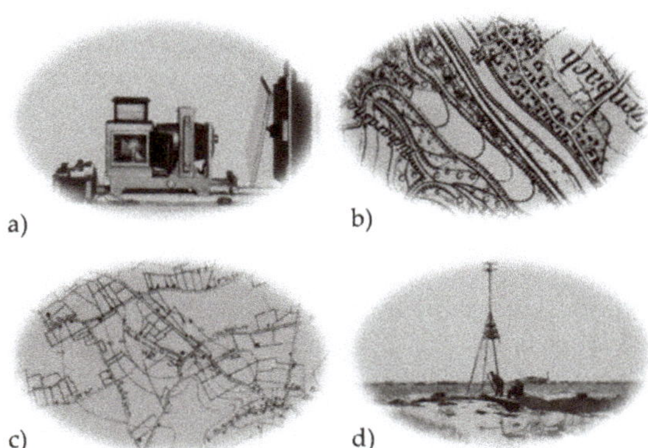

Figure 3. New-era techniques for mapping from aerial photographs. (**a**) Enlarging lantern and tilting copying board for the rectification of aerial photographs; (**b**) extraction of 1:25,000 maps from photos; (**c**) map for revision by plane-table and aerial photograph; (**d**) German triangulation signal (adapted [34,36]).

The role of non-mapping experts undertaking work is also declared: *"Anyone with local knowledge, not necessarily a surveyor, can take this 1/2500 mosaic and go over the ground collecting names, defined plots of ground, wells, etc."*

So positive was the idea that the approach was considered and planned for the entirety of Palestine during the British mandate; however, it was never implemented (see: *Gavish D. An account of an unrealized aerial cadastral survey in Palestine under the British mandate. Geographical Journal. 1987 Mar 1:93-8*). There are, of course, caveats made, including the need for ready access to airplanes, good weather (perhaps humorlessly, the British Isles are mentioned as not being ideal), and the need for touch-up work with the support of a draftsperson. Perhaps predictably, and as was to be the norm in decades to come, these articles elicited conjecture around the accuracy and reliability of the figures relating to cost and time [40]. That said, the notion of using aerial photogrammetric methods in creating large-scale wide-area topographic, and even cadastral, maps was now firmly on the agenda [41].

4.2. 1920s

In the 1920s, recognition continued to increase that beyond military needs, governments should be leading the compilation, and maintenance of, jurisdiction-wide domestic mapping programs, incorporating relevant themes and scales, to support civil governance [42,43]. In the same context, the improving techniques for aerial survey and mapping—and more specific for photogrammetry—were articulated in handbooks and guides [44].

For the case of cadastral maps, the Geographical Journal continued to be a platform of choice for debating the merits of aerial photography applied to mapping. The previously articulated aerial survey approaches, including those estimates of time and cost commitments, would be more thoroughly tested in numerous contexts. Dowson [45], who had acted as Surveyor-General of Egypt, in the previous decade, 1909-19, presented results from that country. Whilst acknowledging the improvements to techniques in that decade, he makes clear that aerial surveys alone cannot replace conventional ground techniques in that context. He cautions against hype: *"Great as is the promise of aeroplane photography as an aid to map-making, there is an obvious danger that too much may be expected of this valuable method of filling in map detail"* and *"So far aeroplane photography has rendered the very great service of enabling a closely accurate record to be instantaneously taken of a considerable block of topographic detail; the accurate assembly of this detail into controlled position in a map is the province of the surveyor, and, so far at least, this requisite control has not been established through the agency of aeroplane photograph"*. Specifically, on cadastral surveys, he goes on to state: *"In sparsely settled, semi-arid areas, where property is held in large units and no great degree of accuracy is required in defining a boundary, it is not usual for the limits of properties to be outlined with sufficient continuity, visibility, and lack of ambiguity for photographic record"* and that for more densely occupied urban and rural areas: *"the accuracy that is obtainable from aerial photographs is so far of a totally different order to that required for a cadastral survey of a fertile and closely-settled land, and it is in fertile and closely-settled lands that cadastral surveys are principally needed"*.

Others, such as Bagley [46], reviewing applications to that point, by the French in Morocco, the English in India, and US Government bureaus, arrive at similar conclusions: the developing technique certainly has merit for difficult to access terrain, but is not appropriate for large-scale detailed topographic or cadastral maps. Bergen [47] similarly argues that aerial survey mapping has limited application for cadastral mapping: its best application is still for large-scale contour maps where there is an abundance of ground control—as found in Europe. Its economical application in the Americas, where ground control is often absent, is questioned. Tuttle [48] appears to counter this view, at least in established urban areas, providing details of the application of aerial mapping to support city planning in New York.

Another development in the 1920s was the necessity for cooperation between surveying and mapping disciplines and the emerging and maturing domain of aeronautics. Burchall [49] explores the administrative necessity and relevant costs of linking the disciplines, and later Durward [50] provides a more matured overview of the integrating disciplines. Winterbotham [51], relaying the status-quo in Canada, remarks of the relationship forming between aeronautics and surveying in that jurisdiction, and sees a move beyond hype and despair, towards productive application: *"There is everywhere the keenest interest in method and instrument and a marked absence of that sloppy over-confidence or wilful pessimism we have seen sometimes elsewhere in airman and surveyor respectively"*.

Perhaps confirming this view, and realizing the need for a 'fit-for-purpose' use of aerial surveying in mapping applications, Fiske [52] puts it best: *"In approaching the subject of the use or aerial photographs the engineer must formulate clearly in his own mind a definite opinion as to what constitutes a map and the purpose it will serve."* ... and what would be the good purpose in ... *"securing data and trying to incorporate it in a map with any higher degree of accuracy that can or will be employed by the user?"*. Herein lies an example of the mindset that would later grow into 'fit for purpose land administration' agenda.

5. Switzerland, Scaling and Spreading (1930 to 1945)

5.1. 1930s

The 1930s began to reveal, for the first time, scaled whole-of-country implementations of aerial survey for cadastral survey, especially pioneered by Switzerland. Spender [53] (Figure 4), writing in 1932, outlines the extensive use of terrestrial photogrammetry in topographic mapping in Switzerland, and perhaps even more remarkably, the almost exclusive use of aerial surveying for cadastral mapping, having commenced in the mid-1920s. Switzerland, having been isolated from the demands of rebuilding post the Great War, and spurred on by scientific demands, matured aerial photographic methods for cadastral surveying. Interestingly, Spender makes it clear that the remarkable speed in uptake—aside from the benefits of swift coverage, addition contextual information, and reduced costs—was a result of private survey firms being primarily responsible for cadastral surveying—and therefore being keen to utilise new technologies, such as stereo-plotting machines, and being prepared to take on the risks to maximise economic gains.

Figure 4. A terrestrial phototheodolite station in Switzerland, although post-1925, cadastral surveys were almost exclusively completed using aerial survey (adapted from [53]).

In terms of the benefits for cost and time: *"It is unprofitable to attempt the aerial survey of a district smaller than 20 sq. km.; a suitable size to treat as a unit is 100 sq. km. The average total cost of the preparation of cadastral map and topographical "Uebersichtsplan" by these methods, including flying costs but excluding the 4th order triangulation, is 800 Swiss francs per sq. km. This is at least 15 percent cheaper than by the use of terrestrial photogrammetry and up to 30 percent cheaper than a plane-table survey. The cost may be taken as less than 1 percent of the value of the property. If the aeroplane makes photographs of 500 sq. km. in the course of the summer and this figure is exceeded without any difficultly, the flying costs, including the crew, fuel, insurance, and sinking fund, represent 10 percent of the total; the marking of points on the ground 3.75 percent; the photographic work 1.25 percent; and the remaining survey and plotting duties undertaken by the private firm represent 85 percent of the total cost"*.

In a review by Ripley and others [54], referring to the work of one Colonel Birdseye, the personnel cost between ground and aerial methods is suggested to be equivalent, but the time commitment is cut to one third when using the aerial approach.

Following the lead of Switzerland, other European countries, including Germany, France, Italy (via outsourcing arrangements with the private sector, on the agreement that costs would be equivalent or lower to ground survey methods [55]), and Spain followed—although, despite some use in India [56], Wolff [57] suggests British colonies lagged in the uptake of this method. Salmon [58] countered this view, recalling the earlier British developments from the Great War: *"I hope Dr Wolff's interesting article will stimulate to action some of those who have not given sufficient attention to air survey as a method of mapping or planning those areas which lend themselves to that method. At the same time, whether so many of us are as conservative as the author appears to think is a matter for doubt, and moreover we do not all look upon air survey as an "innovation"*.

That said, it could hardly be argued that those innovations had translated to use in cadastral mapping in British colonies, and rudimentary understandings of photogrammetric principles were still lacking in the surveyor community [59]. Winterbotham [60] lamented the lack of innovation and updating of British maps themselves, seemingly linking the neglect to the ongoing economic depression. The discourse between Wolff and Salmon was part of the emerging professional dialogue on cadastral surveying, occurring in the recently established Empire Survey Review, itself connected to the first Conference of Empire Survey Offices in 1928 [61].

Meanwhile, technological photogrammetric advances and refinements emerging in France, Germany, Italy, and Switzerland were directly contributing to faster-paced cadastral map production [62]. These initiatives relied upon what was later termed 'analogue photogrammetry' (for a more detailed chronology of 'analogue photogrammetry', see: https://www.asprs.org/wp-content/uploads/pers/1985journal/jul/1985_jul_919-933.pdf, accessed on 18 October 2021), underpinned by earlier-developed stereoscopes, aircraft for aerial surveying, and the methodological refinements of Dr. Carl Pulfrich (later described as the so-called "Father of Stereophotogrammetry"). The United States too was increasingly adopting and using the technological approach, albeit more for topographic applications [54,63]. Similar developments can be observed in Australia [64].

5.2. 1940s

Perhaps predictably, those first scaled applications of photographic methods for land administration appear to have gone into hiatus at the beginning of the 1940s, wholly due to the advent of World War II, at least in terms of reporting. The front-running European nations—Italy, France, and Germany—were either busy with the war effort, or occupied by foreign forces. Those outside Europe, including many colonies of the British Empire, were also equally embroiled in the conflict. This meant that what surveying and mapping capacity was available was almost entirely directed to military mapping. Dick [65], reporting on New Zealand's status in the Empire Survey Review, makes this clear: " … *ordinary routine work had further to be reduced to meet the barest needs of the day and the main activity of the staff has consisted of topographic mapping for military purposes*".

However, whilst there may have been a growing backlog or hiatus of cadastral mapping, in those contexts where there was no conflict on the ground, geodetic work and national mapping were certainly a focus. This was especially in locations strategic to the war effort, and/or where actual conflict was not a day-to-day impediment to survey work [66–68]. Moreover, the war itself spawned photogrammetry and remote sensing innovations, albeit most likely not being openly reported. This was particularly with regard to the use of aerial and aerospace technologies for surveillance and observation [69]. Therefore, it is of no great surprise that by the end of the conflict in 1945, surveyors were already contemplating the tasks of adequately surveying post-war Europe and beyond [70], and utilizing the innovations developed therein [71,72], including the implications for cadastral maps.

6. Going Global (1946 to 1969)

6.1. 1950s

In the later part of the 1940s and into the 1950s, many new case applications would appear from outside Europe: across Africa, Asia, the Middle East, the Americas, and

Oceania, the maturing techniques were gaining widespread interest and application. The backdrop here was an emergent global perspective on the issues of land tenure security and land reform [73,74]. This marked a move beyond the conventional national-level or colonial mindsets, and the beginnings of a more integrated discourse in land economics, land law, and land surveying/mapping. The development of the United Nations and FAO was key here, and the Land Tenure Centre at the University of Wisconsin [74] was central to the discourse, and the amount of published works begins to increase at this point. What follows cannot be considered complete global coverage of all instances of photogrammetric methods applied to land administration tasks; rather, the aim is to provide insights into the breadth and scale of uptake.

In Africa, Dowson [75], countering his own earlier claims against photogrammetric methods, suggests application in the protectorate of Zanzibar, where an increasing number of unstable agricultural small holdings needed recording in a quick and economical manner (Figure 5). The advancement in techniques and the need for speed most likely informing the change of heart. Smith and Whittaker [76] provide a new commentary and comparison of on-ground methods versus aerial survey techniques, based on work undertaken in Kigezi District in Uganda. Menzies [77], providing a broad historical overview, details the uptake in South Africa, and Adams [78] suggests the use of aerial triangulation techniques to support cadastral mapping in Kenya. For the case of Kenya, it can be noted that rectified photography eventually formed the basis for title plans. Whilst initially intended as temporary records, these were in use well into the 2000s (see Section 8.1). However, the approach was ultimately rejected in other African contexts: it required landowners to plant a specific type of tree along boundary lines, and at least anecdotally speaking, this was found as to be too onerous to achieve at scale. These developments in Africa, during this period, gave rise to debates over the legal implications of the photogrammetric method, at least for jurisdictions where cadastres informed legal ownership. These took the form of discussions around 'pegs versus plans', 'measurements versus monuments', 'fixed boundaries versus general boundaries', and the legal responsibilities (and liabilities) of those undertaking both the ground-based and photogrammetric survey work (see Sections 6.1 and 7.3 and [1], where these debates were later documented).

Figure 5. Long before GIS, overlay of cadastral boundaries using photographic techniques was trialled and applied, including in Africa (adapted from [75]).

In the Middle East, Park [79] summarises the effectiveness of the photo mosaic technique to act as a base for resource inventory mapping in the Hashemite Kingdom of Jordan:

even in the featureless flat landscape, the photographic method is shown to have great utility, including the ability for quick training of Jordanian nationals, and is one-tenth of the cost to produce compared to conventional topographic mapping.

In the Americas, where the application had until this point been limited, Van Zandt [80] provides results of scaled application in Utah in the United States (Figure 6). Other experiments demonstrated positive results in the States of Maryland and Vermont [81] and other areas in the United States [82]. Certainly, as reported by McVay [83], the capacity to cover large areas quickly was increasingly recognised as being suited for mapping public/state land tenures, at least by the responsible government agencies. Andrews [84], called for its use in cadastral surveying in Canada, with Slessor [85] delivering the results from an experimental application of photogrammetric methods versus field methods to map an 'Indian reserve'. From a scientific perspective, the results were considered a great success; however, the obtained accuracies were argued as inadequate for practice. In South America, proponents argued for the application of aerial surveying to support cadastral mapping in Peru [86] and Northeast Brazil [87].

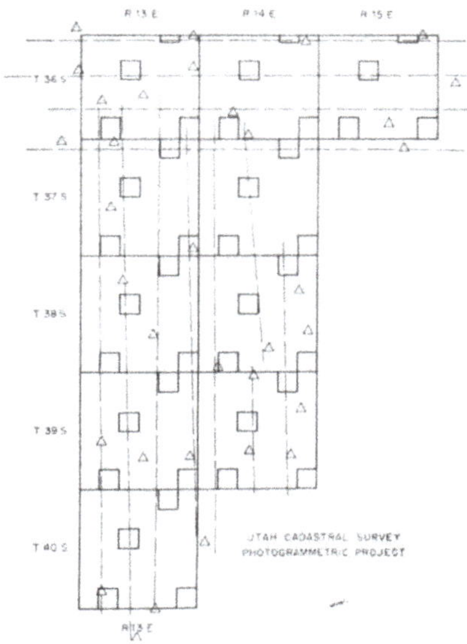

Figure 6. Fusion of photogrammetric and cadastral surveying in Utah, USA: photographic flight lines and control points (adapted from [81]).

In Asia and Oceania, results from applications or experiments in the Philippines [88], Japan [89], and Vietnam [90] (Figure 7)—the latter case having recently had most of its land records destroyed due to conflict—are reported. In the Philippines, this included an initial pilot of 149,000 hectares, 15 municipalities, and 61,000 lots. Plans for a 15-year program to cover the entire country at the cost of USD 6/Ha were put forward. In Australia, there was certainly growing interest, if not a debate about the relevance of these global developments [91,92]. In the state of NSW, Rasseby [93] reports how whilst not used for parcel mapping directly: *"Some significant progress has also been made in the use of photogrammetry to provide control for subdivision of rural land"*.

Figure 7. Conflict-strewn Vietnam demonstrated the potential for photogrammetric methods for fast-tracked in parcel mapping to support land reform (Adapted from [90]).

In terms of technical advances, during this period, there was a first move from analogue towards analytical photogrammetry, enabled by the invention of the analytical plotter in 1957 by Uuno (Uki) Vloho Helava: he used a computer to digitally transfer the coordinates between the image and the map (note: Helava later played a central role in the development of the first digital photogrammetric workstation in 1979). Additionally, there were also new technologies emerging: the use of radar, infrared, colour, and other remote-sensor techniques were developing [94]. There were also iterative improvements and refinements to existing techniques and technologies. As was the case in other fields—such as soil surveys, forestry, and engineering surveys—in cadastral surveying, these were enabled by the closer collaboration of *"surveyors, air navigators, photographers, radar specialists, meteorologists, and instrument designers"* [95]. That is, the technical improvements in measurement and plotting accuracy from the war era, were transferred, via professional collaboration, into the domain. The period also provided the first mention of electronic methods and automation, particularly with regard to providing survey control via automatic registration methods [96,97]—although, the same author laments a lack of investment in technical skills to utilise the method more widely. Later in the period, the concept of the 'numerical cadastre' (based around coordinates), enabling the 'multi-purpose cadastre' and 'integrated survey' concepts [98] came to the fore. As outlined by Basye, the first digital land information system was also developed [99]. These innovations were leveraging and combining the outputs of aerial mapping with other survey data, aiming to deliver one integrated map that, amongst other land management goals, as declared by Bonacci [100]: *"could serve the title examiner, appraiser, land negotiator, the courts, and many others serving this whole complex of land acquisition"*.

6.2. 1960s

Moving into the 1960s, applications in land use planning, and broader land management—an emergent and closely related discipline to land registration and cadastral mapping—gained attention [101–104]. A prime example is provided from Norway [105], where an intended 12-year whole-country mapping program sought to use the technique to produce land use classification maps at scales between 1:5000 and 1:100,000. These would later serve as the basis for property boundary determination. This new focus on land use mapping was driven by the utilitarian desire to manage food production, housing, industrialisation, and economic development more efficiently. The pace of post-war development meant that the speed and extra details provided by photographic methods leant well to the context.

Interestingly, with the planning domain still in its infancy, the relevant works are found in more technical photogrammetric or aeronautic journals.

This period also saw a maturing of land registration and cadastral domains (later to be conceptually merged and known as 'land administration' in 1990) into an aggregated disciplinary body of knowledge. The seminal aggregation [106] and synthesis work of Dowson and Sheppard [107,108] are exemplary here. In the Netherlands, the International Training Centre (ITC) with its dedicated focus skills development in aerial surveying and photogrammetry in developing countries, was established. From early on, the institution considered the specific case of photogrammetry for cadastral surveying [109]. With such developments, aerial surveying and mapping were now truly legitimizing in the field globally. As Hart [110] and Ray [111] both assert, this forced a rethink of what it meant to be a 'surveyor' and how to be trained as one. Certainly, the need for degree programs with a holistic approach to survey and mapping techniques was recognised [112].

That said, the cadastral profession was slow to move in the adoption of photogrammetric methods for cadastral work in many country contexts. Abrams [113] describes how the legal validity of the boundaries produced by photogrammetry is one issue: *"One of the problems of this development has been the admissibility of photogrammetrically prepared exhibits into court and other legal proceedings. Demonstrative evidence, particularly in the form of aerial photographs and other photogrammetrically prepared exhibits, has great potential in the field of eminent domain. Research of the relevant case law indicates that admissibility of demonstrative evidence generally has been a question for the discretion of the trial judge in any given litigation"*.

Irving [114] suggests that it is more to do with the relatively small size of the sector and the lack of competition and drive for innovation: *" ... it may be that, as we are not among the great spenders of public or private funds, there were more important fields for new methods and the doing of more in less time; still again it may be that equipment manufacturers were enjoying the period of repose with us, and not constantly urging us on with new and bright ideas in bronze and glass"*.

Thompson [115] is blunter. Referring to the issue of slow adoption by surveyor community, he blames the surveyors themselves: *"In the earlier period it was claimed that the photograph could not give the necessary accuracy, and in later period, that even if it could, it was uneconomical. Many results in the 1920s and 1930s were interpreted as showing, both at large and medium scales, that air photograph was not good enough."* and ... *"it is now clear that the plane-table and chain were being treated as sacred cows, and it was blasphemy to suggest that maps produced by their means might be inaccurate"*.

At this point, it seems important to make the distinction between the typical work of the government and the private sector with regard to land administration. Typically, it was national or state governments that were concerned with large areas or 'whole of jurisdiction' cadastral coverage, whilst the private (licensed) cadastral surveyor tended to be more orientated towards individual parcels, or small collections thereof. For this reason, given its advantage regarding fast and large-scale coverage, photogrammetric methods tended to be of more interest to government mapping agencies. Most of the students at the abovementioned ITC hailed from these government departments. The method was a 'harder sell' with regard to cost and time benefits for private land surveyors. This was despite the pleas of proponents such as Orvington [116]: *" ... there are many aspects of air survey and photogrammetry which have particular significance for the private surveyor, both for cadastral and topographical surveys and other features of his professional duties"*.

The divergence in perspectives was important. In many jurisdictions, private agents of the state made up a large proportion of the cadastral surveying professional body. The different foci, business models, and financial interests at play can be seen to have underpinned debates on the merits of ground versus aerial methods, long after the technical and accuracy challenges had been surmounted.

7. Space, Cities and Digital Systems (1970 to 1999)

7.1. 1970s

Four pervasive and emergent forces came to the fore commencing in the 1970s: space technologies, digital computing, urban planning, and systems thinking. Whilst the geopolitical 'space race' played out in the decade prior, its fruits were exploited by mappers in the following decade. The new terms 'remote sensing' and 'satellite imagery' are observed at this point. Likewise, whilst digital computing developed in the 1960s, it was the 1970s that saw the technology make its first appearance in innovative national mapping and cadastral agencies [117]. Additionally, as rural populations transformed into urban ones in developing contexts, and city centres began de-industrialising in more developed contexts, urban and city planning grew as a domain, and it too needed its maps. The fourth, perhaps less obvious force, cutting through each of the other three, was 'systems thinking'. Its relevance to cadastral studies was made clear by Dale [118], and the theory would impact greatly in developments in subsequent decades.

Regarding space technologies, the benefits of wide-area coverage, at repeated intervals, with multi-spectral coverage, brought about by satellite-based remote sensing, were being recognised across the domains of forestry, soil science, land use planning [119], and the growing area of environmental protection [120]. The characteristics were also identified as being particularly useful for developing contexts [121], where large parts of the landscape suffered from a lack of adequate or up-to-date topographic and natural sources maps. For cadastres, as with earlier photogrammetric approaches, it was public agencies, typically dealing with larger-sized parcels of public lands, that saw the utilisation. Torbet and Woll [122] describe initial applications in the United States linked to public land, deserts, watersheds, and First Nations lands. Lambert [123] explores the potential impacts in Australian surveying and mapping. Kellie [124] appears to undertake a direct appraisal of the emerging techniques on the domain of cadastre; however, unfortunately, only a citation could be found for this work.

Regarding digital computers, leveraging off the convergence of surveying and mapping professions in the previous decades [125], the 1970s saw the convergence of the numerical cadastre concept based upon coordinates from photogrammetric methods [126]), integrated surveys concept, and multi-purpose cadastre concept, into a matured conceptual design, most notably by McLaughlin [127]. The idea of building an integrated and open land information system, incorporating digital imagery as a key dataset, was taking shape [128,129]. Key here was the 1979 work of Duane Brown(For more detail on Brown's work see: Brown, J., 2005. "Duane C. Brown Memorial Address", Photogrammetric Engineering and Remote Sensing, 71(6):677–681), whose short-arc method of geodesy helped to prepare the way for the integration of photogrammetry into GIS, via the use of reflective targets.

Regarding urban planning, although numerical and computation photogrammetric approaches had been improving since the 1950s, Braasch [130], noting developments out of Hamburg, Germany, still notes: *"The 'graphical cadastre" may be produced by either plane table, or simple photogrammetric methods, but is not recommended"* (for cadastral purposes in urban areas) and that: *"Photogrammetric methods give excessive errors on short lines, but are gaining favor in their economy, especially in rural areas"*.

That is, the growing challenge of urban tenure mapping, via photogrammetric methods, was considered a separate challenge to cadastral mapping in other areas [131,132]. Not only did it require higher accuracies (due to higher values), and therefore higher-grade imagery—urban areas generally changed more rapidly, and whilst the time and costs advantages of photogrammetric approaches over time-consuming ground methods were clearly apparent, the embedded approach of using ground methods for urban cadastres continued to curtail the use of imagery [133].

Meanwhile, alongside the abovementioned digital and remote sensing innovations, applications of more traditional photogrammetric methods continued. Weissman [134] provided an update on the contemporary process being used in Switzerland, combining photogrammetric measurements with ground methods in cadastral survey creation. Bonnell [135] explains the extensive use of photomapping for legal boundary surveys in the contexts of mining and natural resource management. A fascinating account of the extensive use of photogrammetric techniques, including for creating urban cadastral line maps in Saudi Arabia is also provided [136]. Blachut [137] identifies how, for cadastral purposes, winter photographs are highly useful for boundary determination (at least in more temperate climates). Lafferty [138], echoing the much earlier work of the 1910s–20s, seeks to provide an updated cost–benefit analysis for certain terrain types and differing cadastral accuracy requirements of the use of photogrammetric methods. Dale [139] provides a comprehensive overview of the role of photogrammetric methods in cadastral surveys in Commonwealth countries, finding significant use of the techniques, although limited use in areas with high-value land. Meanwhile, Barrie [140] questions outright the need for ground-based parcel surveys altogether, given the recent advances in computation and photogrammetric methods.

7.2. 1980s

Into the 1980s, space technologies and digital technologies continued to converge, with a growing recognition of the impending impact on land surveying and cadastral mapping [141]. New generations of satellite remote sensing technologies were launched (i.e., Landsat 4; SPOT-1): in the Scandinavian context, including Denmark and Sweden, it was argued that for data capture, traditional aerial photography and geodetic methods were already giving way to spatial data acquired from satellites [142]. However, as pointed out by Lodwick and Paine [143], due to the limitations in resolutions with Landsat 1, 2, and 3; challenges with image registration; and issues with handling the large quantities of data, overall, the surveying profession had lagged behind other fields in the application of remotely sensed imagery.

Linked to these space-driven developments were techniques for integrating remotely sensing data into geographic information systems (GIS) [144]. In the context of cadastres, this took on the specialist form of a land information system (LIS), or equivalents variously incorporating 'multipurpose' and other terms [145], and the sub-domain of land information management emerged [146]. Digital techniques for extracting vector data from imagery and enabling its incorporation into LIS were also developed [147], as were procedures for cadastral map renovation based on similarity transformation [148] and digitisation of data from photogrammetric inputs [149,150]. Many countries were at least piloting, for example, in Colombia [151] and Taiwan [152] or were undertaking scaled implementations of these developments [153], as per the case in Canada. This area of cadastral map renovation, upgrading, and updating would be a continued area of focus over subsequent decades, and is more fully unpacked in Bennett et al. [10].

However, it was with conventional photogrammetric methods where most scaled applications continued to occur. Photogrammetric techniques for control network densification were developed [154,155] as were cadastral survey data capture techniques [156]: cost reduction and improved legal certainty were highlighted as key benefits. The application to development projects was also a focus [157], with the World Bank project in Thailand a prominent example [158]. Other experiments also took place, for example, in Zambia [159,160] and Taiwan [153]—but it was still recognised that changeable terrain and tenure systems meant the approach might not be suitable in all locations, for example, in Fiji [161]. The convergence created by the move from analogue photogrammetry in 1960 (e.g., stereo plotters) to analytical/digital ones—in terms of data creation, capture and storage—was again demanding a reappraisal of what constituted a cadastral surveyor [162], of how to offer education and training programs [163,164], of what name to use (e.g., *Geomatics* [165]), and what research programs should constitute [166].

7.3. 1990s

It is tempting to thematically separate the highly digitalised 1990s from the more analogue 1970s and 1980s, given the ubiquity of PCs and scaled uptake of the Internet in that decade; however, ultimately the 1990s capped much of the work of the previous decades. A key development, as explained in Bennett et al. [10], was the internationalisation of the cadastral surveying profession, spurred by the post-Cold War re-establishment of cadastres and land registries in eastern European countries, and the uptake of the unifying term 'land administration'.

In a book of the same name, Dale and McLaughlin [1] provided a synthesis on the debates and options with regard to the use of photogrammetric versus ground-based methods, overall finding that a combination of both is possible. For the specific case of remotely sensed satellite imagery, in agreement with Paulsson and Mundial [167], Dale [168] flags lingering concerns: *"In spite of claims that satellite imagery can be used for cadastral surveying, remote sensing is still too crude a set of tools for such a purpose and, like the use of photogrammetric techniques, addresses only part of the cadastral problem."* He rightly affirms that data capture is but one component of the challenge of getting agreement on boundary locations, which is fundamentally a social process, not only technical one.

That said, Jensen [169], revealing an awareness of emerging higher resolution options, is more positive: *" . . . cadastral (property line) information are best monitored using high spatial resolution panchromatic sensors, including aerial photography (5 0.25 to 1 m) and, possibly, the proposed EOSAT Space Imaging IKONOS (1 by 1 m), Earthwatch Quickbird pan (0.8 by 0.8 m), and Orbview-3 (1 by 1 m) data."* Although it should be noted that he argues that a 0.25 to 0.50 m spatial resolution is acceptable: this is generally (rightly, or wrongly) outside what cadastral surveying professionals (and associated regulations) would deem acceptable.

Rao et al. [170,171] similarly suggest that the Indian remote sensing satellite program will shortly deliver spatial resolutions aligned with cadastral mapping requirements, particularly in rural areas [172]. Gonzalez [173] predicts the new generations of high-resolution satellite imagery could be used for both state-wide and local-level cadastral map production. Jensen [169] also illuminates the issue of invisible boundaries and combined use of ground surveys, ortho photos, and even satellite imagery in the United States: *"In many instances, the fence lines are the cadastral property lines. If the fence lines are not visible or are not truly on the property line, the property lines are located by a surveyor and the information is overlaid onto an orthophotograph or planimetric map database to represent the legal cadastral (property) map. Many municipalities in the United States use high spatial resolution imagery such as this as the source for some of the cadastral information and or as an image back-drop upon which surveyed cadastral and tax information are portrayed"*.

Meanwhile, Schmitt et al. [174] show the application of available high-resolution satellite imagery in identifying settlement structures and changes therein. Leberl et al. [175], commentating on the relevance of remote sensing technologies to the Austrian context, outline the need—if uptake and use are to increase—for the tailoring of imagery products to suit those local and district users who are not looking for nationwide coverage. Moreover, they also make clear the great benefit of repeated capture enabled by satellites, something that is not a given with traditional aerial photogrammetry.

On digital computing, the transition to digital ortho-photo production and use occurred in the 1990s [176], enabling the fusion of GIS/LIS and digital imagery sources [177]. Konecny [178] provides many more scaled examples from German and World Bank donor projects, including Albania (USD 5/parcel) (as do Leke et al. [179]), Georgia, Cambodia, Ethiopia, Argentina, Peru, and Honduras. Holstein [180] adds Brazil to a similar list, but also explains that whilst the use of imagery has its advantages, these techniques require up to an 18-month lead time in terms of flight preparation and base imagery production. For this reason, more flexible methods, including an increased exploration of softcopy digital imagery, were underway. However, here, despite the clear benefits of going digital, issues around poor underlying technology infrastructure and limited capacity were already recognised [181]. Anderson [182] proposes the approach in Mozambique, in alignment with new land laws, and Christensen et al. [183] in Namibia. With regard to the issue of urban cadastral data, Al-garni [184] demonstrated the application using aerial photographs in Riyadh. Harcombe and Williamson [185] show the novel use for low-value lands in the western parts of New South Wales in Australia, making use of helicopter surveys for geodetic control.

New applications of other imagery-based technologies also arrived in the 1990s, for example, historical land record archive scanning, as suggested by Boatta [186]. Mohamed et al. [187] propose the novel use for the identification and demarcation of, until then, unrecorded indigenous lands. Ehlers [188] reveals an imagery-based approach for informal settlement identification and management—including informal land tenure parcel identification. Fourie [181] also highlights the need for systems build around visualisation (i.e., imagery) for these contexts. Onsrud [189], also on informal or unrecorded land tenures, almost harking back to the first terrestrial photographic methods in the late 1800s, but also foreshadowing the pro-poor approaches to come, suggests the incorporation of photos into an integrated data gathering approach, for use by locals within an unmapped community. Similarly, Mason and Fraser [190] look at the issue of informal settlement mapping and propose the use of *"high-resolution satellite imaging, small format digital aerial imagery and digital multispectral video systems"* and *"also discuss the example of automated shack extraction from aerial imagery."*

Bartle et al. [191], with a similar mindset, propose an automated approach for matching field boundaries in Landsat imagery with cadastral boundaries. This appears to be one of the earliest works on cadastral boundary feature extraction: a topic that would garner much interest in the subsequent decades. However, as would be experienced later, Pinz et al. [192] predict active fusion of remotely sensed data and cadastral boundaries would be highly challenging compared to other thematic layers. This is not even to mention, as explained by Okpala [193], how current laws and regulations continued to impede or disallow the use of imagery-based techniques for the generation of crucial nationwide land parcel maps for land management purposes.

8. Deluge of Digital, Drones, Dimensions and Data (2000 to 2021)

8.1. 2000s and 2010s

In the 2000s, a plethora of technological developments created strong momentum for uptake of remote sensing methods in land administration: digital photogrammetry, high-resolution satellite imagery (HRSI); unmanned aerial vehicles (UAVs); lidar; SAR radar; oblique photogrammetry; and pictometry all emerged, or matured, as alternate technological approaches that could support endeavours. This included not only conventional 2D cadastral mapping, but also the move towards new land administration applications, namely 3D cadastres, marine cadastres, and previously unknown or unmapped property rights, restrictions, and responsibilities such as cable networks, biota/carbon rights, and solar rights. Moreover, convergence with the broader establishment of a high-speed Internet infrastructure, cloud-based computer processing, web mapping services, smart mobile devices, and artificial intelligence enabled new ways of creating and sharing imagery-based land administration information.

It needs to be noted that during this period, the quantity of published scientific literature increased significantly across many disciplines, including land administration. The reasons for this are not the focus here, and are briefly unpacked in Bennett et al. [5]. However, in the context of this historical review, unlike the other periods covered thus far, this increase makes it challenging to incorporate all contributions whilst also maintaining the structure of the paper. Opportunely, the increase in scientific contributions also drove the compilation of meta-studies, for example [194–196] (this included detailed historic reviews on photogrammetric technologies. For example, see the 2010 work of Hobbie: https://www.isprs.org/society/history/Hobbie-The-development-of-photogrammetric-instruments-and-methods-at-Carl-Zeiss-in-Oberkochen.pdf, accessed on 18 October 2021), and therefore, in this review, where appropriate, we direct readers to those more detailed reviews of specific topics.

Regarding HRSI, improvements in spatial resolution (i.e., pixel size < 50 cm), making it comparable with aerial orthophotos, helped to curb concerns over property boundary identification and delimitation, particularly for rural lands. Whilst the potential was first recognised in the 1990s [197], a wave of experimentation, piloting, and even scaled use was observed globally in the 2000s. Sahin et al. [198] provide an early, although incomplete analysis of Ikonos imagery for cadastres in Turkey. Likewise, Fraser [199] demonstrated the utility and potential in Bhutan. In nearby Pakistan, Ali et al. [200,201] find that costs and time for cadastral mapping, combined with GNSS positioning, could be cut in half. In India, continuing their work from the 1990s, Rao et al. [202] demonstrate how HRSI is applicable for cadastral boundary determination in India, as do Sapra et al. [203], for the case of heads-up digitisation of HRSI for forested lands. Sengupta et al. [204], in a novel experiment, demonstrate the fusion of HRSI (GeoEye) with older colonial-era parcel maps for cadastral updating purposes. In neighbouring Nepal, Panday [205] successfully trials the use of HRSI preloaded into mobile devices for remote community boundary definitions. Further south, Andri et al. [206] suggest HRSI application for participatory tenure mapping in Indonesia. In Africa, Asiama et al. [207] present positive results from participatory mapping activities in rural Ghana, also with HRSI preloaded into mobile devices. Balas et al. [208] show similar application potential in Mozambique, and importantly, regulations were friendly towards its use. Ondulu et al. [209] consider the use of HRSI an ideal application for undertaking long-overdue updates to what were originally intended as temporary parcel index maps created in Kenya, 50 years earlier (See Section 6.1 for a description of this earlier work). Lengoiboni et al. [210] suggest the approach could be extended for recording the dynamic land tenures of Kenya's nomadic pastoralists. In Iraq, Hassan et al. [211] provide an accuracy assessment for the improvement of historical graphical cadastral maps in Kurkuk City, Iraq. In another post-conflict situation, Jones et al. [212] show the potential in Colombia. HRSI was clearly now proven, if not ubiquitous, in land administration, certainly in terms of R&D, but unfortunately, laws and regulations did not always enable easy or scaled application.

Alongside HRSI, much focus was also afforded to the possibilities brought about by digital camera technology, including automatic orientation, dense image matching, and automated data processing. In terms of image acquisition, film-based cameras were replaced by a variety of active and passive sensors, and a combination of those, mounted on different platforms [213]. These were increasingly integrated with onboard GNSS receivers and inertial measurement units (IMU) [214]. With the significant increase in image quality and quantities of data, attention turned to algorithm development for sensor modelling [215]. Automatic image orientation also gained significant attention, being inspired by the computer vision algorithms such as structure from motion (SfM) and from robotics, simultaneous localisation and mapping (SLAM), new methods were developed, including scale-invariant feature transform (SIFT) [216], and speed up robust feature transformation (SURF) [217]. For detecting blunders, the random sample consensus (RANSAC) [218] algorithm was created. Another focus and advance were thematic information extraction [219]. In this respect, classifications such as support vector machine (SVM) [220] and random forest (RF) [221] have been applied actively. In addition, new change detection techniques were developed [222]. The above-mentioned developments in photogrammetry have been applied in numerous countries. To achieve a digital cadastral database in support of LIS in India, tests for Andhra Pradesh districts, across over 10,000 sq.km were performed in 2011 [223]. Another example is the work on land parcel boundary delineation based on aerial survey in Azerbaijan [224]. In addition, digital aerial images taken over Ghana in 2014 were analysed by Offei et al. [225], aiming to assess the compliance with residential building standards in the context of the local customary land tenure system. Other examples of assessment of digital aerial photogrammetry with small or large format cameras for cadastral applications were tested in Nepal [226], Indonesia [227], Costa Rica [228],

Jordan [229], Turkey [230]. A framework for the automatic characterisation of real property based on aerial photography was proposed by Austrian researchers in [231]. In addition, an interesting exploration estimating the positional accuracy of a parcel boundary dataset based on unrectified aerial images has been done by Siriba in [232].

Regarding UAVs, developments here countered arguments that the one-time collection of imagery was costly and too quickly became outdated: UAVs could, in a cost-effective and as-needed manner, quickly capture a small number of parcels at high accuracy and provide more contextual information than an equivalent ground-based survey. Mumbone [233] trials the application in rural Namibia, in the context of mapping communal villages, for which aerial imagery was explored years earlier [234]. These communities are often separated by large distances making regular aerial photography prohibitive. Ramadhani et al. [235] and Yuwono et al. [236] assess and determine high relevance for the approach in both rural and urban Indonesia, and later Aditya [237] undertakes a larger scaled pilot in the context of participatory tenure mapping. Kurczynski et al. [238] and Cienciala [239] both reveal the potential for UAVs for sporadic cadastral updating in Poland. Other arguments are made for Kenya [240], and trials are undertaken [241] here and in Rwanda [242]. Stocker et al. [243] reveal how for the case of Rwanda, the three different UAV methods could align with administrative requirements, notwithstanding the limits relating to law and capacity. In the same context, Flores et al. [244] consider the governance challenges of UAV introduction. In a related application, Ali [245] demonstrates the potential for land valuation in Zimbabwe. Other investigations include the comparative work of Karatas in Turkey [246], against classical methods, Koeva et al.'s comparative work in Rwanda [247], Mbarga's [248] assessment for application in Cameroon, and perhaps most influentially, Stocker et al.'s [249] recognition that, again, it is regulatory issues—for both UAV usage generally, and specifically for cadastral surveys—that may determine the ultimate update of the technology.

Lidar techniques, both terrestrial and aerial, provided a new means for creating cadastral information. Point cloud data are inherently 3D, and this creates the opportunity to support the growing demand for 3D cadastres, 'indoor' data capture, and marine/littoral zone tenure mapping. Until now, these needs were not well supported by conventional ground-survey methods nor conventional photogrammetric ones. An overview of the main developments on 3D cadastres, including applications of lidar, is provided by Stoter et al. [250], with updates in Van Oosterom et al. [251].

Specifically, on lidar applications outdoors, in an early work, Filin et al. [252] identify methods for fusing lidar point cloud data with cadastral maps. A significant challenge would be creating efficient workflows for extracting simple vector boundaries: automated feature recognition would become a focus not only for lidar data, but also other large, remotely sensed datasets. Whist these automation techniques are covered in more detail below, Kodors et al. [253] and Kumar et al. [254] provide an early method for building and real-estate capture. Meanwhile, others undertook country-specific explorations: Giannaka et al. [255] explore the potential in Greece; Drobez et al. [256] more generally in Slovenia; Luo et al. [257] develop a workflow for Vanuatu (Figure 8); Wierzbicki et al. [258] fuse lidar and orthophoto techniques for cadastral modernisation in Poland; and Griffith Charles et al. [259] trial the approach in low-value informal lands in Trinidad and Tobago. On the latter, the low-cost specifications, whilst not ideal, could support the preparation of spatial data in the context of 3D cadastres. Lubeck [260], in related developments, uses SAR-radar and its application in fence detection to support ground methods in Brazil. Going underground and indoor with lidar, Rajabifard et al. [261] provide full coverage on BIM developments relating to land administration, and provide full coverage for BIM data capture options, including lidar techniques. Koeva et al. [262] provide a novel indoor cadastral data capture solution based on terrestrial scanning. Beida et al. [263] demonstrate the use case for capturing underground 3D objects and converting them into cadastral objects. Yan et al. [264] supplement similar methods with ground-penetrating radar (GPR) to support cadastral object capture. Other novel data capture technologies emerging in this era and applied to land administration, at least conceptually and/or experimentally, included oblique aerial imagery and pictometry (the process of capturing and stitching building façade oblique imagery together), as demonstrated in Kisa et al. [265] and Lemmens et al. [266], respectively.

Figure 8. The 2000s saw convergence in data capture technologies and integrated workflows for map production, as demonstrated by [257], where lidar data, aerial imagery, and open-source software were used to generate a parcel layer (red lines) with limited human input. Note: grey lines are the original cadastral boundaries, used for comparison.

Here, linked to Luo et al. [257] above, it is worth mentioning Luo et al. [267]. Using existing cadastral maps as baseline data, they quantify the overlap between legal/cadastral boundaries and visible features. This is important: remote sensing and photogrammetric methods are premised on the idea that physical boundaries overlap with legal boundaries. The results here tend to confirm the anecdotal notion that 70% of cadastral boundaries were indeed visible or physical, at least in the context studied.

Additionally, during this period, computer processing speeds, networking speeds, and storage capacity exponentially increased. As mentioned above, this reduced image processing times, mosaic creation, and so on, although the amount of imagery data being captured and the density of pixels within these images also exponentially increased. Geocloud platforms combining multiple sources of image data needed for land tenure recording have been developed [268]. The first era of land information systems began to give way to second generation systems, relying on web services for transaction and data delivery [269]. The concept of SDIs fully matured, and alongside cadastral data layers [270], high-resolution georeferenced imagery was often considered a fundamental layer or part of a broader land administration data warehouse [271]: the division between imagery data and land administration data was increasingly blurred. These developments promoted standardisation in the domain, with ISO 19152 Land Administration Domain Model (LADM) [272], a data model standard, being endorsed in 2012. The cadastral and survey data packages within the model were generalisable enough to handle the incorporation of imagery-derived cadastral data and imagery itself as source data.

The final major technological advance in the era was the resurgence of artificial intelligence (AI) and machine learning techniques. For land administration, these offered the opportunity to automate processes for identifying, vectorizing, and validating cadastral boundaries. On this, Commelinck et al. [273] provide a review of these developments up to 2016, albeit primarily focused on UAV imagery. More recently, Bennett et al. [5] provide a review of AI techniques applied to the specific case of land administration maintenance. In Crommelinck et al.'s [273] generalised workflow consisting of preprocessing, image segmentation, line extraction, contour generation and post-processing, an open-source solution is developed. Due to the difficulties in training algorithms, semi-automated methods tend to be more promising [274]. Whilst Masouleh [275] focusses on 3D cadastres, proposing a deep learning methodology to support the reconstruction of buildings from aerial images, most of the work at this point focuses on 2D applications. Indeed, the techniques are seen to offer much hope in developing contexts where the greater majority of land parcels might not be mapped, or at least are very outdated [276]. Wassie et al. [277] (Figure 9) develop an approach using HRSI for rural Ethiopia, finding that regular smallholder parcels lend themselves well to the technique.

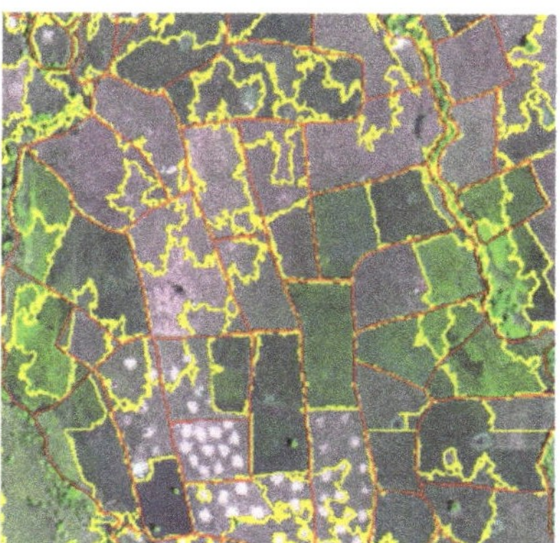

Figure 9. Wassie et al. [277] used open-source software to develop a workflow to extract smallholder parcel boundaries using HRSI in Ethiopia. Note: red = control boundaries; yellow = extracted.

Koeva et al. [278] show how such methods could be integrated with other innovations, including cloud services, UAV usage, and sketch mapping. Meanwhile, Fetai et al. [279] develop an approach using UAV imagery and off-the-shelf feature extraction tools, also with promising results. Park and Song [280] develop an approach for detecting cadastral parcel changes, also using hyperspectral UAV imagery. A key requirement for any AI technique is that the total cost and time for delivering the final boundaries, including pre- and post-processing and editing, should not be higher than the cost of non-automated techniques. This remains a challenge: even if 60–70% of boundaries can be extracted, the editing work involved still often takes the total cost over that of manual methods. For the case of land administration, unlike other thematic geospatial layers, it is generally an expectation to have 100% accuracy (or very close to it). Therefore, comparative work, between manual and automated methods, undertaken by Nyandwai et al. [281] (Figure 10), continues to be important: current rates are too low to be brought into production in many contexts, but could act as a 'first cut' cadastre in some contexts. Most recently, Xia et al. [282] used a convolutional neural network to improve extraction quality further (Figure 11).

Perhaps most promisingly in this era was the fact that the digital advances from previous decades became affordable and accessible in most contexts globally: mobile communications, smart devices, and high-speed Internet were not only the domain of developed contexts. Moreover, high-resolution satellite imagery covered the majority of the Earth's surface. This motivated the concepts of crowdsourced cadastres, participatory land administration, pro-poor land recordation [283], and more broadly, fit for purpose land administration (FFPLA) [284], all of which, learning from the lessons of development projects in the previous decades, and ongoing ones in the early 2000s, such as Cambodia [285], heavily advocated for the use of remotely sensed imagery, in all forms, to support data capture, and as Bennett et al. [5] explain, were ultimately endorsed in the Framework for Effective Land Administration (FELA) of the United Nations Committee of Experts on Global Geospatial Information Management. As always, this was understood to include ground visits, for sensitisation, demarcation, or validation, especially in the initial registration projects [286]: imagery alone was not enough [287]. Whilst many in the land sector argued that FFPLA, and associated terms, were nothing new (indeed, as shown in the preceding sections, imagery-intensive methods in land administration projects dated back decades), the branding was used to frame and propel experimentation with new technologies, pilots, and scaled work. Regarding the latter, the Rwandan case of mapping +10M parcels, with imagery, over a 3–5-year period was oft-cited [284]. Other explorations, including participatory methods and/or imagery-based approaches, were undertaken in Greece [288], Namibia, Ghana, and Kenya, as described in Chigbu et al. [289] and Koeva et al. [290]. The South African context is also demonstrated by Williams-Wynn [291] as being ready for FFPLA, based upon imagery, and that the legislative basis is already supportive. Zein et al. [292] provided an updated comparison of imagery sources—HRSI, UAVs and digital orthophoto techniques—in the context of FFPLA. They find that a cost of USD 7/parcel is achievable and that the selection of the most appropriate source will depend on the context. Simplifying things even further, the point cadastre concept was revisited [293], where a single point (rather than complex to capture polygons), overlaid on high-resolution imagery, could be used to identify rights.

Figure 10. Nyandwai et al.'s method [281] created a 'first cut' cadastre (yellow) that could be taken to the field for validation.

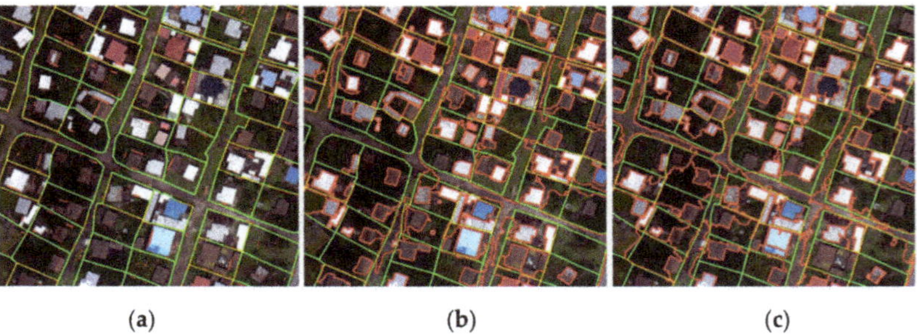

(a) (b) (c)

Figure 11. Xue et al. [282] compared three emerging feature extraction approaches for parcel boundaries: (**a**) fully convolutional networks (FCNs); (**b**) globalised probability of boundary (gPb); (**c**) multi-resolution segmentation (MRS). Note: Yellow lines are 'true positive'; red lines are 'false positive'; and green lines *are 'false negative.'*.

Meanwhile, beyond all the innovations around digital photogrammetry, HRSI, UAVs, AI, lidar and FFPLA, work continued, on the many decades long, surveyor-realisation that photogrammetric methods could and ought to be used to support formal and conventional cadastral surveying tasks. Here, use cases including hilly areas in Nepal [226], forest land in Greece [294], cadastral updating and illegal building detection in Turkey [230,295], urban mapping based on satellite data in Bulgaria [296], digital aerial photogrammetric building footprint additions to cadastres in Poland [297], and the use of historic imagery in the Slovak Republic for updating [298] are observed.

9. Discussion

This section does not seek to revisit the minutia in the developments, debates and discourse outlined above. Instead, it focuses on: (i) providing a concise synthesis of the key periods, drives, developments, and cases from the review; (ii) confirming the overarching hypothesis that photogrammetric and remote sensing methods have a strong historical and contemporary presence in land administration practice; (iii) providing a conclusive statement on the various cost–benefit analyses covered in this review; (iv) showing the limitations, at least in the contemporary era, in framing data capture methods in land administration as a dichotomous issue; (v) providing an important reminder of the issue of 'invisible boundaries' in the context of remote sensing techniques; (vi) highlighting legal and regulatory constraints; (vii) making mention of the need to consider the broader land management domain (versus land administration); and (viii) briefly casting forward to hypothesise emerging approaches.

First, the major findings from the review are presented both thematically (Figure 12) and geographically (Figure 13). For the thematic depiction, these are organised by the chronological periods identified in the review process. In addition, the key drivers, technological developments and illustrative cases are also depicted.

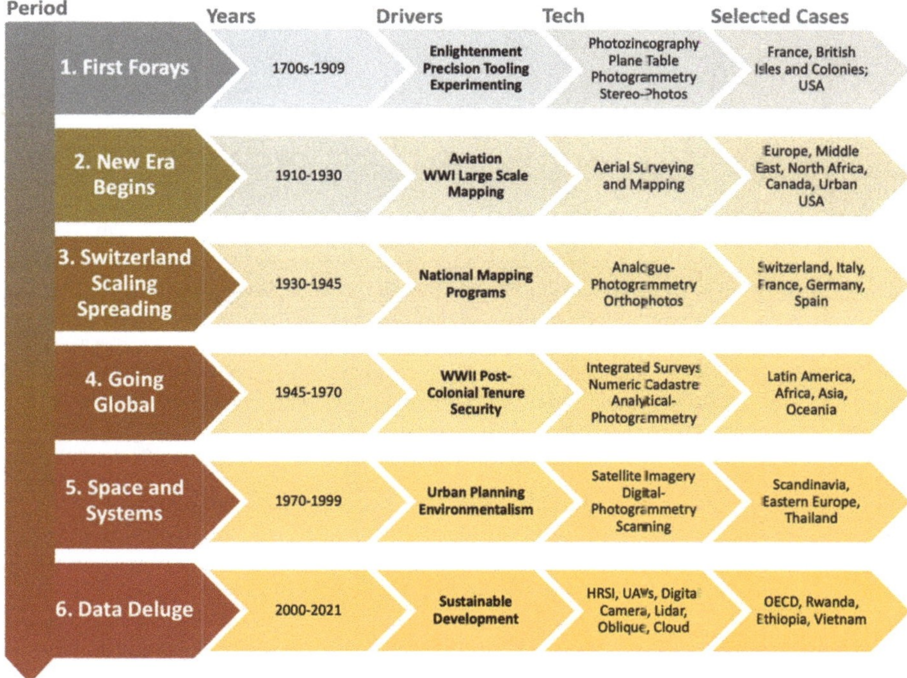

Figure 12. Origins and developments of photogrammetry and remote sensing applied in land administration.

Second, the overarching message is clear enough in the evidence. That is, almost one hundred years after European countries demonstrated the ability to use photogrammetric methods to produce high-quality and comprehensive cadastral coverage—with far more rudimentary technologies than have since developed—any remnant arguments on the use, and apparent limitations, of photogrammetric methods and remote sensing applied to land administration can hardly be sustained. This is not to say that ground methods have become redundant; on the contrary, ground methods continue to dominate in many jurisdictions. Whether this is to do with regulatory inertia, sector self-interest, or driven by considered cost–benefit analyses can be debated, but really ought not to be. What is more certain is that the surveying community, regardless of the jurisdiction in which they operate, owe it to its citizenry to ascertain how best to incorporate imagery-driven cadastral mapping approaches, at least in conjunction with ground-based methods, into land administration functions. Arguments around cost, time, and accuracy for capture would appear very hard to sustain; and cloud computing and high-speed internet overcome the issue of transferring and processing large amounts of remotely sensed data between stakeholders.

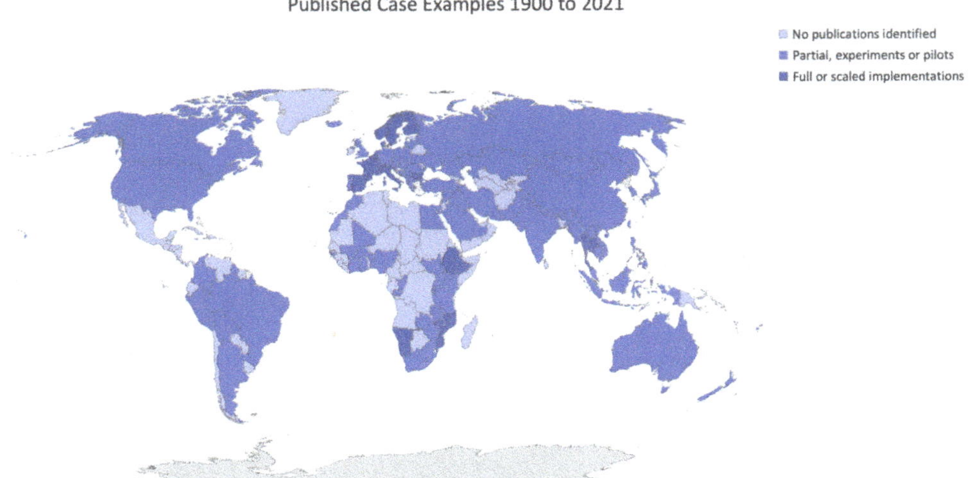

Figure 13. Remote sensing and photogrammetry applied to land administration: review results depicted geographically.

Third, this review did not provide a conclusive endpoint or structured comparison of the cost–benefit studies completed, as they emerged over the decades, with regard to photogrammetric methods (including remote sensing) and ground-based methods. In general, most of these studies sought to demonstrate the efficiencies that could be gained by aerial or image-based methods, at a particular point in time. If these were countered, it was usually with regard to what costs were not being included or excluded in the analyses. Here, the authors agree that cost–benefit analyses can be open to bias or manipulation, and without the means to test the claims in several papers, we simply presented the claims and rebuttals of both sides. That said, through the significant body of works provided, counter claims that imagery-based methods are not suitable in land administration tended to be accompanied with less empirical work.

Fourth, the more recent eras demonstrate that convergence of ground-methods and photogrammetric/remote-sensing methods is increasingly the norm: in practice, thanks to digitalisation, the dichotomy between ground and air is harder to ascertain—that is, the tools, the techniques, the resulting data and maps, and the underlying training programs are increasingly intertwined.

Fifth, noting the positive developments above, it is important to always keep in mind the issue of 'invisible boundaries'—that is, those boundaries that exist purely as social (and potentially legal) concepts in the minds of parties, and do not have a physical presence. Surprisingly, this issue was only explored more empirically in the final era; however, for imagery-based approaches, it is key issue in terms of achieving completeness and coverage, where such approaches are said to be superior. Whilst many cadastral boundaries have a physical presence, being able to be sensed remotely, many do not [299], and as shown by Luo et al. [267], whilst up to 70% could be sensed, other studies show [300,301] that in some contexts, physical boundaries may be absent altogether. That said, this still does not negate the use of imagery in those contexts: imagery, by its nature, provides contextual information that can be extremely supportive of land tenure, land value, and land use planning activities. Moreover, land administration as a domain has long recognised the different manifestations, perspectives, or representations that cadastral boundaries have, variously including combinations of physical natural features, person-made features (stakes or monuments), legal authority, social recognition, textual descriptions (metes and bounds), graphical depictions (be they scaled accurately scale, or not), numerical or coordinated descriptions, and more recently, digital representations. No technological approach can cover all of these aspects; however, remote sensing and photogrammetry can certainly support in some of them.

Sixth, mainly due to the historic nature of the land administration profession being more technology-oriented, the works reviewed here tended to have an overtly technological bias, with an overt focus on the spatial accuracy debate. For example, many works focused on 'how to' apply technologies—or, as in other examples, the cost–benefit analyses tended to focus on the costs for data capture, rather than taking a broader view on legal implications, staff retraining costs, awareness raising in beneficiary communities, or governance costs. However, these 'other' issues are most likely where the major blockers for the uptake of remote sensing technologies in land administration occur. In particular, laws and regulations around what tools and techniques can be used to create cadastral surveys, the legal responsibilities or mandate (e.g., licences) of those completing the surveying (whether ground-based or photogrammetric), or even the more philosophical debate on what can or should constitute a cadastral boundary (see [302] and [303], for example), are of considerable importance, but, apart from the work of Stocker et al. [249], far fewer dedicated works on these issues, directly relating to remote sensing and land photogrammetric methods, were found in the review.

Seventh, it needs to be recognised that this review focused on 'land administration': the broader area of land management cannot be said to have been fully covered. Most definitions of land administration, and certainly land management, would incorporate land tenure, land value, land use planning, and land development. Although these other functions are certainly covered variously in this review, primarily, the review here systematically concentrated on land tenure, linked to registration and cadastres. This limitation was somewhat intentional, given that the scope of work would have been too large. Nonetheless, it is noted, and moreover, it is suggested that the literature from these related areas would reveal a similar trajectory in terms of imagery-based technologies and techniques with regard to application, although this cannot be said for sure. Likewise, the limitation of focusing only on English literature is again noted. Where deemed necessary, it is encouraged for others to undertake similar studies of French, German and Spanish works—noting that many of the English works cited in this work refer directly to developments from contexts using those languages.

Eighth, looking ahead, whilst past developments cannot necessarily be used to predict future progress, it appears quite certain that emerging remote sensing technologies will continue to be experimented on within land administration. The period for scaled diffusion and uptake of those innovations, however, may be more rapid than in the past. The era of digital transformation makes it harder to sustain legal and institutional barriers to change: starts-ups and alternate land administration service providers can now more easily enter the market. In terms of immediate developments, AI and feature extraction techniques will continue to garner attention and will likely be fused with other data sources, both statutory and non-statutory, and social and environmental, to create more intelligent land boundary recognition algorithms. Regardless of this automation, it seems likely that human mediation will remain in some form for the foreseeable future. Integration of the use cases of land use planning, land valuation, land development, marine environment, underground, indoor and 3D, more generally, will continue to drive developments in practice and training courses.

10. Conclusions

This paper began from the premise that, at least conventionally in many countries (although, not all), land administration used ground-based survey and methods: the application of photogrammetry and remote sensing was said to be far more contemporary, if not considered inappropriate by some practitioners, only commencing later into the 20th century. This paper sought to counter this prevailing view, and contended that the use of remote sensing and photogrammetry to support land administration was far from a recent addition to the land administration toolkit: scaled implementation dated back much earlier.

Using now more accessible historical works, made available through archive digitisation, this paper presented an enriched and more complete synthesis of the developments of photogrammetric methods and remote sensing applied to the domain of land administration. Developments from early phototopography and aerial surveys, through to numeric photogrammetric methods, the emergence of satellite remote sensing, digital computing, and later lidar surveys, UAVs, and artificial intelligence were covered. That said, the review has limitations in terms of relevant languages covered (e.g., German was not included), and being based upon the available literature. It is encouraged for others to undertake similar studies, where deemed necessary, of other language groups, and to undertake a more complete country-level comparison of remote sensing and photogrammetric techniques, and related laws, applied in land administration.

The synthesis illustrated how declarations of the benefits of the technique are hardly new—and neither are well-meaning, though oft-flawed, comparative analyses based on time, cost, coverage, and quality. The historic case for, and application of, photogrammetric and remote sensing methods, in land administration is undisputable. Alongside this key finding, this review also identified other recurring challenges, in selected country contexts, throughout the decades, including: the problem of land sector inertia, conservatism and legal constraints when it comes to imagery-based approaches; the recognition that invisible boundaries will always mean some boundaries can only be identified through human interaction; and more recently the increasing irrelevance of the distinction between ground and aerial survey methods: technology convergence is driving fusion in practice and educational programs.

Apart from providing this more holistic view and a timely and important reminder of previous pioneering work, the paper brought contemporary practical value in further demonstrating to land administration practitioners that aerial and remote methods of data capture, and subsequent map production, are an entirely legitimate, if not essential part of the domain. Any contemporary arguments that the tools and approaches do not bring adequate quality for land administration purposes cannot be sustained. Indeed, these arguments tend to undermine what should be essential characteristics of the land surveying profession—pragmatic and pioneering mindsets. That said, it is left to land administration practitioners to assess whether the available methods are suitable for a given jurisdiction, and also whether cadastral laws and standards require revisiting.

Author Contributions: Conceptualisation, R.M.B.; methodology, R.M.B.; validation, M.K., K.A.; formal analysis, R.M.B.; investigation, R.M.B., M.K.; resources, R.M.B., M.K.; data curation, R.M.B. and M.K.; writing—original draft preparation, R.M.B.; writing—review and editing, M.K. and K.A.; visualisation, M.K. and R.B; All authors have read and agreed to the published version of the manuscript.

Funding: This research received no external funding.

Institutional Review Board Statement: Not applicable.

Informed Consent Statement: Not applicable.

Data Availability Statement: Not applicable.

Acknowledgments: In this section, you can acknowledge any support given which is not covered by the author contribution or funding sections. This may include administrative and technical support, or donations in kind (e.g., materials used for experiments).

Conflicts of Interest: The authors declare no conflict of interest.

References

1. Dale, P.; McLaughlin, J. *Land Administration*; Oxford University Press: Oxford, UK, 2000.
2. Henssen, J. Land Registration and Cadastre Systems: Principles and Related Issues. In *Lecture Notes of Technische Universität München*; Technische Universität München: München, Germany, 2010.
3. Konecny, G. *Geoinformation: Remote Sensing, Photogrammetry and Geographic Information Systems*; CRC Press: Boca Raton, FL, USA, 2014.
4. Toth, C.; Jóźków, G. Remote sensing platforms and sensors: A survey. *ISPRS J. Photogramm. Remote Sens.* **2016**, *115*, 22–36. [CrossRef]
5. Bennett, R.; Oosterom, P.v.; Lemmen, C.; Koeva, M. Remote Sensing for Land Administration. *Remote Sens.* **2020**, *12*, 2497. [CrossRef]
6. Enemark, S.; Bell, K.C.; Lemmen, C.H.; McLaren, R. *Fit-for-Purpose Land Administration*; International Federation of Surveyors (FIG): Copenhagen, Denmark, 2014.
7. Rossiter, D.G. *Research Concepts & Skills Volume 1: Concepts*; ITC Faculty, The University of Twente: Enschede, The Netherlands, 2011; p. 35.
8. Kaushik, V.; Walsh, C.A. Pragmatism as a Research Paradigm and Its Implications for Social Work Research. *Soc. Sci.* **2019**, *8*, 255. [CrossRef]
9. Asiama, K.; Bennett, R.; Zevenbergen, J. Towards Responsible Consolidation of Customary Lands: A Research Synthesis. *Land* **2019**, *8*, 161. [CrossRef]
10. Bennett, R.; Unger, E.-M.; Lemmen, C.; Dijkstra, P. Land Administration Maintenance: A Review of the Persistent Problem and Emerging Fit-for-Purpose Solutions. *Land* **2021**, *10*, 509. [CrossRef]
11. Badampudi, D.; Wohlin, C.; Petersen, K. Experiences from using snowballing and database searches in systematic literature studies. In Proceedings of the 19th International Conference on Evaluation and Assessment in Software Engineering, Nanjing, China, 27 April 2015; pp. 1–10.
12. Martindale, A. *The Country-Survey-Book: Or Land-Meters Vade-Mecum. Wherein the Principles and Practical Rules for Surveying of Land, Are . . . Delivered . . . With an Appendix, Containing Twelve Problems Touching Compound Interest and Annuities . . . Illustrated with Copper Plates*. Clavel, R., Sawbridge, T., Eds.; 1702. Available online: https://quod.lib.umich.edu/e/eebo/A52120.0001.001?view=toc (accessed on 14 October 2021).
13. Love, J. *Geodæsia: Or, The Art of Surveying and Measuring of Land, Made Easie*. 1731. Available online: https://quod.lib.umich.edu/e/eebo2/A49269.0001.001?view=toc (accessed on 14 October 2021).
14. Breaks, T. *A Complete System of Land-surveying: Both in Theory and Practice: Containing the Best, the Most Accurate, and Commodious Methods of Surveying and Planning of Ground by All the Instruments Now in Use*; Forgotten Books: London, UK, 1771.
15. Leybourn, W. *The Compleat Surveyor: Or, The Whole Art of Surveying of Land: By a New Instrument Lately Invented; as Also by the Plain Table, Circumferentor, the Theodolite as Now Improv'd, Or by the Chain Only. Samuel Ballard . . . , and Aaron Ward . . . , and Tho. Woodward*. 1722. Available online: https://books.googleusercontent.com/books/content?req=AKW5Qac9auCDImyAs4FEX4GemK0WOiz2hn-bElmFE26UpCSPyzG-0DCtQW2aUpJbjG_uFPA7rzE-WK_kLVlZlJRCsHr4lOj5dBpveWEtjjGYkQFbyDxorqr2BsfuD68o8nCi40i6P98lnsI0qYiXvgVzXo79WjljSIF9yKnnSuAO6Rz0sx3EfYkdegH-hHIGHd1gRMgJliz4ce72uAXQuzN4K-YDjlMP6 (accessed on 20 October 2021).
16. Ainslie, J. *Comprehensive Treatise on Land Surveying*; Doig, S., Stirling, A., Eds.; Nabu Press: Charleston, SC, USA, 1812.
17. Thornthwaite, W.H. *A Guide to Photography*; Nabu Press: Charleston, SC, USA, 1845.
18. Lerebours, N.P. *A Treatise on Photography: Containing the Latest Discoveries and Improvements Appertaining to the Daguerreotype*; Longman, Brown, Green & Longmans: London, UK, 1843.
19. Tissandier, G. *A History and Handbook of Photography*; Sampson, Low, Marston, Low & Searle: London, UK, 1877.
20. Reed, H.A. *Photography Applied to Surveying*; Wiley: New York, NY, USA, 1889.
21. Thomson, J. Photography and Exploration. In Proceedings of the Royal Geographical Society and Monthly Record of Geography; Royal Geographical Society (with the Institute of British Geographers). Wiley: New York, NY, USA, 1891; Volume 13, pp. 669–675.
22. Deville, E. *Photographic Surveying: Including the Elements of Descriptive Geometry and Perspective*; Government Printing Bureau: Tokyo, Japan, 1895.
23. Wilson, C. Ordnance survey, methods and processes of the. *RSA J.* **1890**, *39*, 258. [CrossRef]
24. Palmer, H.S. *The Ordnance Survey of the Kingdom: Its Objects, Mode of Execution, History, and Present Condition*; Stanford, E., Ed.; Forgotten Books: London, UK, 1873.
25. James, C. On the Ordnance Survey. *R. United Serv. Inst. J.* **1859**, *3*, 28–38. [CrossRef]
26. Black, C.E. The Survey of India, 1892–1893. *Geogr. J.* **1894**, *4*, 31–33. [CrossRef]

27. Comstock, C.B. *Notes on European Surveys*. 1876. Available online: https://books.google.co.jp/books?hl=en&lr=&id=cvwZUbHoPKMC&oi=fnd&pg=PA1&dq=Notes+on+European+Surveys.+1876.+&ots=WdTJ4jqjTO&sig=Cn7TBbChzZPhQczyUsQ27CGseDM&redir_esc=y#v=onepage&q=Notes%20on%20European%20Surveys.%201876.&f=false (accessed on 18 October 2021).
28. Flemer, J.A. Phototopography. *Science* **1895**, *2*, 152–154. [CrossRef] [PubMed]
29. Thompson, F.V. Stereo-photo surveying. *Geogr. J.* **1908**, *31*, 534–549. [CrossRef]
30. Atkinson, K.B. Vivian Thompson (1880–1917): Not only an officer of the Royal Engineers. *Photogramm. Rec.* **1980**, *10*, 5–38. [CrossRef]
31. Gill, D.; Hills, M.; Close, M. Stereo-Photo Surveying: Discussion. *Geogr. J.* **1908**, *31*, 549–551. [CrossRef]
32. Johnston, D. Surveys and Maps. *Bull. Am. Geogr. Soc.* **1909**, *41*, 751–754. [CrossRef]
33. Johnston, D. The Survey and Mapping of New Areas. *Geogr. J.* **1909**, *34*, 423–431. [CrossRef]
34. MacLeod, M.N. Mapping from air photographs. *Geogr. J.* **1919**, *53*, 382–396. [CrossRef]
35. Hinks, A.R. German war maps and survey. *Geogr. J.* **1919**, *53*, 30–40. [CrossRef]
36. Winterbotham, H.S. Geographical work with the Army in France. *Geogr. J.* **1919**, *54*, 12–23. [CrossRef]
37. Winterbotham, H.S. British survey on the Western Front. *Geogr. J.* **1919**, *53*, 253–271. [CrossRef]
38. Whitlock, G.F.; Newcombe, L.C.; Salmon, L.C.; Brock, M.; Holdich, T.; Hinks, M.; Hardy, G.; MacLeod, M.N. Mapping from Air Photographs: Discussion. *Geogr. J.* **1919**, *53*, 396–403. [CrossRef]
39. Newcombe, S.F. The practical limits of aeroplane photography for mapping. *Geogr. J.* **1920**, *56*, 201–206. [CrossRef]
40. Winterbotham, H.S. The Economic Limits of Aeroplane Photography for Mapping, and Its Applicability to Cadastral Plans. *Geogr. J.* **1920**, *56*, 481–483. [CrossRef]
41. Thomas, H.H. Aircraft photography in war and peace. Lecture III. Aeroplane photography in time of peace. *J. R. Soc. Arts* **1920**, *68*, 777–781.
42. Dodds, J.S. The Government Mapping Program in a Map-Minded Age. *Science* **1930**, *71*, 471–474. [CrossRef]
43. Australian Survey Committee. Report on the Need for a Geodetic and Topographical Survey of Australia. *Aust. Surv.* **1929**, *2*, 6–19. [CrossRef]
44. Winterbotham, H.S. General principles of photographic surveying. *Trans. Opt. Soc.* **1925**, *27*, 65. [CrossRef]
45. Dowson, E.M. Further notes on aeroplane photography in the Near East. *Geogr. J.* **1921**, *58*, 359–370. [CrossRef]
46. Bagley, J.W. Concerning aerial photographic mapping: A review. *Geogr. Rev.* **1922**, *12*, 628–635. [CrossRef]
47. Bergen, G.T. Closure to "Bergen on Aeroplane Topographic Surveys". *Trans. Am. Soc. Civ. Eng.* **1927**, *90*, 672–679. [CrossRef]
48. Tuttle, A.S.; Olmsted, F.L.; Green, C.N.; Ripley, T.M. Discussion of "Tuttle on Aerial Surveys for City Planning". *Trans. Am. Soc. Civ. Eng.* **1927**, *91*, 326–331. [CrossRef]
49. Burchall, P.R. An investigation of the possibilities attaching to aerial co-operation with survey, map-making and exploring expeditions. *R. United Serv. Institution. J.* **1922**, *67*, 112–127. [CrossRef]
50. Durward, J. Air Photography Surveys. *Aeronaut. J.* **1930**, *34*, 344–358. [CrossRef]
51. Winterbotham, H.S. The Surveys of Canada. *Geogr. J.* **1926**, *67*, 403–416. [CrossRef]
52. Fiske, H.C.; Davis, A.P.; Faison, H.R.; Matthes, G.H. Discussion of "Fiske on Aeroplane Topographic Surveys". *Trans. Am. Soc. Civ. Eng.* **1927**, *90*, 656–672. [CrossRef]
53. Spender, M. The New Photographic Survey of Switzerland. *Geogr. J.* **1932**, *79*, 383–397. [CrossRef]
54. Ripley, T.M.; Reading, O.S.; Stewart, L.O.; Peters, F.H.; Crosson, W.H.; Johns, D.F.; Nelles, D.H.; Ballester, R.E.; Pendleton, T.P.; Lemberger, O.; et al. Ripley on Stereo-Topographic Mapping. *Trans. Am. Soc. Civ. Eng.* **1933**, *98*, 795–822. [CrossRef]
55. Le Divelec, G.P. Aerophotogrammetry applied to the survey of large areas at mean scale. *Photogrammetria* **1950**, *7*, 40–43. [CrossRef]
56. Anderson, D.J. Aerial surveying. (Includes bibliography, photographs and plates). *Sel. Eng. Pap.* **1932**, *1*. [CrossRef]
57. Wolff, N. Air Survey and Colonial Cadastral Mapping. *Emp. Surv. Rev.* **1938**, *4*, 281–290. [CrossRef]
58. Salmon, F.J. Cadastral Air Survey. *Emp. Surv. Rev.* **1938**, *4*, 334–338. [CrossRef]
59. Eden, J.A. Air Survey and the Photograph. *Emp. Surv. Rev.* **1933**, *2*, 105–108. [CrossRef]
60. Winterbotham, H.S. Mapping of the colonial empire. *Scott. Geogr. Mag.* **1936**, *52*, 289–299. [CrossRef]
61. HLC; JEEC; EMD. Conference of empire survey officers 1931: Report of proceedings. *Emp. Surv. Rev.* **1933**, *2*, 108–120. [CrossRef]
62. Hinks, A.R. The Fifth International Congress of Photogrammetry, Rome, 1938. *Geogr. J.* **1939**, 240–246.
63. Birdseye, C.H. Stereoscopic phototopographic mapping. *Ann. Assoc. Am. Geogr.* **1940**, *30*, 1–24. [CrossRef]
64. Follet, F.W. Aerial photography and its application to surveying. *Aust. Surv.* **1938**, *7*, 37–41. [CrossRef]
65. None. New Zealand, department of lands and survey, annual report on surveys, 1943–1944. *Emp. Surv. Rev.* **1945**, *8*, 115–118.
66. None. Report of the survey of jamaica for 1941–42. *Emp. Surv. Rev.* **1943**, *7*, 168–182.
67. Vance, T.A. Mapping a Continent. *Aust. Surv.* **1940**, *8*, 148–156. [CrossRef]
68. Marschner, F.J. Maps and a Mapping Program for the United States. *Ann. Assoc. Am. Geogr.* **1943**, *33*, 199–219. [CrossRef]
69. None. Military Surveys. *Emp. Surv. Rev.* **1941**, *6*, 96–101.
70. Jones, S.B. The description of international boundaries. *Ann. Assoc. Am. Geogr.* **1943**, *33*, 99–117. [CrossRef]
71. Winterbotham, H.S. The international boundaries of Europe. *Emp. Surv. Rev.* **1945**, *8*, 133–137. [CrossRef]
72. Cheetham, G. The post-war programme of the ordnance survey of Great Britain. *Emp. Surv. Rev.* **1945**, *8*, 93–102. [CrossRef]

73. El-Ricaby, A.; Velmonte, J.E.; Costa, A.; Dantwala, M.L.; Romero, C. First World Land Tenure Problems Conference and Report of Its Steering Committee. *Land Econ.* **1952**, *28*, 75–81. [CrossRef]
74. Thome, J.R. The process of land reform in Latin America. *Wis. L. Rev.* **1968**, *9*. Available online: https://heinonline.org/HOL/LandingPage?handle=hein.journals/wlr1968amp;div=10amp;id=amp;page= (accessed on 18 October 2021).
75. Dowson, E.M. Direct use of air photographs for cadastral purposes in Zanzibar. *Emp. Surv. Rev.* **1947**, *9*, 2–14. [CrossRef]
76. Smith, W.P.; Whittaker, B.B. Photogrammetry and land tenure surveys with particular reference to Uganda. *Photogramm. Rec.* **1959**, *3*, 42–54. [CrossRef]
77. Menzies, G.H. Monthly Notes of the Astronomical Society of South Africa. *Land Surv.* **1950**, *9*, 24.
78. Adams, L.P. The Computation of Aerial Triangulation for the Control of Cadastral Mapping in High Density Agricultural Areas. Ph.D. Thesis, University of East Africa, Kampala, Uganda, 1969.
79. Park, B.C. Use of Photo Mosaics as a Base for Range Resource Iventory in the Hashemite Kingdom of the Jordan. *Rangel. Ecol. Manag. /J. Range Manag. Arch.* **1955**, *8*, 257–260.
80. Van Zandt, F.K. A Photogrammetric Cadastral Survey in Utah. *Photogramm. Eng.* **1959**, *23*, 493.
81. Loelkes, G.L., Jr. Orthophotography as a Data Base for Land Descriptions. *Can. Surv.* **1969**, *23*, 54–60. [CrossRef]
82. Steiner, D. Use of air photographs for interpreting and mapping rural land use in the United States. *Photogrammetria* **1965**, *20*, 65–80. [CrossRef]
83. McVay, D.M. Cadastral Surveys by Photogrammetry. Highway Research Record. 1967. Available online: https://onlinepubs.trb.org/Onlinepubs/hrr/1967/201/201-004.pdf (accessed on 18 October 2021).
84. Andrews, G.S. Some statutory aspects in cadastral use of photogrammetry. *Can. Surv.* **1960**, *15*, 309–316. [CrossRef]
85. Slessor, D.R. Use of Photogrammetry on a Legal Survey. *Can. Surv.* **1959**, *14*, 330–336. [CrossRef]
86. Fitchett, D.A. Cadastral Systems on the Northern Coast of Peru: Some Problems and Proposals. *J. Inter-Am. Stud.* **1964**, *6*, 537–547. [CrossRef]
87. Osterhoudt, F. Land Titles in Northeast Brazil; The Use of Aerial Photography. *Land Econ.* **1965**, *41*, 387–392. [CrossRef]
88. Marzan, G.T.; Umadhay, G.; Jimenez, T.C. Philippine-Numerical Photogrammetric Cadastre. *Photogramm. Eng.* **1964**, *30*, 278–283.
89. Oshima, T. Photogrammetry on Japan National Report of Japan. *J. Jpn. Soc. Photogramm.* **1973**, *11*, 61–77. [CrossRef]
90. Koffman, L.A. Photogrammetry for Land Reform, Vietnam. *Mil. Eng.* **1970**, *62*, 188–191.
91. Eekhout, L. Photogrammetry and the cadastral system: A paper presented to the 9th Survey Congress, Perth, April, 1966. *Aust. Surv.* **1966**, *21*, 909–924. [CrossRef]
92. Lee, B.J. Application of photogrammetry to cadastral surveying. *Aust. Surv.* **1965**, *20*, 515–518. [CrossRef]
93. Rassaby, H.S. Some applications of photogrammetry in engineering and cadastral surveying in New South Wales. *Cartography* **1960**, *3*, 145–151. [CrossRef]
94. Whitmore, G.D. Fifty Years in Surveying–Mapping–and the Future. *J. Surv. Mapp. Div.* **1969**, *95*, 143–150. [CrossRef]
95. Hart, C.A. Air Survey: The Modern Aspect. *Geogr. J.* **1946**, *108*, 179–198. [CrossRef]
96. Van der Weele, A.J. Graphical or numerical photogrammetry. *Photogrammetria* **1959**, *16*, 90–96. [CrossRef]
97. Petrie, G. The President's Prize Essay. *Photogramm. Rec.* **1959**, *3*, 125–138. [CrossRef]
98. Holden, G.J. Integrated surveys and large scale mapping. *Aust. Surv.* **1970**, *23*, 15–20. [CrossRef]
99. Basye, A.; Ul, A. *Procedures and Standards for a Multipurpose Cadastre. By the Panel on a Multipurpose Cadastre, Committee on Geodesy*; Commission on Physical Sciences, Mathematics, and Resources, National Research Council; National Academy Press: Washington, DC, USA, 1983.
100. Bonacci, F. Problems in Property Surveys and Right-of-Way Maps. *J. Surv. Mapp. Div.* **1963**, *89*, 91–112. [CrossRef]
101. Schermerhorn, W. Planning in modern aerial survey. *Photogrammetria* **1960**, *17*, 7–17. [CrossRef]
102. Weatherhead, T.D. The Application of Air Survey to the Economic Development of a Country. *Aeronaut. J.* **1955**, *59*, 682–689. [CrossRef]
103. Robertson, V.C. Aerial photography and proper land utilisation. *Photogramm. Rec.* **1955**, *1*, 5–12. [CrossRef]
104. Biesheuvel, H. Maps and land use. *Emp. Surv. Rev.* **1956**, *13*, 342–353. [CrossRef]
105. Einevoll, O. Land classification maps of areas basic to agricultural production. *Nor. J. Geogr.* **1968**, *22*, 4. [CrossRef]
106. EMD; VLOS. A bibliography of cadastral survey and land records. *Emp. Surv. Rev.* **1946**, *8*, 210–214. [CrossRef]
107. Dowson, E.M.; Sheppard, V.L. Evolution of land records. *Emp. Surv. Rev.* **1948**, *9*, 295–311. [CrossRef]
108. HBT; BW; EAM; PNR. Land registration. *Emp. Surv. Rev.* **1953**, *12*, 87–95. [CrossRef]
109. Schermerhorn, W.; Witt, G.F. Photogrammetry for cadastral survey. *Photogrammetria* **1953**, *10*, 45–57. [CrossRef]
110. Hart, C.A. Modern Influences on the University Aspect of Professional Training in Surveying. *Emp. Surv. Rev.* **1948**, *9*, 282–295. [CrossRef]
111. Ray, P.N. Surveying instruction at the university. *Emp. Surv. Rev.* **1953**, *12*, 104–110. [CrossRef]
112. Angus-Leppan, P.V. University education and the modern surveyor: A paper read at the Annual Congress of the Institution of Surveyors of Australia, Perth, April, 1966. *Aust. Surv.* **1966**, *21*, 833–846. [CrossRef]
113. Abrams, M.M. Photogrammetry and Highway Law. *J. Surv. Mapp. Div.* **1964**, *90*, 153–168. [CrossRef]
114. Irving, G.C. Photogrammetric techniques in land title boundary surveys: An address given to the NSW Division on 13th November, 1959. *Aust. Surv.* **1960**, *18*, 154–160. [CrossRef]
115. Thompson, E.H. The prospect for British photogrammetry. *Photogramm. Rec.* **1958**, *2*, 355–362. [CrossRef]

116. Ovington, J.J. Photogrammetry and the private surveyor: A paper given to the NSW Division by JJ Ovington, Assoc. IS Aust., MAIC on 10th July, 1959. *Aust. Surv.* **1960**, *18*, 53–57. [CrossRef]
117. Bress, D.L. Computers and Cartography. *Computer* **1972**, *5*, 44–47. [CrossRef]
118. Dale, P.F. A Systems view of the Cadastre. *Surv. Rev.* **1979**, *25*, 28–32. [CrossRef]
119. Hardy, E.E.; Anderson, J.R. A Land Use Classification System for Use with Remote-Sensor Data. Available online: https://pubs.usgs.gov/pp/0964/report.pdf (accessed on 18 October 2021).
120. Mullens, R.H.; Senger, L.W.; Thrower, N.J.; Walton, K.J. *Satellite Photography as a Geographic Tool for Land Use Mapping of the Southwestern United States Technical Report, 1 July 1968–31 January 1970*; United States Department of Interior, Geological Survey, for NASA: Washington, DC, USA, 1970.
121. Kio, P.R. Developing Countries and the new science of remote sensing. *Commonw. For. Rev.* **1974**, *53*, 137–145.
122. Torbert, G.B.; Woll, A.M. Remote sensing on Indian and public lands. In NASA. Manned Spacecraft Center 4 th Ann. Earth Resources Program Rev. 1972; 2. Available online: https://ntrs.nasa.gov/citations/19720021712 (accessed on 20 October 2021).
123. Lambert, B.P. The impact of satellites on mapping. *Aust. Surv.* **1973**, *25*, 303–315. [CrossRef]
124. Kellie, A.C.; AC, K. Evaluation of Remote Sensing Imagery for Cadastral Mapping. Available online: http://pascal-francis.inist.fr/vibad/index.php?action=getRecordDetail&idt=PASCAL7930209581 (accessed on 18 October 2021).
125. McLaughlin, J. The Cadastral Surveying Challenge. *Can. Surv.* **1975**, *29*, 131–136. [CrossRef]
126. Harley, I.A. The determination of XYZ coordinates using numerical photogrammetry. *Aust. Surv.* **1973**, *25*, 89–108. [CrossRef]
127. McLaughlin, J.D. *The Nature, Function and Design Concepts of Multi-Purpose Cadastres*; The University of Wisconsin-Madison: Madison, WI, USA, 1975.
128. Cook, R.N. Land Data Systems: The Next Steps. *U. Cin. L. Rev.* **1974**, *43*, 527.
129. Smith, W. *The Presentation of Spatial Information. InNatural Resources Forum*; Blackwell Publishing Ltd.: Oxford, UK, 1977; Volume 1, pp. 203–213.
130. Braasch, H.W. The arrangement of numerical cadastral data in a modern cadastre of land holdings. *Can. Surv.* **1975**, *29*, 39–48. [CrossRef]
131. Gilliam, J.J. Aerial Photography and Related Products | Aids in Expediting the Construction and Development of Urban Land-Use Maps. Available online: https://scholarworks.umt.edu/cgi/viewcontent.cgi?article=2501&context=etd (accessed on 19 October 2021).
132. Blachut, T.J.; Chrzanowski, A.; Saastamoinen, J.H. Use of Photogrammetry in Urban Areas. In *Urban Surveying and Mapping*; Springer: New York, NY, USA, 1979; pp. 246–329.
133. Jaksic, Z. Photogrammetric Data in Urban Information Systems. *Can. Surv.* **1972**, *26*, 558–566. [CrossRef]
134. Weissmann, K. Photogrammetry applied to cadastral survey in Switzerland. *Photogramm. Rec.* **1971**, *7*, 5–15. [CrossRef]
135. Bonnell, C. Photomapping and Its Application to Legal Surveys. *Can. Surv.* **1977**, *31*, 331–346. [CrossRef]
136. Leatherdale, J.; Kennedy, R. Mapping Arabia. *Geogr. J.* **1975**, *141*, 240–251. [CrossRef]
137. Blachut, T.J. Winter Photographs in Cadastral Surveying (A Suggestion). *Can. Surv.* **1971**, *25*, 603–612. [CrossRef]
138. Lafferty, M.E. Accuracy/Costs with Analytics. *Photogramm. Eng.* **1973**, *39*, 507–514.
139. Dale, P.F. Cadastres and Cadastral maps. *Cartogr. J.* **1977**, *14*, 44–48. [CrossRef]
140. Barrie, J.K. Land registration and boundary surveys. *Aust. Surv.* **1977**, *28*, 256–262. [CrossRef]
141. Forster, B.C. An introduction to modern Remote Sensing techniques and their implication for surveying practice. *Aust. Surv.* **1989**, *34*, 763–779. [CrossRef]
142. Andersson, U.; Rystedt, B. Scandinavian Activities in the LIS/GIS Area. *Photogramm. Eng. Remote Sens.* **1988**, *54*, 201–204.
143. Lodwick, G.D.; Paine, S.H. Satellite remote sensing in surveying present opportunities, future possibilities. *Can. Surv.* **1986**, *40*, 315–326. [CrossRef]
144. Zhou, Q. A method for integrating remote sensing and geographic information systems. *Photogramm. Eng. Remote Sens.* **1989**, *55*, 591–596.
145. Cooperative, G.I.; Collins, F. The unique qualities of a geographic information system: A commentary. *Photogramm. Eng. Remote Sens.* **1988**, *54*, 1547–1549.
146. Dale, P.F.; McLaughlin, J.D. *Land Information Management*; Oxford University Press: Oxford, UK, 1988.
147. Muzakidis, P.D. *Photogrammetric Mapping for Cadastral Land and Information Systems*; University of London, University College London: London, UK, 1990.
148. Gagnon, R. Cadastral plotting by similarity transformation. *CISM J.* **1988**, *42*, 121–125. [CrossRef]
149. Visser, J. The European Organisation for Experimental Photogrammetric Research (OEEPE). *Photogramm. Rec.* **1982**, *10*, 655–668. [CrossRef]
150. Walker, A.S. A review of map revision by photogrammetry. *Photogramm. Rec.* **1984**, *11*, 395–405. [CrossRef]
151. Or, K. The Prototype Land Information System for the Cadastral Pilot Project in Colombia. In *Auto-Carto Six: Automated Cartography: International Perspectives on Achievements and Challenges: Proceedings of the Sixth International Symposium on Automated Cartography, October 16–21, National Capital Region of Canada, Canada 1983*; Steering Committee for the Sixth International Symposium on Automated Cartography= Comité organisateur pour le sixième Symposium sur la Cartographie Automatisée; American Congress on Surveying and Mapping: Frederick, MD, USA, 1983; Volume 1, p. 128.
152. Clerici, E.; Walker, E. Photogrammetric cadastral mapping in rural Taïwan. *Aust. Surv.* **1987**, *33*, 469–479. [CrossRef]

153. Cremont, D. Application of Computer Technology in Processing of Cadastral Surveying and Mapping Data. *Can. Surv.* **1980**, *34*, 21–40. [CrossRef]
154. Smith, G.L.; Nisbet, K.A. Geodetic network densification by analytical aerial triangulation. *Aust. Surv.* **1985**, *32*, 644–657. [CrossRef]
155. Ziemann, H. High Accuracy Photogrammetric Determinations Using Image Deformation Corrections. *Can. Surv.* **1980**, *34*, 65–74. [CrossRef]
156. Karns, D.O. Photogrammetric cadastral surveys and GLO corner restoration. *Photogramm. Eng. Remote Sens.* **1981**, *47*, 193–198.
157. Blachut, T.J. Cadastre for developing countries based on orthophoto techniques. *Can. Surv.* **1985**, *39*, 31–43. [CrossRef]
158. Williamson, I.P. Cadastral survey techniques in developing countries—with particular reference to Thailand. *Aust. Surv.* **1983**, *31*, 496–512. [CrossRef]
159. Bujakiewicz, A. Simple Photogrammetric Methods for Registration of rural land in African countries. In *Technical Commission IV: Cartographic and Data Bank Application of Photogrammetry and Remote Sensing*; ISPRS: Kyoto, Japan, 1–10 July 1988.
160. Van Loenen, B. *Land Tenure in Zambia*. 1999. Available online: http://citeseerx.ist.psu.edu/viewdoc/download?doi=10.1.1.460.7561&rep=rep1&type=pdf (accessed on 18 October 2021).
161. Williamson, I.P. The cadastral survey requirements of developing countries in the pacific regionwith particular reference to fiji. *Surv. Rev.* **1982**, *26*, 355–366. [CrossRef]
162. Hannigan, B.J. The role of surveying in society—has it changed? *Aust. Surv.* **1990**, *35*, 209–228. [CrossRef]
163. Gracie, G. Restructuring Photogrammetry and Remote Sensing Education for the Future. *Can. Surv.* **1985**, *39*, 338–344. [CrossRef]
164. Forster, B.C.; Williamson, I.P. Past and Future Trends of Surveying Education in Australia. *Can. Surv.* **1985**, *39*, 427–435. [CrossRef]
165. Bédard, Y.; Gagnon, P.; Gagnon, P.A. Modernizing surveying and mapping education: The programs in geomatics at laval university. *CISM J.* **1988**, *42*, 105–114. [CrossRef]
166. Dale, P.F. Evolution and developments in cadastral studies. *Can. Surv.* **1985**, *39*, 353–362. [CrossRef]
167. Paulsson, B.; Mundial, B. *Urban Applications of Satellite Remote Sensing and GIS Analysis*; World Bank: Washington, DC, USA, 1992.
168. Dale, P. Is technology a blessing or a curse in land administration. In Proceedings of the UN-FIG Conference on Land Tenure and Cadastral Infrastructure for Sustainable Development, Melbourne, Australia, 25 October 1999; pp. 25–27.
169. Jensen, J.R.; Cowen, D.C. Remote sensing of urban/suburban infrastructure and socio-economic attributes. *Photogramm. Eng. Remote Sens.* **1999**, *65*, 611–622.
170. Rao, M.; Krishnamurthy, J.; Raj, U.; Patan, S.K.; Ragavaswamy, V.; Jayaraman, V. Classification of high resolution satellite imagery—The experience from IRS-1C/1D. In Proceedings of the IAF, International Astronautical Congress 49th, Melbourne, Australia, 2 October 1998.
171. Rao, D.P.; Navalgund, R.R.; Murthy, Y.K. Cadastral applications using IRS-1C data–Some case studies. *Curr. Sci.* **1996**, *70*, 624–628.
172. Das, R.K.; Ghosh, S.; Rajesh, K.; Prithviraj, M. Cadastral map overlaying upon irs-ic. 8ybrid image: A critical analysis. *Geogr. Environ.* **1997**, *2*, 53–58.
173. González, A.R. Horizontal Accuracy Assessment of the New Generation of High Resolution Satellite Imagery for Mapping Purposes. Master's Thesis, Ohio State University, Columbus, OH, USA, 1998.
174. Schmitt, U.; Sulzer, W.; Schardt, M. Analysis of settlement structure by means of high resolution satellite imagery. *Int. Arch. Photogramm. Remote Sens.* **1998**, *32*, 557–561.
175. Leberl, F.; Kalliany, R. Earth Observation Data Services for Users-an Austrian Perspective. In ESRIN EEOS Workshop on Networks. In Proceedings of the ESRIN EEOS Workshop on Networks, Frascati, Italy, 13–15 December 1994; pp. 461–469.
176. Baltsavias, E.P. Digital ortho-images—A powerful tool for the extraction of spatial-and geo-information. *ISPRS J. Photogramm. Remote Sens.* **1996**, *51*, 63–77. [CrossRef]
177. Chagarlamudi, P.; Plunkett, G.W. Mapping applications for low-cost remote sensing and geographic information systems. *Int. J. Remote Sens.* **1993**, *14*, 3181–3190. [CrossRef]
178. Konecny, G. International Technical Cooperation in the Geoinformatics Field. In Proceedings of the EARSeL Workshop on Remote Sensing in the Developing Countries, Gent, Belgium, 19 September 2000.
179. Leka, E.; Gjika, M. Aerial Photography and Parcel Mapping for Immovable Property Registration in Albania. Available online: https://minds.wisconsin.edu/bitstream/handle/1793/69877/wp10.pdf?sequence=1 (accessed on 18 October 2021).
180. Holstein, L. Towards best practice from World Bank experience in land titling and registration. In Proceedings of the International Conference on Land Tenure and Administration, November 1996; pp. 1–26. Available online: https://citeseerx.ist.psu.edu/viewdoc/download?doi=10.1.1.472.1914&rep=rep1&type=pdf (accessed on 18 October 2021).
181. Fourie, C.; Nino-Fluck, O. Cadastre and land information systems for decision makers in the developing world. *Geomatica* **2000**, *54*, 335–343.
182. Anderson, P.S. Mapping land rights in Mozambique. *Photogramm. Eng. Remote Sens.* **2000**, *66*, 769–776.
183. Christensen, S.F.; Werner, W.; Højgaard, P.D. *Innovative land surveying and land registration in Namibia*; University College London: London, UK, 1999.
184. Al-garni, A.M. Urban photogrammetric data base for multi-purpose cadastral-based information systems: The Riyadh city case. *ISPRS J. Photogramm. Remote Sens.* **1996**, *51*, 28–38. [CrossRef]
185. Harcombe, P.R.; Williamson, I.P. A cadastral model for low value lands: The NSW western lands experience. In Proceedings of the FIG XXI International Congress, Brighton, UK, 19–25 July 1998.

186. Boatto, L.; Consorti, V.; Del Buono, M.; Di Zenzo, S.; Eramo, V.; Esposito, A.; Melcarne, F.; Meucci, M.; Morelli, A.; Mosciatti, M.; et al. An interpretation system for land register maps. *Computer* **1992**, *25*, 25–33. [CrossRef]
187. Mohamed, M.A.; Ventura, S.J. Use of geomatics for mapping and documenting indigenous tenure systems. *Soc. Nat. Resour.* **2000**, *13*, 223–236. [CrossRef]
188. Ehlers, M. *Remote Sensing and Geographic Information Systems: Image-Integrated Geographic Information Systems. In Geographic Information Systems (GIS) and Mapping—Practices and Standards 1992 Jan*; ASTM International: West Conshohocken, PA, USA, 1992.
189. Onsrud, H.J. Integrated cadastral technologies field system (ictfs) for documenting title and boundary evidence. *Geomatica* **1998**, *52*, 25–35.
190. Mason, S.O.; Fraser, C.S. Image sources for informal settlement management. *Photogramm. Rec.* **1998**, *16*, 313–330. [CrossRef]
191. Bartl, R.; Petrou, M.; Christmas, W.J.; Palmer, P.L. Automatic registration of cadastral maps and Landsat TM images. In *Image and Signal Processing for Remote Sensing III*; International Society for Optics and Photonics: Bellingham, WA, USA, 1996; Volume 2955, pp. 9–20.
192. Pinz, A.J.; Prantl, M. Active fusion for remote sensing image understanding. In *Image and Signal Processing for Remote Sensing II*; International Society for Optics and Photonics: Bellingham, WA, USA, 1995; Volume 2579, pp. 67–77.
193. Okpala, D.C. Land survey and parcel identification: Data for effective land management. *Land Use Policy* **1992**, *9*, 92–98. [CrossRef]
194. Zevenbergen, J.; De Vries, W.; Bennett, R.M. (Eds.) *Advances in Responsible Land Administration*; CRC Press: Boca Raton, FL, USA, 2015.
195. Polat, Z.A. Evolution and future trends in global research on cadastre: A bibliometric analysis. *GeoJournal* **2019**, *84*, 1121–1134. [CrossRef]
196. Reis, S.; Torun, A.T.; Bilgilioğlu, B.B. Investigation of Availability of Remote Sensed Data in Cadastral Works. In *Cadastre: Geo-Information Innovations in Land Administration*; Springer: Cham, Switzerland, 2017; pp. 63–76.
197. Jarica, C.C. *Commercialization of High-Resolution Earth Observation Satellite Remote Sensing*; Florida Atlantic University: Boca Raton, FL, USA, 1996.
198. Şahin, N.; Bakıcı, S.; Erkek, B. An Investigation on High Resolution IKONOS Satellite Images for Cadastral Applications. Available online: https://cartesia.org/geodoc/isprs2004/comm7/papers/222.pdf (accessed on 18 October 2021).
199. Fraser, C.; Tshering, D.; Grün, A. Satellite Mapping in Bhutan. *GIM Int.* **2008**, *22*, 18–24.
200. Ali, Z. Assessing Usefulness of High-Resolution Satellite Imagery (HRSI) in GIS-based Cadastral Land Information System. *J. Settl. Spat. Plan.* **2012**, *3*, 93–96.
201. Ali, Z.; Tuladhar, A.; Zevenbergen, J. An integrated approach for updating cadastral maps in Pakistan using satellite remote sensing data. *Int. J. Appl. Earth Obs. Geoinf.* **2012**, *18*, 386–398. [CrossRef]
202. Rao, S.S.; Sharma, J.R.; Rajasekhar, S.S.; Rao, D.S.; Arepalli, A.; Arora, V.; Singh, R.P.; Kanaparthi, M. Assessing usefulness of High-Resolution Satellite Imagery (HRSI) for re-survey of cadastral maps. *ISPRS Ann. Photogramm. Remote Sens. Spat. Inf. Sci.* **2014**, *2*, 133–143. [CrossRef]
203. Kumar, K.E.M.; Singh, S.; Attri, P.; Kumar, R.; Kumar, A.; Hooda, R.S.; Sapra, R.K.; Garg, V.; Kumar, V.; Sarika; et al. GIS based Cadastral level Forest Information System using World View-II data in Bir Hisar (Haryana). *ISPRS Int. Arch. Photogramm. Remote Sens. Spat. Inf. Sci.* **2014**, *XL-8*, 605–612. [CrossRef]
204. Sengupta, A.; Lemmen, C.; Devos, W.; Bandyopadhyay, D.; Van der Veen, A. Constructing a seamless digital cadastral database using colonial cadastral maps and VHR imagery–an Indian perspective. *Surv. Rev.* **2016**, *48*, 258–268. [CrossRef]
205. Panday, U.S.; Chhatkuli, R.R.; Joshi, J.R.; Deuja, J.; Antonio, D.; Enemark, S. Securing Land Rights for All through Fit-for-Purpose Land Administration Approach: The Case of Nepal. *Land* **2021**, *10*, 744. [CrossRef]
206. Andri, H.; Sella, N.; Alfita, P.; Putri, R.; Winna, P.P.; Lasmi, R.; Ratri, W.; Rani, A. Incremental Improvement of High resolution Satellite Imagery For Participatory Mapping in Land Registration. *IOP Conf. Ser. Earth Environ. Sci.* **2019**, *280*, 012035. [CrossRef]
207. Asiama, K.O.; Bennett, R.M.; Zevenbergen, J.A. Participatory Land Administration on Customary Lands: A Practical VGI Experiment in Nanton, Ghana. *ISPRS Int. J. Geo-Inf.* **2017**, *6*, 186. [CrossRef]
208. Balas, M.; Carrilho, J.; Lemmen, C. The Fit for Purpose Land Administration Approach-Connecting People, Processes and Technology in Mozambique. *Land* **2021**, *10*, 818. [CrossRef]
209. Ondulo, J.D.; Kalande, W. High spatial resolution satellite imagery for Pid improvement in Kenya. In Proceedings of the FIG Congress: Shaping the Change, Munich, Germany, 8 October 2006; pp. 8–13.
210. Lengoiboni, M.; Bregt, A.; van der Molen, P. Pastoralism within land administration in Kenya—The missing link. *Land Use Policy* **2010**, *27*, 579–588. [CrossRef]
211. Hassan, N.D.; Noori, A.M.; Hasan, S.F.; Shareef, M.A.; Ajaj, Q.M. Cadastral Mapping Accuracy Assessment Using Various Surveying Techniques and High-Resolution Satellites Images. In Proceedings of the 2019 2nd International Conference on Electrical, Communication, Computer, Power and Control Engi-neering (ICECCPCE), Mosal, Iraq, 13–14 February 2019; pp. 182–187.
212. Jones, B.; Lemmen, C.H.; Molendijk, M. Low Cost, Post Conflict Cadastre with Modern Technology. In Proceedings of the Responsible Land Governance, Towards and Evidence Based Approach, Washington, DC, USA, 20–24 March 2017; pp. 20–24.
213. Chen, J.; Dowman, I.; Li, S.; Li, Z.; Madden, M.; Mills, J.; Paparoditis, N.; Rottensteiner, F.; Sester, M.; Toth, C.; et al. Information from imagery: ISPRS scientific vision and research agenda. *ISPRS J. Photogramm. Remote Sens.* **2016**, *115*, 3–21. [CrossRef]
214. Cramer, M. *Digital Camera Calibration*; EuroSDR no 55; Gopher Amsterdam: Amsterdam, The Netherlands, 2011; 262p.

215. Remondino, F.; Fraser, C. Digital camera calibration methods: Considerations and comparisons. In: The International Archives of Photogrammetry. *Remote Sens. Spat. Inf. Sci* **2006**, *XXXVI-5*, 266–272.
216. Lowe, D.G. Distinctive Image Features from Scale-Invariant Keypoints. *Int. J. Comput. Vis.* **2004**, *60*, 91–110. [CrossRef]
217. Bay, H.; Ess, A.; Tuytelaars, T.; van Goal, L. SURF: Speeded up robust features. *Comput. Vis. Image Und.* **2008**, *110*, 346–359. [CrossRef]
218. Fischler, M.A.; Bolles, R.C. Random Sample Consensus: A Paradigm for Model Fitting with Applications to Image Analysis and Automated Cartography. *Commun. ACM* **1987**, 726–740. [CrossRef]
219. Vosselman, G. Advanced point cloud processing. In *Photogrammetric Week*; Wichmann Heidelberg, Germany, 2009; pp. 137–146.
220. Mountrakis, G.; Im, J.; Ogole, C. Support vector machines in remote sensing: A review. *ISPRS J. Photogramm. Remote Sens.* **2011**, *66*, 247–259. [CrossRef]
221. Gislason, P.O.; Benediktsson, J.A.; Sveinsson, J.R. Random Forests for land cover classification. *Pattern Recognit. Lett.* **2006**, *27*, 294–300. [CrossRef]
222. Lu, D.; Li, G.; Moran, E. Current situation and needs of change detection techniques. *Int. J. Image Data Fusion* **2014**, *5*, 13–38. [CrossRef]
223. Srinivas, P.; Venkataraman, V.R.; Jayalakshmi, I. Digital Aerial Orthobase for Cadastral Mapping. *J. Indian Soc. Remote Sens.* **2011**, *40*, 497–506. [CrossRef]
224. Ahn, K.; Song, Y. Digital Photogrammetry for Land Registration in Developing Countries Digital Photogrammetry for Land Registration in Developing Countries. In Proceedings of the FIG Working Week 2011 Bridging the Gap between Cultures, Marrakech, Morocco, 18–22 May 2011.
225. Offei, E.; Lengoiboni, M.; Koeva, M. Compliance with Residential Building Standards in the Context of Customary Land Tenure System in Ghana. *Planext Next Gener. Plan.* **2018**, *6*, 25–45. [CrossRef]
226. Tamrakar, R.M. A Prospect of Digital Airborne Photogrammetry Approach for Cadastral Mapping in Nepal. *J. Geoinformatics Nepal* **2012**, *11*, 1–6. [CrossRef]
227. Harintaka, S.; Susanto, A. Assessment of Low Cost Small Format Aerial Photogrammetry for Cadastral Mapping (Case Study in Klaten Regency, Central Java, Indonesia). In Proceedings of the Spatial Data Serving People. Land Governance and the Environment-Building the Capacity, Hanoi, Vietnam, 19–22 October 2009.
228. Burgos, A.S. *Digital Mapping for Cadastral Purposes*; ASPRS/MAPPS: Sanantonio, TX, USA, 2009.
229. Al-Ruzouq, R.; Dimitrova, P. 2006 Photogrammetric Techniques for Cadastral Map Renewal. In Proceedings of the XXIII FIG Congress, Munich, Germany, 8–13 October 2006.
230. Alkan, M.; Solak, Y. An investigation of 1: 5000 scale photogrammetric data for cadastral mapping uses: A case study of Kastamonu-Taskopru. *Afr. J. Agric. Res.* **2010**, *5*, 2576–2588.
231. Meixner, P.; Leberl, F. From aerial images to a description of real properties—A framework. In *Proceedings of the International Conference on Computer Vision Theory and Applications—Volume 2*; VISAPP: Angers, France; SCITEPRESS—Science and Technology Publications: Setúbal, Portugal, 2010; pp. 283–291.
232. Siriba, D. Positional Accuracy Assessment of a Cadastral Dataset based on the Knowledge of the Process Steps used. In Proceedings of the 12th AGILE Conference on GIScience. Leibniz Universität Hannover, Germany. Available online: https://www.springer.com/gp/book/9783642003172 (accessed on 18 October 2021).
233. Mumbone, M. Innovations in Boundary Mapping: Namibia, Customary Land and UAV's. Master's Thesis, University of Twente, Enschede, The Netherlands, 2015.
234. Meijs, M.G.; Kapitango, D.; Witmer, R. Land Registration using aerial photography in Namibia: Costs and lessons. In Proceedings of the FIG–World Bank Conference on Land Governance in Support of the MDGs: Responding to New Challenges, Washington DC, USA, 9–10 March 2009.
235. Ramadhani, S.A.; Bennett, R.M.; Nex, F.C. Exploring UAV in Indonesian cadastral boundary data acquisition. *Earth Sci. Inform.* **2018**, *11*, 129–146. [CrossRef]
236. Yuwono, B.D.; Suprayogi, A.; Azeriansyah, R.; Nukita, D. UAV Photogrammetry Implementation Based on GNSS CORS UDIP to Enhance Cadastral Surveying and Monitoring Urban Development (Case Study: Ngresep Semarang). *IOP Conf. Ser. Earth Environ. Sci.* **2018**, *165*, 012031. [CrossRef]
237. Aditya, T.; Maria-Unger, E.; Berg, C.V.; Bennett, R.; Saers, P.; Syahid, H.L.; Erwan, D.; Wits, T.; Widjajanti, N.; Santosa, P.B.; et al. Participatory Land Administration in Indonesia: Quality and Usability Assessment. *Land* **2020**, *9*, 79. [CrossRef]
238. Kurczynski, Z.; Bakuła, K.; Karabin, M.; Kowalczyk, M.; Markiewicz, J.S.; Ostrowski, W.; Podlasiak, P.; Zawieska, D. The possibility of using images obtained from the uas in cadastral works. In Proceedings of the International Archives of the Photogrammetry, Remote Sensing & Spatial Information Sciences, Prague, Czech Republic, 2–19 July 2016.
239. Cienciała, A.; Sobolewska-Mikulska, K.; Sobura, S. Credibility of the cadastral data on land use and the methodology for their verification and update. *Land Use Policy* **2021**, *102*, 105204. [CrossRef]
240. Kameri-Mbote, P.; Muriungi, M. Potential contribution of drones to reliability of Kenya's land information system. *Afr. J. Inf. Commun.* **2017**, *20*, 159–169. [CrossRef]
241. Wayumba, R.; Mwangi, P.; Chege, P. Application of unmanned aerial vehicles in improving land registration in Kenya. *Int. J. Res. Eng. Sci.* **2017**, *5*, 5–11.

242. Koeva, M.; Muneza, M.; Gevaert, C.; Gerke, M.; Nex, F. Using UAVs for map creation and updating. A case study in Rwanda. *Surv. Rev.* **2018**, *50*, 312–325. [CrossRef]
243. Stöcker, C.; Ho, S.; Nkerabigwi, P.; Schmidt, C.; Koeva, M.; Bennett, R.; Zevenbergen, J. Unmanned Aerial System Imagery, Land Data and User Needs: A Socio-Technical Assessment in Rwanda. *Remote Sens.* **2019**, *11*, 1035. [CrossRef]
244. Flores, C.C.; Tan, E.; Buntinx, I.; Crompvoets, J.; Stöcker, C.; Zevenbergen, J. Governance assessment of the UAVs implementation in Rwanda under the fit-for-purpose land administration approach. *Land Use Policy* **2020**, *99*, 104725. [CrossRef]
245. Ali, F. Fit-for-Purpose Boundary Mapping and Valuation of Agricultural Land Using UAVs: The Case of a1 Farms in Zimbabwe. Master's Thesis, University of Twente, Enschede, The Netherlands, 2017.
246. Karataş, K.; Altinişik, N.S. The Effect of UAV Usage on Detail Points in Cadastre Update Studies: Çorum-Karaköy Case Study. *Int. J. Environ. Geoinformatics* **2020**, *7*, 140–146. [CrossRef]
247. Koeva, M.; Gasuku, O.; Lengoiboni, M.; Asiama, K.; Bennett, R.M.; Potel, J.; Zevenbergen, J. Remote Sensing for Property Valuation: A Data Source Comparison in Support of Fair Land Taxation in Rwanda. *Remote Sens.* **2021**, *13*, 3563. [CrossRef]
248. Mbarga, T.C. Advantages of a Digital Cadastre Using an Unmanned Aerial Vehicle (UAV) Tool to Support Better Governance and Land Administration in Cameroon: An Exploratory Study. Available online: https://fig.net/resources/proceedings/fig_proceedings/fig2020/papers/ts01e/TS01E_tobie_camille_vivian_et_al_10715.pdf (accessed on 18 October 2021).
249. Stöcker, C.; Bennett, R.; Nex, F.; Gerke, M.; Zevenbergen, J. Review of the Current State of UAV Regulations. *Remote Sens.* **2017**, *9*, 459. [CrossRef]
250. Stoter, J.E.; van Oosterom, P. *3D Cadastre in an International Context: Legal, Organizational, and Technological Aspects*; CRC Press: London, UK, 2006.
251. Van Oosterom, P.; Bennett, R.; Koeva, M.; Lemmen, C. 3D land administration for 3D land uses. *Land Use Policy* **2020**, *98*, 104665. [CrossRef]
252. Filin, S.; Borka, A.; Doytsher, Y. From 2D to 3D Land Parcelation: Fusion of LiDAR Data and Cadastral Maps. *Surv. Land Inf. Sci.* **2008**, *68*, 81–91.
253. Kodors, S.; Ratkevics, A.; Rausis, A.; Buls, J. Building Recognition Using LiDAR and Energy Minimization Approach. *Procedia Comput. Sci.* **2015**, *43*, 109–117. [CrossRef]
254. Kumar, P.; Rahman, A.A.; Buyuksalih, G. Automated Extraction of Buildings from Aerial Lidar Point Cloud and Digital Imaging Datasets for 3D Cadastre—Preliminary Results. In *Cadastre: Geo-Information Innovations in Land Administration*; Springer: Cham, Switzerland, 2017; pp. 159–165.
255. Giannaka, O.; Dimopoulou, E.; Georgopoulos, A. Investigation on the contribution of LiDAR data in 3D cadastre. Paphos, Cyprus. (RSCy2014). 2014, Volume 9229, p. 922905. Available online: https://www.researchgate.net/publication/269320030_Investigation_on_the_contribution_of_LiDAR_data_in_3D_Cadastre (accessed on 18 October 2021).
256. Drobež, P.; Grigillo, D.; Lisec, A.; Fras, M.K. Remote sensing data as a potential source for establishment of the 3D cadastre in Slovenia. *Géod. Vestn.* **2016**, *60*. [CrossRef]
257. Luo, X.; Bennett, R.M.; Koeva, M.; Lemmen, C. Investigating Semi-Automated Cadastral Boundaries Extraction from Airborne Laser Scanned Data. *Land* **2017**, *6*, 60. [CrossRef]
258. Wierzbicki, D.; Matuk, O.; Bielecka, E. Polish Cadastre Modernization with Remotely Extracted Buildings from High-Resolution Aerial Orthoimagery and Airborne LiDAR. *Remote Sens.* **2021**, *13*, 611. [CrossRef]
259. Griffith-Charles, C.; Sutherland, M. 3D cadastres for densely occupied informal situations: Necessity and possibility. *Land Use Policy* **2020**, *98*, 104372. [CrossRef]
260. Lubeck, D. Airborne Dual-band Radar for Cadastre. *Gim Int. Worldw. Mag. Geomat.* **2016**, *30*, 26–27.
261. Rajabifard, A.; Atazadeh, B.; Kalantari, M. *BIM and Urban Land Administration*; CRC Press: London, UK, 2019.
262. Koeva, M.; Nikoohemat, S.; Elberink, S.O.; Morales, J.; Lemmen, C.; Zevenbergen, J. Towards 3D Indoor Cadastre Based on Change Detection from Point Clouds. *Remote Sens.* **2019**, *11*, 1972. [CrossRef]
263. Bieda, A.; Bydłosz, J.; Warchoł, A.; Balawejder, M. Historical Underground Structures as 3D Cadastral Objects. *Remote Sens.* **2020**, *12*, 1547. [CrossRef]
264. Yan, J.; Jaw, S.W.; Soon, K.H.; Wieser, A.; Schrotter, G. Towards an Underground Utilities 3D Data Model for Land Administration. *Remote Sens.* **2019**, *11*, 1957. [CrossRef]
265. Kisa, A.; Ozmus, L.; Erkek, B.; Ates, H.B.; Bakici, S. Oblique photogrammetry and usage on land administration. *ISPRS Int. Arch. Photogramm. Remote Sens. Spat. Inf. Sci.* **2013**, *XL-2/W2*, 161–165. [CrossRef]
266. Lemmens, M.; Lemmen, C.; Wubbe, M. Pictometry: Potentials for land administration. In Proceedings of the 6th FIG Regional Conference, San José, Costa Rica, 12–15 November 2007.
267. Luo, X.; Bennett, R.; Koeva, M.; Lemmen, C.; Quadros, N. Quantifying the Overlap between Cadastral and Visual Boundaries: A Case Study from Vanuatu. *Urban Sci.* **2017**, *1*, 32. [CrossRef]
268. Koeva, M.; Humayun, M.; Timm, C.; Stöcker, C.; Crommelinck, S.; Chipofya, M.; Bennett, R.; Zevenbergen, J. Geospatial Tool and Geocloud Platform Innovations: A Fit-for-Purpose Land Administration Assessment. *Land* **2021**, *10*, 557. [CrossRef]
269. Bennett, R.M.; Pickering, M.; Sargent, J. Transformations, transitions, or tall tales? A global review of the uptake and impact of NoSQL, blockchain, and big data analytics on the land administration sector. *Land Use Policy* **2019**, *83*, 435–448. [CrossRef]
270. Van Oosterom, P.; Groothedde, A.; Lemmen, C.; van der Molen, P.; Uitermark, H. Land administration as a cornerstone in the global spatial information infrastructure. *Int. J. Spat. Data Infrastruct. Res.* **2009**, *4*, 298–331.

271. Roić, M.; Vranić, S.; Stančić, B.; Kliment, T.; Tomić, H. Development of Multipurpose Land Administration Warehouse. In Proceedings of the FIG Working Week, Helsinki, Finland, 29 May–2 June 2017.
272. Lemmen, C.; Van Oosterom, P.; Bennett, R. The land administration domain model. *Land Use Policy* **2015**, *49*, 535–545. [CrossRef]
273. Crommelinck, S.; Bennett, R.; Gerke, M.; Nex, F.; Yang, M.Y.; Vosselman, G. Review of Automatic Feature Extraction from High-Resolution Optical Sensor Data for UAV-Based Cadastral Mapping. *Remote Sens.* **2016**, *8*, 689. [CrossRef]
274. Davidse, J. *Semi-Automatic Detection of Field Boundaries from High-Resolution Satellite Imagery*; Wageningen University: Wageningen, The Netherlands, 2015.
275. Masouleh, M.K.; Sadeghian, S. Deep learning-based method for reconstructing three-dimensional building cadastre models from aerial images. *J. Appl. Remote Sens.* **2019**, *13*, 024508. [CrossRef]
276. Pichel, F. Faster cadastre. *RICS Land J.* **2018**, *2*, 12–13.
277. Wassie, Y.A.; Koeva, M.; Bennett, R.; Lemmen, C. A procedure for semi-automated cadastral boundary feature extraction from high-resolution satellite imagery. *J. Spat. Sci.* **2017**, *63*, 75–92. [CrossRef]
278. Koeva, M.; Bennett, R.; Gerke, M.; Crommelinck, S.; Stöcker, C.; Crompvoets, J.; Ho, S.; Schwering, A.; Chipofya, M.; Schultz, C.; et al. Towards innovative geospatial tools for fit-for-purpose land rights mapping. *ISPRS Int. Arch. Photogramm. Remote Sens. Spat. Inf. Sci.* **2017**, *XLII-2/W7*, 37–43. [CrossRef]
279. Fetai, B.; Oštir, K.; Kosmatin Fras, M.; Lisec, A. Extraction of Visible Boundaries for Cadastral Mapping Based on UAV Imagery. *Remote Sens.* **2019**, *11*, 1510. [CrossRef]
280. Park, S.; Song, A. Discrepancy analysis for detecting candidate parcels requiring update of land category in cadastral map using hyperspectral UAV Images: A case study in Jeonju, South Korea. *Remote Sens.* **2020**, *12*, 354. [CrossRef]
281. Nyandwi, E.; Koeva, M.; Kohli, D.; Bennett, R. Comparing Human Versus Machine-Driven Cadastral Boundary Feature Extraction. *Remote Sens.* **2019**, *11*, 1662. [CrossRef]
282. Xia, X.; Persello, C.; Koeva, M. Deep Fully Convolutional Networks for Cadastral Boundary Detection from UAV Images. *Remote Sens.* **2019**, *11*, 1725. [CrossRef]
283. Zevenbergen, J.; Augustinus, C.; Antonio, D.; Bennett, R. Pro-poor land administration: Principles for recording the land rights of the underrepresented. *Land Use Policy* **2013**, *31*, 595–604. [CrossRef]
284. Enemark, S.; McLaren, R.; Lemmen, C. Fit-for-Purpose Land Administration—Providing Secure Land Rights at Scale. *Land* **2021**, *10*, 972. [CrossRef]
285. Törhönen, M.-P. Developing land administration in Cambodia. *Comput. Environ. Urban Syst.* **2001**, *25*, 407–428. [CrossRef]
286. .Burns, T. *International experience with land administration projects: A framework for monitoring of pilots. In National Workshop on Land Policies and Administration for Accelerated Growth and Poverty Reduction in the 21st Century*; World Bank: Washington, DC, USA; Available online: https://www.researchgate.net/publication/228744867_International_experience_with_land_administration_projects_A_framework_for_monitoring_of_pilots (accessed on 18 October 2021).
287. Arruñada, B. Evolving practice in land demarcation. *Land Use Policy* **2018**, *77*, 661–675. [CrossRef]
288. Mourafetis, G.; Apostolopoulos, K.; Potsiou, C.; Ioannidis, C. Enhancing cadastral surveys by facilitating the participation of owners. *Surv. Rev.* **2015**, *47*, 316–324. [CrossRef]
289. Chigbu, U.; Bendzko, T.; Mabakeng, M.; Kuusaana, E.; Tutu, D. Fit-for-Purpose Land Administration from Theory to Practice: Three Demonstrative Case Studies of Local Land Administration Initiatives in Africa. *Land* **2021**, *10*, 476. [CrossRef]
290. Koeva, M.; Stöcker, C.; Crommelinck, S.; Ho, S.; Chipofya, M.; Sahib, J.; Bennett, R.; Zevenbergen, J.; Vosselman, G.; Lemmen, C.; et al. Innovative Remote Sensing Methodologies for Kenyan Land Tenure Mapping. *Remote Sens.* **2020**, *12*, 273. [CrossRef]
291. Williams-Wynn, C. Applying the Fit-for-Purpose Land Administration Concept to South Africa. *Land* **2021**, *10*, 602. [CrossRef]
292. Zein, T.A. Fit-For-Purpose Land Administration: An implementation model for cadastre and land administration systems. In Proceedings of the Land and Poverty Conference, Washington, DC, USA, 14–18 March 2016.
293. Hackman-Antwi, R.; Bennett, R.; de Vries, W.; Lemmen, C.; Meijer, C. The point cadastre requirement revisited. *Surv. Rev.* **2013**, *45*, 239–247. [CrossRef]
294. Vogiatzis, M. Cadastral Mapping of Forestlands in Greece. *Photogramm. Eng. Remote Sens.* **2008**, *74*, 39–46. [CrossRef]
295. Köktürk, E.; Köktürk, A.P. The Role of Photogrammetry and Remote Sensing on Determining the Forest Boundaries and Unauthorized Buildings in Turkey (A Sample Area: Beykoz (İstanbul)). Available online: https://citeseerx.ist.psu.edu/viewdoc/download?doi=10.1.1.184.1578&rep=rep1&type=pdf (accessed on 18 October 2021).
296. Madzharova, T.; Petrova, V.; Ivanova, K.; Koeva, M. Mapping from high resolution data in GIS SOFIA Ltd. In Proceedings of the XXI Congress: Silk Road for Information from Imagery: The International Society for Photogrammetry and Remote Sensing, Beijing, China, 3–11 July 2008; pp. 3–11.
297. Busko, M. Evaluation of the Possibilities to use the Photogrammetric Method to Determine the Course of Boundaries of Cadastral Parcels during the Modernization of the Cadastre. In Proceedings of the 10th International Conference "Environmental Engineering", Vilnius, Lithuania, 27–28 April 2017; Volume 10, pp. 1–8.
298. Chromčák, J.; Šafář, V. The use of aerial photogrammetry in cadastre of real estates. *Int. Multidiscip. Sci. GeoConference SGEM* **2016**, *2*, 1035–1041.

299. Kohli, D.; Bennett, R.; Lemmen, C.; Asiama, K.; Zevenbergen, J. A Quantitative Comparison of Completely Visible Cadastral Parcels Using Satellite Images: A Step towards Automation. In Proceedings of the FIG Working Week, Helsinki, Finland, 29 May–2 June 2017; pp. 1–14.
300. Kohli, D.; Unger, E.M.; Lemmen, C.H.; Bennett, R.M.; Koeva, M.N.; Friss, J.; Bhandari, B. Validation of a cadastral map created using satellite imagery and automated feature extraction techniques: A case of Nepal. In *XXVI FIG Congress 2018: Embracing Our Smart World Where the Continents Connect: Enhancing the Geospatial Maturity of Societies*; International Federation of Surveyors (FIG): Istanbul, Turkey, 2018.
301. Crommelinck, S.; Lemmen, C.; Kohli, D.; Bennett, R.; Koeva, M. Object-based image analysis for cadastral mapping using satellite images. *Image Signal Process. Remote Sens. XXIII* **2017**, *10427*. [CrossRef]
302. Grant, D.; Enemark, S.; Zevenbergen, J.; Mitchell, D.; McCamley, G. The Cadastral triangular model. *Land Use Policy* **2020**, *97*, 104758. [CrossRef]
303. Bennett, R.; Kitchingman, A.; Leach, J. On the nature and utility of natural boundaries for land and marine administration. *Land Use Policy* **2010**, *27*, 772–779. [CrossRef]

Article

High-Quality UAV-Based Orthophotos for Cadastral Mapping: Guidance for Optimal Flight Configurations

Claudia Stöcker [1,*], Francesco Nex [1], Mila Koeva [1] and Markus Gerke [2]

[1] Faculty of Geo-Information Science and Earth Observation, University of Twente, 7514 AE Enschede, The Netherlands; f.nex@utwente.nl (F.N.); m.n.koeva@utwente.nl (M.K.)
[2] Institute of Geodesy and Photogrammetry, Technische Universität Braunschweig, 38106 Braunschweig, Germany; m.gerke@tu-bs.de
* Correspondence: e.c.stocker@utwente.nl; Tel.: +49-1573-701-6558

Received: 13 October 2020; Accepted: 2 November 2020; Published: 4 November 2020

Abstract: During the past years, unmanned aerial vehicles (UAVs) gained importance as a tool to quickly collect high-resolution imagery as base data for cadastral mapping. However, the fact that UAV-derived geospatial information supports decision-making processes involving people's land rights ultimately raises questions about data quality and accuracy. In this vein, this paper investigates different flight configurations to give guidance for efficient and reliable UAV data acquisition. Imagery from six study areas across Europe and Africa provide the basis for an integrated quality assessment including three main aspects: (1) the impact of land cover on the number of tie-points as an indication on how well bundle block adjustment can be performed, (2) the impact of the number of ground control points (GCPs) on the final geometric accuracy, and (3) the impact of different flight plans on the extractability of cadastral features. The results suggest that scene context, flight configuration, and GCP setup significantly impact the final data quality and subsequent automatic delineation of visual cadastral boundaries. Moreover, even though the root mean square error of checkpoint residuals as a commonly accepted error measure is within a range of few centimeters in all datasets, this study reveals large discrepancies of the accuracy and the completeness of automatically detected cadastral features for orthophotos generated from different flight plans. With its unique combination of methods and integration of various study sites, the results and recommendations presented in this paper can help land professionals and bottom-up initiatives alike to optimize existing and future UAV data collection workflows.

Keywords: UAV; cadastral mapping; data quality; geometric accuracy; impact assessment; ground control points; feature extraction; flight plan

1. Introduction

Harnessing disruptive technologies is crucial to achieving the Sustainable Development Goals. Amongst others, unmanned aerial vehicles (UAVs) play a significant role in the so-called Fourth Industrial Revolution. They are being referred to as mature technologies for remote delivery, geospatial mapping, and land use detection and management [1]. In the domain of land administration, UAV technology gained in importance as a promising technique that can bridge the gap between time-consuming but accurate field surveys and the fast pace of conventional aerial surveys [2,3]. Various publications tested UAV-based workflows for cadastral applications, covering formal cadastral systems such as in Albania [4] Poland [5], The Netherlands [6], or Switzerland [7] as well as less formal systems in Namibia [8], Kenya [9], or Rwanda [10]. Findings examine vast opportunities, especially with the additional information of textured 3D models and high-resolution orthophotos that ease public participation in boundary delineation [4,8,11]. Benefitting from the advantages of UAV data, various authors utilized approaches in artificial intelligence and developed (semi-) automatic scene understanding procedures to extract cadastral boundaries [12–15].

The fact that UAV-derived geospatial information can support decision-making processes involving people's land rights raises questions about the quality of UAV data. In this context, the concept of quality is closely linked to spatial accuracy, which can be defined as absolute (external) or relative (internal) accuracy. According to [16], absolute accuracy refers to the closeness of reported coordinate values to values accepted as or being true. In contrast, relative accuracy describes the similarity of relative positions of features in the scope to their respective relative positions accepted as or being true. Both measures are equally crucial in land administration contexts (c.f. [17]), firstly, the correct representation of image objects such as houses or walls (relative accuracy) as well as the correct position of corner points (absolute accuracy) [16]. Generally speaking, the spatial accuracy depends on configurations of the UAV flight mission such as sensor specifications, UAV itself, mode of georeferencing, flight pattern, flight height, photogrammetric processing, image overlap, but also on external factors such as weather, illumination, or terrain.

During the past decades, remote sensing, as well as computer vision communities alike, studied those impacting parameters emphasizing image matching algorithms, different means of georeferencing, and various flight planning parameters, among others. Finding accurate and reliable image correspondences is the basis for a successful image-based 3D reconstruction. Numerous authors investigated this fundamental part of the photogrammetric pipeline while trying to increase the precision of image correspondences and to optimize computational costs [18–21]. The quantity of tie-points derived during feature matching mainly depends on the type and the content of the image signal. Deficient success rates negatively impact the spatial accuracy and overall reliability of the 3D reconstruction and ultimately worsen the quality of the digital surface model (DSM) and orthophoto [22].

Next to the aspect of feature matching, georeferencing refers to one of the most practice-relevant yet most discussed topics when utilizing UAV imagery for surveying and mapping applications. More than 60 studies examined various methods of sensor orientation for terrestrial applications, as outlined by [23]. The choice for a georeferencing approach typically represents trade-offs between spatial accuracy and operational efficiency [24]. Even though direct sensor orientation or integrated sensor orientation brings significant time-savings for the data collection operation, planimetric accuracies usually range between 0.5 and 1 m due to the low accuracy and reliability of directly measured attitude and positional parameters by onboard navigational units without a reference station [25–27]. Due to inaccurate scale estimation of those insufficient methods, not only the absolute but also the relative accuracy might be not suitable for a particular application. In contrast, the use of real-time kinematic (RTK) or post-processing kinematic (PPK) enabled GNSS devices allows to improve the spatial accuracy to a range of several centimeters [28–32]. However, issues of sensor synchronization, as well as insufficient lever-arm and boresight calibration, remain challenging [29,33], particularly for off-the-shelf UAVs.

In addition to positional or full aerial control, integrated sensor orientation offers the option to include ground observations, known as ground control points (GCPs). This has proven to be beneficial to mitigate systematic lateral and vertical deformations in the resulting data products [34]. Various studies addressed the impact of the survey design of GCPs in terms of quantity and distribution. In their meta-study, [23] did not find a clear relationship between the number of GCPs and the size of the study area, but investigated a weak negative relationship between statistics of the residuals and the number of GCPs collected per hectare. Data from several sources confirm that the distribution of GCPs strongly impacts the spatial accuracy, and an equal distribution is recommended [35–37]. However, looking at the results of the optimal number of GCPs, different conclusions are evident. Results from relatively small study sites suggest that the vertical error stabilizes after 5 or 6 GCPs [35,38] and the horizontal error after 5 GCPs [35,36]. In contrast, [39] obtained a low spatial quality with 5 GCP and recommended to use a medium to a high number of GCPs to reconstruct large image blocks accurately. In [40,41] the authors achieved similar results with a concluding recommendation to integrate 15 or 20 GCPs in the image processing workflow, respectively. Aside from GCPs, higher spatial accuracy can be achieved by additionally including oblique imagery [42] or perpendicular flight strips [30]. In most cases, checkpoint residuals were measured in the point cloud or obtained directly after the bundle block adjustment and, thus, do not necessarily represent the displacement of image points in the final data product, as potential offsets during the orthophoto generation were not taken into consideration. However, particularly for the application in cadastral mapping, the correct estimation of the spatial accuracy is of vital importance.

Even though weak dependencies between several impacting factors on the data quality are evident, the results of existing studies are very heterogeneous. Furthermore, most studies remain narrow in focus, dealing mainly with only one study site situated in non-populated areas, and it is questionable whether recommendations can be transferred to the cadastral context. To the best of the authors' knowledge, existing studies on UAV-based cadastral mapping only highlight the usability of UAVs without assessing different flight configurations or the impact on the final absolute or relative accuracy. To this end, a comprehensive analysis of varying data quality measures should provide a factual basis for clear recommendations that ensure data quality for UAV-based cadastral mapping. Thus, this paper seeks to conclude on best practice guidance for optimal flight configurations by integrating results of a detailed quality assessment including three main aspects: (1) feature matching, (2) ground-truthing, and (3) reconstruction of cadastral features. Whereas the first two approaches target the evaluation of the data quality during and after photogrammetric processing, the latter method focusses on the implications of different orthophoto qualities for the automated extraction of cadastral features. Similar to diverse practices of quality assessment, research data are also manifold and are drawn from six study sites located in Africa and Europe.

In many low- and middle-income countries, conditions for flying, controlling, and referencing respective data are more complex than in Western-oriented countries, a situation which is often underestimated. Primarily spatial and radiometric accuracy can be negatively influenced by poor flight planning and adverse meteorological conditions. Moreover, ground control measurements can be problematic due to a lack of reference stations, the availability of professional surveying equipment, or capacity. In the field of land administration in general and cadastral mapping in particular, incorrect geometries of the orthophoto might cause negative consequences to civil society as the subject deals with a spatial representation of land parcels and attached rights and responsibilities. As an example, erroneous localization and estimations of parcel sizes might imply inadequate tax charges, problems with land compensation funds, or challenges to merge existing databases spatially. With its unique combination of methods and integration of various study sites, it is hoped that the results and recommendations presented in this paper help land administration professionals and bottom-up initiatives alike to optimize existing and future data collection workflows.

The remainder of the paper is structured as follows. Section two provides background information on data collection, data processing, and quality assessment methods. The results section is divided into three separate subsections with (1) findings showing the impact of land use on the number of tie-points, (2) a comprehensive comparison of different ground control setups and its effect on the final absolute accuracy, and (3) an evaluation of qualitative and quantitative characteristics of extracted cadastral features. The discussion critically reflects on the results based on existing literature and outlines best practice guidance for UAV-based data collection workflows in land administration contexts.

2. Materials and Methods

The study setup foresaw three different means of quality assessment targeting at absolute as well as relative accuracy as outlined in the conceptual framework in Figure 1. Well-known methods as the statistical evaluation of checkpoint residuals were combined with quantitative measures of image matching results as well as characteristics of automatically delineated cadastral features. Different clues on the spatial accuracy substantiate the results to provide best practice guidance. Detailed workflows and specifications of the analysis are outlined below.

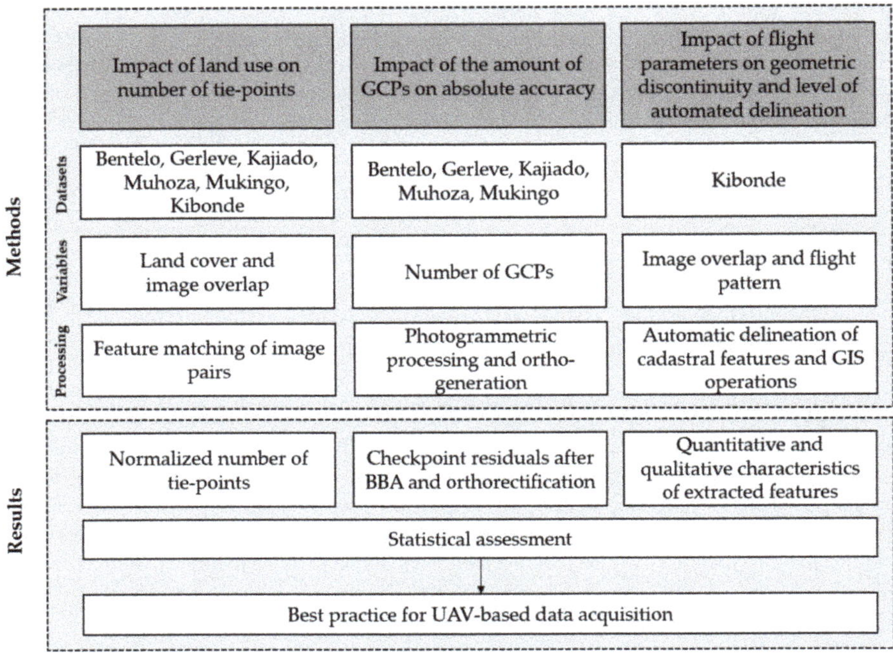

Figure 1. Conceptual framework.

2.1. UAV and GNSS Data Collection

To test the transferability of the findings and to ultimately claim best-practice recommendations, methods were applied to different datasets collected with diverse UAVs and sensor equipment. This includes, in total, six study areas across Europe (Gerleve, Bentelo) and Africa (Kajiado, Kibonde, Muhoza, Mukingo) ranging from 0.14 to 8.7 km^2 (Figure 2). UAV equipment as well as sensor specifications are outlined in Table 1 and included two fixed-wing UAVs (Ebee Plus, DT18), one hybrid UAV (FireFly6), and two rotary-wing UAV (DJI Inspire 2, DJI Phantom 4) equipped with an RGB sensor. Two out of the five UAVs worked with a PPK. Prices for the platforms and sensors range from 1000 to 40,000 €. Flights in Gerleve, Bentelo, Muhoza, Kajiado, and Mukingo were carried out according to a classical flight pattern without cross-flights and an overlap of 80% forward overlap and 70% side lap for all datasets. Additionally, the study in Kibonde foresaw several flights that were repeatedly carried out with varying image overlap (60%, 70%, and 80% side lap) to assess the impact of flight parameters on the characteristics of extracted cadastral features. Following existing literature that proves the benefit of cross flight patterns [16], three perpendicular strips in a different flight height were added to the regular flight and are part of the accuracy evaluation in Kibonde as well.

Figure 2. Overview of all datasets presented as orthomosaics (**a**) Bentelo, (**b**) Gerleve, (**c**) Mukingo, (**d**) Kajiado, (**e**) Kibonde, and (**f**) Muhoza (scales vary).

Table 1. Unmanned Aerial Vehicles (UAVs) and technical specifications of the sensor. GSD refers to ground sampling distance.

Dataset	Area (km^2)	GSD (cm)	UAV	Camera	Sensor Size (mm)	Resolution (MP)
Muhoza	0.98	2.1	BirdEyeView FireFLY6	SONY ILCE-6000	13.50 × 15.60	24.00
Mukingo	0.50	2.2	DJI Inspire 2	DJI FC652	13.00 × 17.30	20.89
Kajiado	8.70	5.8	DJI Phantom 4	DJI FC330	06.20 × 04.65	19.96
Kibonde	0.15	3.0	SenseFly Ebee Plus	SenseFly S.O.D.A.	12.70 × 08.50	19.96
Gerleve	1.10	2.8	DelairTech DT18	DT 3Bands	08.45 × 07.07	5.00
Bentelo	0.14	2.7	DJI Phantom 4	DJI FC330	06.20 × 04.65	11.94

To allow the inclusion of external reference points into the bundle block adjustment (BBA) as well as for means of independent quality assessment, GCPs were deployed. Due to different contexts and time delays between marking and the data collection flights [17], different shapes and methods to mark control points were used. In Musanze, Mukingo, Bentelo, and Kibonde quadratic plastic tiles with two equally sized black and white squares were fixed with iron pegs. Crosses marked with permanent white paint were used in Kajiado, as the flight missions took several days. For Gerleve, white sprayed Compact Disks were deployed and fixed with survey pins. Three-dimensional coordinates of the central point were determined with survey-grade GNSS devices. As Continuous Operating Reference Stations (CORS) are only available at a few locations in Africa, different modes were used to achieve a measurement accuracy of less than 2 cm. Real-time CORS corrections could be harnessed in Europe, while a base-rover setting over a known survey point and either radio-transmitted real-time corrections or a classical post-processing approach was the preferred surveying operation for the African missions. All GCPs were measured twice, before and after the UAV flight. The average of both measurements was converted from the local geodetic datum to WGS84 or ETRF89. A detailed list of specifications about the GNSS device, number of measured control points, as well as original and target geodetic datums are given in Table 2.

Table 2. Specifications of GCP measurements.

Dataset	GNSS Device	Measured Points	Original Datum	Target Datum
Muhoza	Leica CS10	17	ITRF 2005	WGS84 UTM35S
Mukingo	Leica CS10	19	ITRF 2005	WGS84 UTM35S
Kajiado	CHC X900+	16	Cassini	WGS84 UTM37S
Kibonde	Sokkia Stratus	11	Arc1960	WGS84 UTM37S
Gerleve	Trimble	22	ECEF	ETRS89 UTM32N
Bentelo	Leica GS14	18	Amersfoort	WGS84 UTM32N

2.2. Estimating the Impact of Land Cover on the Number of Automatic Tie Points

The establishment of image correspondences is a crucial component of image orientation. In the first step, primitives are extracted and defined by a unique description. Secondly, the descriptors of overlapping pictures are compared, and correspondences determined. With a low number of automatic tie-points, the image orientation is less reliable and negatively impacts the quality of subsequent image matching processes. Different land use classes were defined (cf. Table 3) to evaluate the impact of land cover on the number of automatic tie-points. If a particular land use was present in a dataset, representative image pairs were manually selected and processed as described below.

Table 3. Land use classes and representation in datasets (Bentelo, Gerleve, Kajiado, Kibonde, Muhoza, Mukingo). Digits indicate the number of image pairs used for the experiment. Percentage, as outlined in the definition, refers to pixel representing specific land cover.

Land Use Class	Definition	Ben	Ger	Kaj	Kib	Muh	Muk
Forest	>70% covered by trees	4	5				
Agriculture (cropland)	>70% cultivated agricultural fields	5					5
Agriculture (grassland or uncovered soil)	>70% bare soil or sparse grass vegetation	5	5	5	5		5
Rural context	<20% structures, a predominance of agricultural activities			5	5		5
Peri-urban context	20–70% structures			5	5	5	5
Urban context	>70% structures, densely populated			5		5	

Most commercial photogrammetric software packages do not provide information on their image matching techniques, and respective code might be subject to frequent changes. Instead of using such a black-box software, we chose three state-of-the-art feature matching approaches which were selected, reflecting the variety of blob and corner detectors with binary and string descriptors: SIFT [43], SURF [44], and AKAZE [45]. The open-source photogrammetric software PhotoMatch [46] was utilized to carry out the tests. Before the feature matching process, all images were pre-processed by a contrast-preserving decolorization tool [47], maintaining the full image resolution. The feature matching was conducted with a brute-force method and supported by RANSAC for filtering wrong matches. Thus, image correspondences are searched by comparing each key-point with all key-points in the overlapping image. Settings for feature extraction and description were kept to default values as this analysis is meant to detect relative changes of feature matching rates according to the type of land cover instead of performance evaluation of different approaches. Resulting tie-points (i.e., inlier of key-point matches) were normalized according to the image resolution to reach comparability between various sensor specifications within one land use class. To enable an evaluation of matching quality and variation in different land use classes and feature extraction/matching technique, the number of matches per image pair was normalized with respect to the number of matches within a specific matching algorithm, see equation below for the so-called *z*-score. Here *ATP* indicates the normalized number of automatic tie-points per image pair, $\overline{ATP}$ the mean of all matches of the respective feature extraction approach, and σ *ATP* the standard deviation of all matches of the respective matching approach. The *z*-score provides insights on how many standard deviations below or above the mean the quantity of tie-points in comparison to the other algorithms, within a land use class, is.

$$Z\ score = \frac{\left(ATP - \overline{ATP}\right)}{\sigma\ ATP}$$

To also visualize absolute quantities, the mean value of *ATP* for all different datasets in the same land use class was calculated. Furthermore, the overlap of image pairs was added as an additional variable. For this analysis, data from Kibonde and Bentelo served as input image pairs as both datasets offered various overlap configurations.

2.3. Estimating the Impact of the Number of GCPs on the Final Geometric Accuracy

All images were processed using Pix4D, keeping the original image resolution. Point clouds were created with an optimal point density, and DSMs as well as the orthomosaics were produced with a resolution of 1 GSD. To allow the comparability of the spatial accuracy of different datasets, uniformly distributed GCPs were included in the processing pipeline according to a standard pattern (Figure 3). Ground markers were identified and linked to at least six images. Depending on the specific number of GCPs (0–10), the remaining points were used as independent checkpoints to estimate the vertical and horizontal accuracy of the final data products.

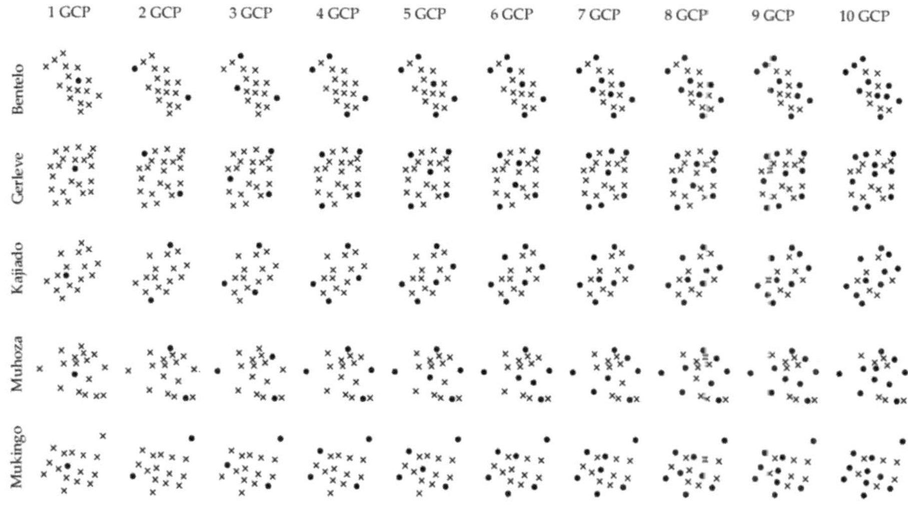

Figure 3. Distribution of GCPs for experimental assessment of the spatial accuracy.

The spatial accuracy was calculated at two different stages of the photogrammetric processing. Firstly, the geometric error was determined after the BBA, as outlined in the quality report of Pix4D. The horizontal error of a checkpoint was calculated using the Euclidean distance of the residuals in X and Y directions. The residuals of the Z coordinate represented the vertical offset. Secondly, this study also foresaw an accuracy assessment of checkpoint residuals in the final data product as the absolute accuracy of points in the orthophoto is of vital importance for cadastral surveying. This measure reveals information about displacement errors introduced during the orthorectification. The center of checkpoints was visually identified and marked in the orthomosaic using QGIS. Horizontal errors were derived by X and Y residuals, whereas the vertical error was extracted based on the raster value of the DSM. To describe the overall planimetric and vertical error of a particular processing scenario, the root mean square error (RMSE) was calculated following the ISO standard [16]. In this context, the GNSS measurement of the checkpoint coordinate was treated as true value and the extracted coordinates from the orthophoto as the predicted value.

2.4. Estimating the Impact of Different Flight Plans on the Characteristics of Extracted Cadastral Features

In contrast to the other two methods, the third quality evaluation utilizes only data from one regional context. Following basic photogrammetric principles, it is clear that the amount of image overlap significantly impacts the quality of the reconstructed scene. Thus, different flight pattern (with and without cross-flight), as well as multiple image overlap (50% and 75% forward overlap as well as 60%, 70%, and 80% side lap) configurations, were exemplified for the study area Kibonde to ultimately show the impact of various flight configurations on the reconstruction quality of cadastral features and subsequent automatic delineation results. Orthophotos were processed. Subsequently, a quadratic shape of 500 × 500 m was extracted as required by the image segmentation algorithm [48]. To ultimately analyze geometric features and line discontinuities, this paper foresaw a workflow including image segmentation algorithms as well as raster and vector operations, as shown in Figures 4 and 5. The first step was the establishment of reference lines and the creation of a mask to clip all candidate lines subject to this analysis. Reference lines were based on independently captured UAV images (80% forward overlap and side lap, cross-flight at a different altitude) and a resulting orthomosaic with 1.5 cm resolution. Two distinct features, namely building rooftops and concrete walls, were selected as representative visible objects that are important for cadastral applications. Both feature types were manually digitized and served as reference lines for subsequent analyses.

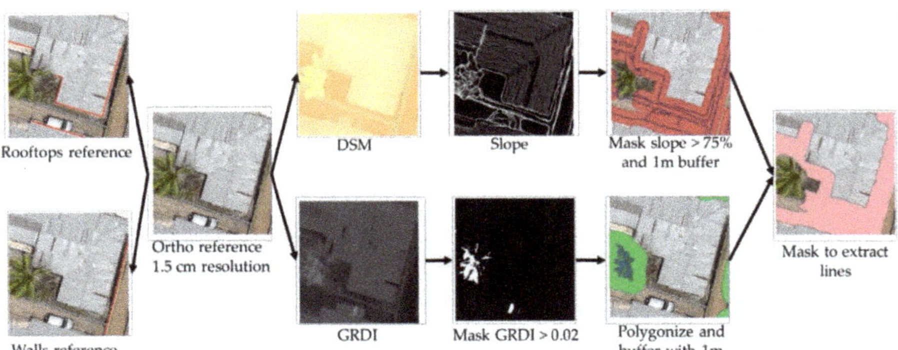

Figure 4. Workflow to define reference lines and a search mask for lines representing concrete walls and rooftops.

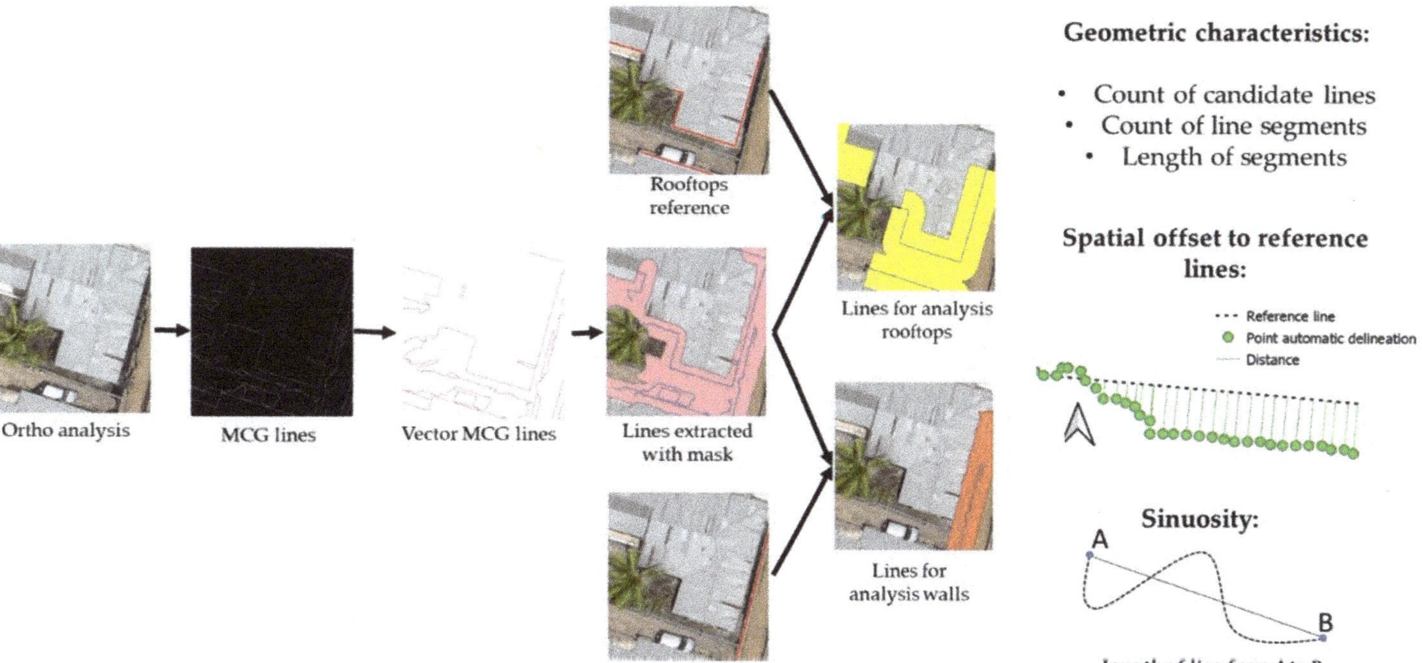

Figure 5. Workflow to compute and select MCG lines representing rooftops or walls and analytical tools to describe geometric characteristics of selected MCG lines.

A uniform vector mask representing the vicinity of concrete walls and rooftops was created to minimize the number of candidate boundary lines. A slope layer served as the basis to select a 1 m buffer of all raster cells of the DSM representing >75% of the height gradient. Additionally, a vegetation mask was created to remove vegetated areas as those would negatively impact the straightness of selected cadastral features independent of the quality of the orthomosaic and thus would introduce unintended noise to the analysis of geometric discontinuities. The vegetation mask was based on the Gree-Red Difference Index (GRDI). Raster cells above a GRDI of 0.02 were classified as vegetation and polygonized to calculate a buffer of 1 m. Finally, the slope-based mask was clipped with the buffer of the GRDI to exclude vegetation from the samples.

In the second step, multiscale combinatorial grouping (MCG) [49] was applied to all orthophotos to ultimately derive closed contour lines of visible objects, as suggested by [50]. The segmentation threshold was set to k = 0.6 as this has proven to limit over-segmentation while still maintaining relevant cadastral objects in the context of this study. As shown in Figure 5, resulting lines were polygonized and simplified according to [48]. Once the lines were clipped with the reference mask, several geometric and spatial characteristics were queried (c.f. Figure 5). Candidate lines were selected by overlaying the MCG lines with a 0.5 m buffer of reference lines. From those candidate lines, actual lines representing rooftops and walls were chosen manually. To calculate the correspondence as well as the spatial difference to reference lines, the MCG lines representing walls and rooftops were split to segments of 10 cm and subsequently converted to points. Afterwards the distance from each point in the MCG line to the closest point of the reference line was calculated to derive statistical values for the spatial offset. To describe the amount of MCG lines that could automatically be extracted (i.e., correspondence with reference lines), a neighborhood analysis was carried out to estimate the percentage of reference lines that could be reproduced by the MCG algorithm. As a last characteristic, this study calculated the sinuosity as a measure of the straightness of MCG lines to reflect on inconsistencies of critical features in the orthomosaics. Similar to the spatial offset, the sinuosity was calculated based on summed length of the MCG lines for one object in relation to the length of a virtual straight line (Figure 5).

3. Results

3.1. Image Matching: Image Correspondences

The number of pairwise image correspondences was derived from comparing feature matching success rates representing certain land use classes prevalent in the images. The diagram in Figure 6 depicts standardized z-scores as well as mean values of automatic tie-points for SIFT, SURF, and AKAZE. At first glance, the results of various matching algorithms demonstrate a similar distribution, whereas apparent differences between land use classes are evident.

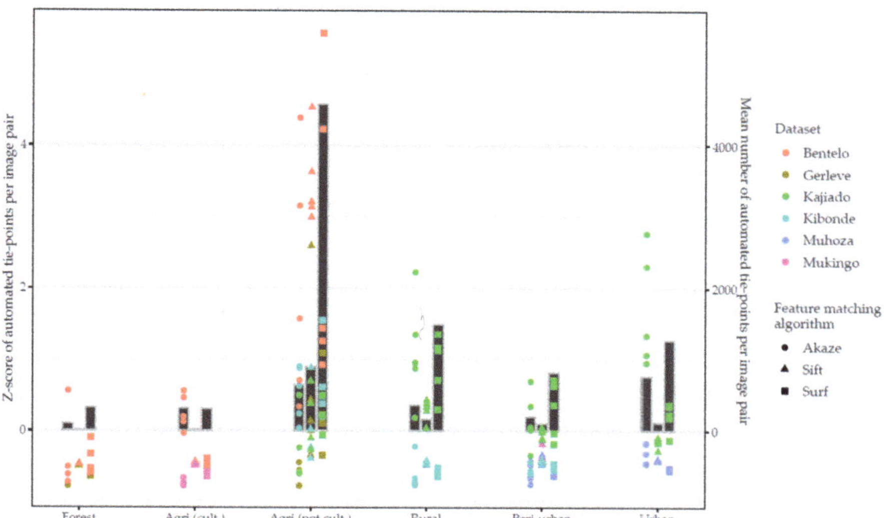

Figure 6. Standardized values of automatic tie-points using SIFT, AKAZE, and SURF as feature extraction, detection, and matching algorithm. The mean number of automatic tie-points per algorithm and land use class is reflected as bars. The x-axis represents land use classes as defined in Table 3.

Image pairs characterized by forest and cultivated agricultural fields show significantly low numbers of automatic tie-points. In some cases, no matches could be found. Images, displaying non-cultivated agricultural field plots stick out by a broad range of images correspondences for all three feature matching approaches. Here, the dataset Bentelo reaches the highest z-scores and results are multiple standard deviations above the mean. However, also insufficient numbers of automatic tie-points are evident in this land use class, particularly for Gerleve. This can be ascribed to poor illumination conditions and little contrast in the images. The remaining datasets are clustered in a range between −0.5 and 1.5 of the z-score.

For image scenes showing human-made structures, two different trends are visible. The first trend describes the following correlation: on average Kibonde, Muhoza, and Mukingo indicate more key-point matches if less vegetation and more structures are prevalent. Thus, for Kibonde and Mukingo, a higher z-score was achieved with the peri-urban scene context compared to the rural context. The same applies to Muhoza with the land uses peri-urban and urban, respectively. In contrast, Kajiado does not follow this trend and represents the dataset with the highest z-scores for all three land use classes (rural, peri-urban, urban). The same applies for all three image matching algorithms. A possible explanation for this may be the climate zone. As indicated above, high vegetation presents an adverse condition for finding tie-points. In contrast to the humid climate in Kibonde, Mukingo, and Muhoza, Kajiado is located in a semiarid region characterized by a sparse shrub and bush vegetation. Thus, the impact of vegetation is almost not visible and rural as well as urban scenes achieve similar z-scores. Secondly, next to the climate zone, also the GSD might have an impact on the above-average z-score of Kajiado for the rural, peri-urban, and urban land use class.

Looking at the impact of image overlap on the automatic tie-points in Table 4, it becomes clear that the poor feature matching results of forest can only be overcome with 90% image overlap while the other land use classes already show sufficient matches with less overlap. Similar to Figure 6, non-cultivated agricultural areas present the highest rate of image correspondences for all image overlap scenarios. Two adverse conditions could explain the low rate of automatic tie-points in the forest. Firstly, although the flight is configured with a high image overlap, the difference in the viewing angle is larger between image points showing the crown of the tree than for image objects on the ground. Thus, we observe that key-points show insufficient similarity to be determined as image correspondence. This challenge can only be overcome by 80–90% image overlap. However, at the same time, the descriptors of leaves could also be too similar, leading to ambiguities during the feature matching process. Both effects are visible and could explain the comparatively low number of automatic tie-points for all four image overlap configurations. In addition, and more or less independently from that, high vegetation cannot be regarded "static", which is, however, an indispensable requirement for mono-camera bundle adjustment. It should be emphasized that those results were derived with single image pairs. It is expected that a priori location and alignment information of images in an image block ease the feature matching process compared to the brute-force approach used in this analysis.

Table 4. Mean of automatic tie-points of image pairs using SURF showing different land use classes and overlap.

	60% Overlap	70% Overlap	80% Overlap	90% Overlap
Agriculture (not cultivated)	289	666	2519	n/a
Rural	83	116	291	n/a
Peri-urban	18	302	326	n/a
Forest	0	5	6	50

3.2. Absolute Accuracy: Checkpoint Residuals in DSM and Orthophotos

The absolute accuracy was determined after the bundle block adjustment as well as after the orthophoto generation. Figure 7 presents the RMSE of horizontal and vertical checkpoint residuals of all datasets. Looking at the results, it is evident that, in general, all datasets show a similar pattern. For photogrammetric processing with less than 5 GCPs, resulting RMSE of the datasets differ widely, whereas, for results with more than 5 GCPs, the final RMSE seems to stabilize at a certain level.

Looking at the horizontal RMSE, the large variance of the datasets for processing scenarios from 0 to 5 GCPs can be explained by the different quality of positional sensors. If no ground truth is included (0 GCPs), the BBA solely uses image geotags to estimate the absolute position of the reconstructed scene. Here, Gerleve was the only dataset with a professional PPK enabled GNSS device and attained the lowest RMSE (10 GSD) for all datasets processed with 0 GCP. In contrast, Kajiado was flown with a consumer-grade UAV showing a large horizontal offset of more than 200 GSD. Bentelo, Mukingo, and Muhoza achieve an RMSE between 50 and 100 GSD without GCPs, which is considered a typical error range of GNSS positioning without enhancement methods. Except for the dataset Mukingo, the RMSE drops significantly with including 1 GCP which corrects systematic lateral shifts. For the scenario with 3 GCP, all datasets achieve a horizontal RMSE between 10 and 20 cm. Gerleve and Bentelo reach an RMSE of less than 10 cm after 6 GCPs and are followed by Kajiado and Muhoza after 7 GCPs. Subsequently, almost all datasets keep the same level alternating within a range of 1 GSD. In this aspect, Mukingo achieves the most accurate results with less than 5 cm RMSE after 5 GCPs. Muhoza is the only dataset which nearly improves its RMSE for each scenario that adds one more GCP.

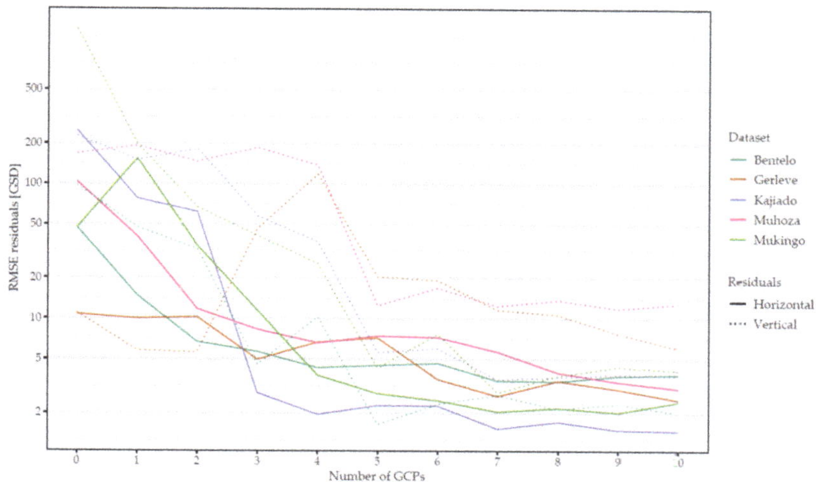

Figure 7. RMSE of checkpoint residuals measured in the DSM (vertical) and orthophoto (horizontal).

Looking at the vertical residuals, Figure 7 suggests a higher dynamic compared to horizontal residuals. In general, residuals are larger than the values of the horizontal RMSE and start to level only after 7 GCPs. With a height offset of more than 1000 GSD, which corresponds to approximately 30 m, the dataset Mukingo shows the maximum value without including GCPs. This can be attributed to a general definition problem of the height model used by DJI and can be corrected by adding at least 1 GCP. Similar to the horizontal residuals, Gerleve achieves the highest accuracy with an RMSE of only 10 GSD. However, after 2 GCPs, the height residuals abruptly increase before decreasing again after 4 GCP, indicating that this dataset requires a checkpoint in the center of the scene to correct severe height deformations. At 5 GCPs, all datasets demonstrate a significant improvement of the vertical RMSE. Independent from the size of the area, five evenly distributed GCPs can be considered as the minimum number of GCPs which efficiently fixes cushion and dome deformations during scene reconstruction. After 7 GCPs the vertical residuals of Bentelo, Kajiado, and Mukingo stabilize within the range of 1 GSD whereas Muhoza and Gerleve continue to lower its RMSE.

Additional to the absolute accuracy, the difference of the RMSE after BBA to the RMSE after DSM and orthophoto generation are shown in Table 5. The presented values reveal insights about the share of the overall error, which accumulates after the BBA during the 3D-reconstruction and ortho-generation process, independent of horizontal or vertical displacement indicated during the BBA. Negative values suggest that the RMSE after the BBA is higher than the RMSE of the residuals taken from the DSM/orthophoto. On average, variations between the error measures remain very low (below 1 GSD) and do not show a clear trend of an overestimation of one or the other, as well as no relation to the number of GCPs. However, for Gerleve and Muhoza, horizontal residuals range up to 3 GSD, and for vertical residuals we observe differences up to 5 GSD in two cases. For both datasets, significantly higher differences in the RMSE of checkpoint residuals could be explained by the challenging conditions for the 3D-reconstruction and orthophoto-generation processes. For Muhoza, difficulties could arise from considerable height (i.e., land surface) dynamics of the densely populated urbanized center. Gerleve stands out for its poor illumination conditions and subsequent problems to reliably reconstruct the image scenes.

Table 5. Differences of RMSE of checkpoint residuals measured after the BBA and in the orthophoto/DSM. Values are normalized, according to GSD. Horizontal (h) and vertical (v) errors are treated separately. Differences >1 GSD are indicated bold.

	Bentelo h/v (GSD)	Gerleve h/v (GSD)	Kajiado h/v (GSD)	Muhoza h/v (GSD)	Mukingo h/v (GSD)
0 GCP	0.39/**−1.96**	−0.03/−0.84	0.05/−0.45	0.29/0.69	0.38/0.22
1 GCP	−0.11/−0.08	0.05/0.73	−0.09/−0.10	−0.01/**3.79**	−0.31/−0.02
2 GCP	0.37/0.07	−0.02/−0.11	−0.24/−0.18	0.01/**2.72**	−0.28/−0.22
3 GCP	0.05/−0.28	−0.19/0.20	0.18/−0.57	0.72/**1.07**	0.24/0.16
4 GCP	0.15/−0.52	**2.11**/−0.72	0.23/0.94	0.60/**−4.12**	−0.27/0.49
5 GCP	0.10/0.01	**2.17/−1.09**	0.44/−0.24	**2.81**/0.18	0.15/−0.34
6 GCP	0.15/0.05	−0.15/−0.55	0.58/−0.37	**1.60**/0.18	0.19/0.16
7 GCP	−0.13/−0.35	−0.05/0.57	0.24/−0.70	**1.42**/−0.12	0.23/−0.23
8 GCP	−0.08/−0.22	−0.03/**1.63**	0.35/−0.37	−0.81/0.41	0.27/−0.07
9 GCP	−0.22/−0.31	**−1.55**/−1.02	0.29/0.39	−0.50/**−4.89**	0.11/−0.66
10 GCP	−0.11/−0.31	0.42/0.68	0.17/−0.02	−0.24/**−2.60**	0.28/−0.05

3.3. Relative Accuracy: Characteristics of Automatically Extracted Cadastral Features

Various line geometry measures present the quality of the scene reconstruction and subsequent feature extraction. For the chosen quadratic scene in the center of the Kibonde dataset, houses are predominantly covered by corrugated iron roofs and parcels are usually separated by concrete walls or bushes. To minimize external noise to our statistical assessment, only walls and rooftops without the interference of vegetation were delineated as reference (Figure 8). This adds up to a total of 692.3 m of lines referring to rooftops and 196.4 m of lines representing walls. As presented in Table 6, this relation is also expressed by candidate lines counted in a 0.5 m buffer of all reference lines. Interestingly, the impact of the flight pattern (cross-flight or no cross-flight) is more evident for rooftops than for walls, shown by the difference of line counts for different flight pattern scenarios. Concerning reference walls, marginally (within 10% range) fewer candidate lines were selected compared to the same scenario without a cross-flight pattern. In contrast, for rooftops, differences range from 10% to 40%.

Figure 8. Selected reference lines representing rooftops (green) and walls (red) for the area of interest in Kibonde.

Table 6. Qualitative and quantitative characteristics of line geometries representing rooftops (R) and walls (W) separated according to flight configuration (forward overlap (f), side lap (s)) and flight pattern (CF = cross flight pattern, no CF = no cross-flight pattern). Minimum and maximum values are presented in bold.

Image Overlap (f/s)		**50%/60%**		**50%/70%**		**50%/80%**		**75%/60%**		**75%/70%**		**75%/80%**	
		no CF	CF	no CF	CF	no CF	CF	no CF	CF	no CF	CF	no CF	CF
Candidate lines 0.5 m buffer (count)	W	144	146	**121**	133	**177**	157	158	122	165	133	129	134
	R	333	273	410	285	**505**	310	402	**271**	366	295	444	369
Selected line segments (count)	W	73	**42**	76	63	67	50	**78**	44	75	57	54	40
	R	209	177	233	180	**256**	180	243	**161**	189	168	220	173
Mean length of line segments (m)	W	3.22	4.68	3.50	3.39	3.32	4.96	3.31	3.71	**3.05**	3.86	4.37	**5.47**
	R	3.65	4.37	3.34	4.28	**3.01**	4.21	3.24	**4.87**	4.14	4.54	4.40	4.50
Correspondence with reference (%)	W	**71.5**	85.8	79.0	90.1	82.5	87.8	88.5	83.6	**93.0**	87.6	91.9	**93.0**
	R	**94.9**	95.8	94.8	95.1	96.6	95.6	96.3	95.9	95.1	95.7	97.2	**97.8**
Sinuosity	W	1.77	**1.78**	1.76	1.70	1.68	1.65	1.66	1.62	1.74	1.62	1.66	**1.58**
	R	**1.58**	1.59	**1.58**	1.60	1.59	1.60	1.60	1.61	**1.58**	1.59	**1.61**	**1.58**

Looking at the count of selected line segments, a more homogenous picture can be drawn. In all cases, the line count for the cross-flight pattern is lower than for the same image overlap scenario without a cross-flight. The mean length of line segments shows no significant difference between walls and rooftops. However, an important observation can be made concerning the image overlap. On average, line segments are shorter for scenarios with only 50% forward overlap compared to flight plans with 75% overlap. The combination of a higher count of line segments and a smaller average line length proves a higher fragmentation of boundary features for orthophotos without a cross flight pattern, as well as for lower image overlap scenarios. This result becomes even more apparent concerning the correlation of selected MCG lines with the reference dataset. Here, the improvement of the correlation with reference lines is more significant for walls than for rooftops. In this aspect, walls demonstrate a range between 71.5% and 93% and steadily increase with higher image overlap (both, forward and side lap). This means, the MCG algorithm applied to the orthophoto generated with a poor flight plan, produces contours for only 71.5% of the walls. In contrast, an orthophoto based on a favorable flight plan achieves an object detection rate of 93%. Hence, the detection range of contour lines for rooftops is comparatively small with maximal 2.9% variance between different flight plan scenarios.

A similar observation is evident for the sinuosity. Here, rooftops do not differ much, and lines of rooftops are on average 1.5 times longer than a perfectly straight line from the start to the endpoint. MCG lines representing walls are on average more curved and show a clear trend concerning the flight parameters reaching a minimal curviness with a cross flight pattern and 75% forward overlap and 80% side lap. Particularly for lines representing rooftops, it should be noted that the sinuosity values are relatively high due to the origin of the MCG lines, which were created based on a raster dataset and consequently still show undulations at the pixel level.

Aside from line feature characteristics, the spatial correlation was also investigated in terms of distance measurements of MCG lines to reference lines. Figure 9 visualizes the results and exemplifies the spatial correlation with a small sample of the entire dataset. Rooftops are mainly delineated close to the reference line, whereas walls show considerable variability. As an example, we included the orthophoto generated with the poorest image overlap at the bottom of Figure 9. The visual interpretation reveals a significant deformation and poor orthorectification of the wall, which ultimately leads to the displacement of MCG lines for the dataset with 50%/60% overlap and without a cross-flight.

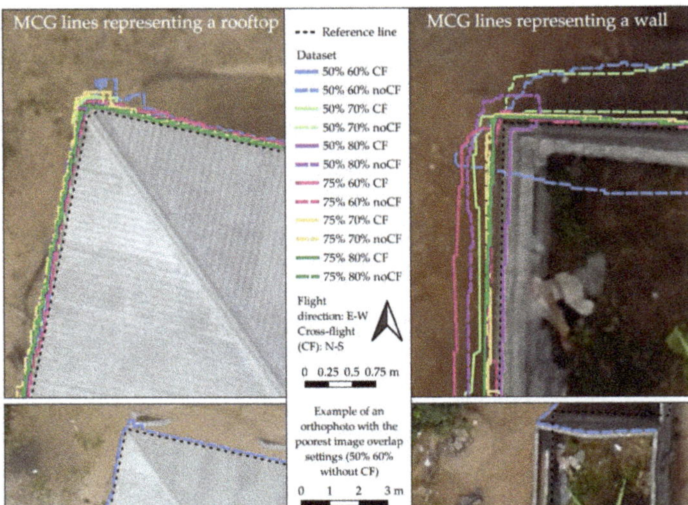

Figure 9. Example showing the differences of automatically extracted rooftops and walls separated according to flight configuration (forward overlap (%), side lap (%)) and flight pattern (CF = cross flight, noCF = no cross flight).

This variability is also apparent in the statistics of the point-to-line distances, presented as box-whisker plots in Figure 10. The interquartile range of rooftops is significantly smaller than the one of the walls. It should be noted that the distances of reference walls are subject to a systematic offset of 15 cm as the reference line was placed in the center of the wall, whereas the MCG algorithm produced lines on the right or left edge.

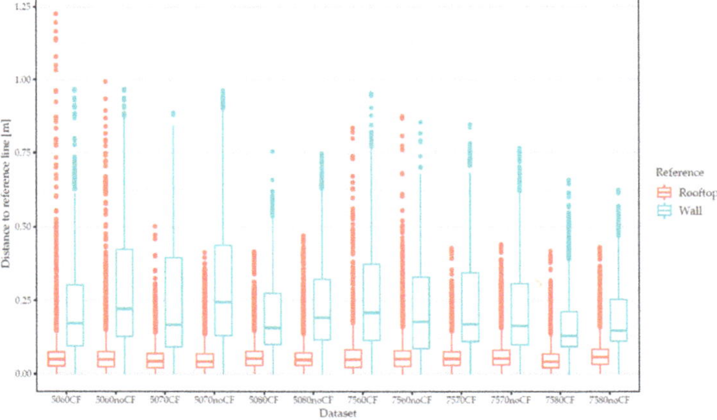

Figure 10. Box-whisker plot of point distances to reference lines separated according to the reference wall and rooftop. Box represents the interquartile range (IQR) with the median; whisker represent 1.5 IQR, points represent outliers. *x*-axes label refers to flight parameter, e.g., 5060CF means 50% forward overlap, 60% side lap and cross-flight (CF) pattern. Distances reflect the length of perpendicular lines from points to reference lines. Points were created every 10 cm from a line geometry that was derived by feature extraction with the MCG algorithm.

For two flight scenarios with low overlap, outliers of point distances of rooftops exceed the outliers of walls. In general, the share of outliers is higher for rooftops than for walls indicating that almost all rooftops are delineated in a range of approximately 20 cm with a few extreme variations. For wall features, the statistical analysis confirms the observations from the line characteristics, showing that the overall quality of delineated walls differs highly with respect to the image overlap and flight plan settings. Best results represented by the lowest five-number values of the box-whisker plot were returned for flight scenarios with 75% forward overlap, 80% side lap, and a cross flight pattern.

As evident in Figure 11, the RMSE of horizontal checkpoint residuals of the orthophoto stays between 0.8 and 1.5 GSD for all flight configurations, which corresponds to 2.5–4 cm. Similar to Figure 9, the statistics of the offset of detected line features show a noticeable discrepancy between rooftops (<10 cm) and walls (20–40 cm).

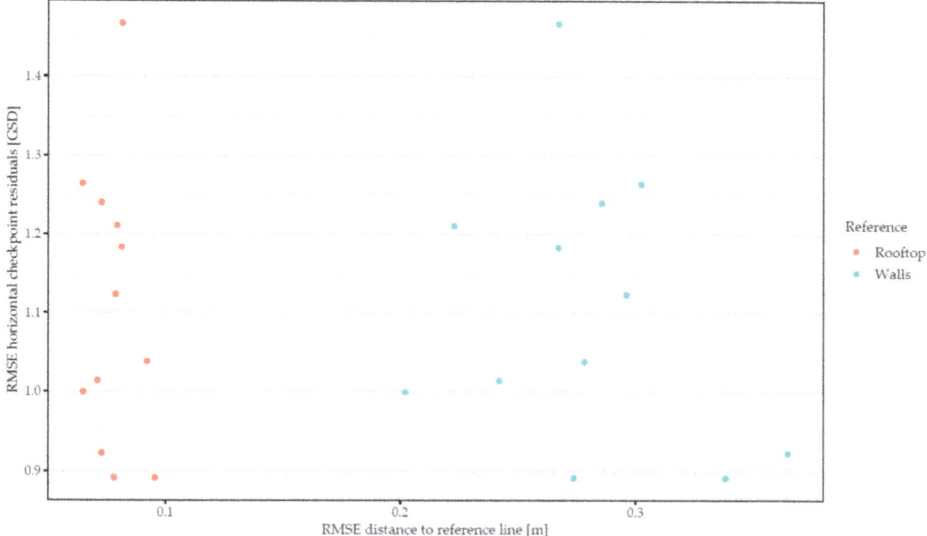

Figure 11. Scatterplot of error metrics for delineated rooftops and walls of orthophotos captured with different flight configurations. Absolute accuracy of the orthophoto is given on the y-axis with the RMSE of horizontal checkpoint residuals. Relative accuracy is shown on the x-axis displayed by the RMSE of point distances to reference lines. Note that both axes have different scales.

In contrast to checkpoint residuals of the orthophoto, which do not show a correlation with the error metrics, Figure 12 reveals that the flight pattern has implications on the relative accuracy of extracted features. Walls directed perpendicular to the flight direction show almost the same statistics for both scenarios, with a cross-flight or without a cross-flight pattern. However, for walls parallel to the flight direction, a cross-flight pattern improves the results indicated by a lower median and a smaller IQR. This result could be attributed to the fact that geometries of features parallel to epipolar lines imply more challenges to correctly estimate the 3D position and subsequent image matching and ortho-generation.

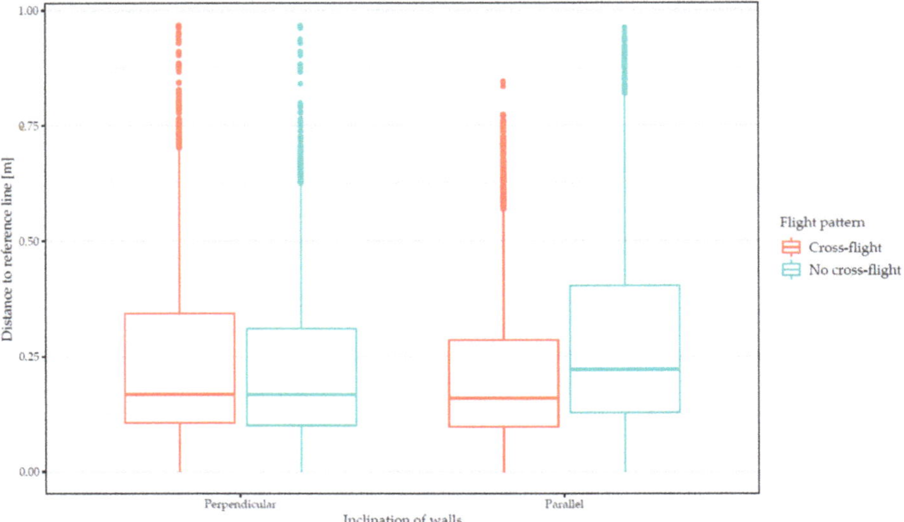

Figure 12. Box-whisker plot of distances to reference lines separated according to the direction of walls (parallel or perpendicular to the flight direction). Box represents the interquartile range (IQR) with the median; whisker represent 1.5 IQR, points represent outliers.

4. Discussion

Even though UAV can collect images with a resolution of a few centimeters, results in this paper show that the absolute and relative accuracy can differ from some centimeters up to several meters depending on the chosen flight configuration. To exploit the full potential of UAV-based workflows for land administration tasks, careful decisions on efficient mission planning are essential. This holds true for both sides: collecting as many images and GCPs as needed to meet the expected survey accuracy, but also for collecting just as many images and GCPs as necessary to minimize computational costs in favor of time constraints or potential hardware limitations.

Several reports have shown that the quantity of automatic tie-points impacts the quality of the photogrammetric 3D reconstruction as image correspondences are fundamental for the correct estimation of image orientation parameters. Even though different sensors, UAV, scenes, and flight conditions were analyzed in this paper, a homogenous picture can be drawn when looking at generated tie-points in relation to land use classes. In image scenes showing trees or crops significantly lower rates of tie-points could be extracted compared to scenes with human-made structures or grassland. In the former case, only a high image overlap of at least 80–90% is sufficient to achieve an adequate number of image correspondences. These results match those reported in [51]. In tropical or subtropical regions, most rural and peri-urban scenes are also characterized by vegetated areas subject to subsistence farming side-by-side to residential buildings. Thus, an optimal flight mission might need to be configured with higher image overlap compared to a flight mission in arid or semiarid regions.

Although a clear correlation of generated tie-points and land use can be shown, the results suggest that the optimal number of GCPs seems to be independent of the climate zone or land cover, as all datasets in this analysis reveal a similar pattern and indicate no significant changes of the RMSE after seven equally distributed GCPs. This result reflects those of [36,38,40] who also observed no significant differences in the final vertical or horizontal RMSE after 5 or 6 GCPs, respectively. In contrast to earlier findings [39], no evidence of the impact of the GCPs density was detected. In terms of GSD and despite considerably different extents of the study areas, all datasets in this analysis reach a similar error level with 10 GCPs, 2–3 GSD for the horizontal accuracy, and 2–4 GSD for the vertical accuracy.

Thus, the number and distribution of GCPs might play a more critical role than the density of GCPs. This is particularly interesting for the mission planning and calculation of costs as placing, marking, and measuring of GCPs is one of the most time-consuming and consequently, most costly aspects of the entire UAV-based data collection campaign. In the case of Mukingo, we observed a substantial offset of the vertical error in the scenario without GCPs. This magnitude of height offset was already reported before [35] and seemed to be specific to DJI UAV.

Contrary to most other studies that investigate checkpoint residuals, this analysis presents the absolute accuracy with regard to the residuals after the BBA and in the final DSM and orthophoto. For two out of six datasets, our results show significant discrepancies between the checkpoint residuals with a magnitude of up to 5 GSD. In both cases, challenging conditions were present, i.e., poor illumination conditions for Gerleve and a densely populated built-up area in Muhoza. We observe that particular the height component could be strongly impacted. Consequently, the consideration of checkpoint residuals measured in the orthophoto is indispensable for the evaluation of the final accuracy, as additional offsets might be introduced during the 3D reconstruction and orthophoto-generation process.

As a third central aspect, this study reveals yet another perspective on the orthophoto quality: success rates of the automatic extraction of cadastral features. Here, our findings point on a clear difference between the delineation of rooftops and walls. Whereas various flight configurations showed less impact on the extractability of rooftops, the automatic extraction of walls achieves more accurate and complete lines with large image overlaps and a cross-flight pattern. Even though the absolute difference of the correlation seems minor in our example, values of either 70% correlation or 93% correlation with reference lines are significant for scaled applications. A smaller percentage would entail a lot more manual work of delineating respective walls that were not represented by MCG lines. Furthermore, the MCG algorithm applied on an orthophoto of a weak image block—as described by lower image overlap—produces shorter line segments which also implies more manual effort to receive a complete delineation finally. Thus, thoughts should also be given to characteristics of extractable features when designing a UAV flight mission. Our findings suggest that planar cadastral features are less sensitive to differences in flight configurations than thin image objects such as walls or fences. Consequently, the latter necessitates a higher percentage of image overlap to be reliably reconstructed and detectable during subsequent automatic delineation processing. Additionally, when thin cadastral objects are oriented towards different cardinal directions, a cross-flight pattern is clearly recommendable.

In combination, the results are significant in at least two aspects. Firstly, although this study investigated very different study sites, common trends are evident. Thus, some general recommendations can be drawn. Independent of the sensor or feature matching algorithm, vegetated spaces, and forests or cultivated agricultural areas, still present challenges to the establishment of image correspondences. However, findings of checkpoint residuals suggest that the impact on the overall accuracy is only marginal when looking at scenes with multiple land use classes.

Secondly, the research investigations reveal large discrepancies between the spatial accuracy and the completeness of automatically detected cadastral features, even though the RMSE of the orthophoto as commonly accepted error measure is low. According to these data, we can infer that the flight configurations play a crucial role in achieving high data quality, particularly for cadastral features characterized by height differences and thin shape as exemplified for concrete walls. Moreover, in most cases, checkpoints are put out in open and visible spaces which do not necessarily reflect objects subject to manual or automatic cadastral delineation. Consequently, it should be emphasized that context-driven error analysis is essential to assess the overall accuracy of UAV-based data products.

Finally, a multifactorial analysis, as presented in this paper includes shortcomings on various ends. The study design foresaw various UAV and sensor configurations to mimic a variety of different contexts and real-world applications, focusing on the quality of final data products. Despite various camera specifications, common trends and characteristics are evident throughout all datasets. However, as a limitation of this study, the impact of hardware differences on the final data quality could not be estimated. Next to this, it cannot be ruled out that the GSD affects the quantity of generated tie-points. Further investigations are needed to evaluate this nexus. Lastly, it should be emphasized that all datasets were collected following flight plans as specified in the methods of this paper and the transferability of our findings to other flight configurations or other contexts cannot be guaranteed.

5. Conclusions

This paper provides recommendations on optimal UAV data collection workflows for cadastral mapping based on a comprehensive analysis of data quality measures applied to numerous orthophotos generated from various flight configurations. Methods covered several aspects ranging from statistics of automatic tie-points and an evaluation of the geometric accuracy to characteristics of automatically delineated cadastral features. The results highlight that scene context, flight configuration, and GCP setup significantly impact the final data quality of resulting orthophotos and subsequent automatic extraction of relevant cadastral features.

In a nutshell, the following recommendations can be drawn:

- Land use has a significant impact on the generation of tie-points. Image scenes characterized by a high percentage of vegetated areas and especially trees or forest require image overlap settings of at least 80–90% to establish sufficient image correspondences.
- Independent of the size of the study area, the error level of planimetric and vertical residuals remains steady after seven equally distributed GCPs (according to the scheme presented in Figure 3), given at least 70% forward overlap and 70% side lap. As the absolute accuracy does not increase significantly with adding more GCPs, 7 GCPs can be recommended as optimal survey design.
- The quality of reconstructed thin cadastral objects, as exemplified for concrete walls, is highly variable to the flight configuration. A large image overlap, as well as a cross-flight pattern, has proven to enhance the reliability of the generated orthophoto as quantified by the increased accuracy and completeness of automatically delineated walls. In contrast, the delineation results of rooftops showed less sensitivity to the flight configuration.
- Even though checkpoint residuals indicate high absolute accuracy of an orthophoto, the reliability of reconstructed scene objects could vary, particularly in adverse conditions with large variations in the height component. We furthermore recommend measuring checkpoint residuals in the generated orthophoto in addition to after the BBA.

Generally, these findings have important implications for developing UAV-based workflows for land administration tasks. This fact that the data quality can significantly change depending on the flight configurations involves risks and opportunities. The risk is that UAVs are used as off the shelf products with little knowledge of photogrammetric principles and options to customize flight configurations. Consequently, even though the end-product appears to be of good quality, spatial offsets, deformations, or poor reconstruction results of relevant features might be present but remain undetected. However, at the same time, we also realize immense opportunities in the customization of UAV workflows. The results in this analysis show that different flight configurations and various ground-truthing measures offer a wide range of options to tailor the data collection task to financial, personnel, and time capacities and optimally align it to customer needs and requirements in the land sector. This equips UAV workflows as a viable and sustainable tool to deliver reliable and cost-efficient information to cope with current and future cadastral challenges.

Future research building upon our results could follow different pathways. Firstly, although our study foresaw six different contexts, the terrain was mostly flat or slightly undulated and showed only minor surface variations. It would be interesting to explore if data of hilly and larger study areas could substantiate our recommendations. Secondly, this study neglects the ground sampling distance as a variable in our assessment. It is expected that next to the orthophoto quality also variations in the resolution might impact the feature matching process as well as completeness and accuracy of automatically extracted line geometries. Clues on this correlation could expand best-practice examples by adding recommendations on camera specifications and flight heights.

Author Contributions: Conceptualization, C.S., F.N., M.G., M.K.; Methodology, C.S., F.N., M.G.; Data Curation, C.S.; Writing—Original Draft Preparation, C.S.; Writing—Review and Editing, C.S., F.N., M.G., M.K.; Visualization, C.S.; Supervision, F.N., M.K.; Project Administration, M.K. All authors have read and agreed to the published version of the manuscript.

Funding: The research described in this paper was funded by the research project "its4land", which is part of the Horizon 2020 program of the European Union, project number 687828.

Acknowledgments: We acknowledge the support of all its4land project partners and local stakeholders in facilitating the UAV flight missions in Europe and Africa.

Conflicts of Interest: The authors declare no conflict of interest.

References

1. World Economic Forum. *Unlocking Technology for the Global Goals*; World Economic Forum: Geneva, Switzerland, 2020.
2. Jazayeri, I.; Rajabifard, A.; Kalantari, M. A geometric and semantic evaluation of 3D data sourcing methods for land and property information. *Land Use Policy* **2014**, *36*, 219–230. [CrossRef]
3. Shakhatreh, H.; Sawalmeh, A.H.; Al-Fuqaha, A.; Dou, Z.; Almaita, E.; Khalil, I.; Othman, N.S.; Khreishah, A.; Guizani, M. Unmanned Aerial Vehicles (UAVs): A Survey on civil applications and key research challenges. *IEEE Access* **2019**, *7*, 48572–48634. [CrossRef]
4. Barnes, G.; Volkmann, W. High-resolution mapping with unmanned aerial systems. *Surv. L. Inf. Sci.* **2015**, *74*, 5–13.
5. Kurczynski, Z.; Bakuła, K.; Karabin, M.; Markiewicz, J.S.; Ostrowski, W.; Podlasiak, P.; Zawieska, D. The possibility of using images obtained from the uas in cadastral works. *Int. Arch. Photogramm. Remote Sens. Spat. Inf. Sci.-ISPRS Arch.* **2016**, *41*, 909–915. [CrossRef]
6. Rijsdijk, M.; Van Hinsbergh, W.H.M.; Witteveen, W.; Buuren, G.H.M.; Schakelaar, G.A.; Poppinga, G.; Van Persie, M.; Ladiges, R. Unmanned aerial systems in the process of juridical verification of cadastral border. *Int. Arch. Photogramm. Remote Sens.* **2013**, *40*, 4–6. [CrossRef]
7. Manyoky, M.; Theiler, P.; Steudler, D.; Eisenbeiss, H. Unmanned aerial vehicle in cadastral applications. *ISPRS-Int. Arch. Photogramm. Remote Sens. Spat. Inf. Sci.* **2012**, *38*, 57–62. [CrossRef]
8. Mumbone, M.; Bennett, R.M.; Gerke, M.; Volkmann, W. Innovations in boundary mapping: Namibia, customary lands and UAVs. In Proceedings of the Land and Poverty Conference 2015: Linking Land Tenure and Use for Shared Prosperity, Washington, DC, USA, 23–27 March 2015.
9. Koeva, M.; Stöcker, C.; Crommelinck, S.; Ho, S.; Chipofya, M.; Sahib, J.; Bennett, R.; Zevenbergen, J.; Vosselman, G.; Lemmen, C.; et al. Innovative remote sensing methodologies for Kenyan land tenure mapping. *Remote Sens.* **2020**, *12*, 273. [CrossRef]
10. Koeva, M.; Muneza, M.; Gevaert, C.; Gerke, M.; Nex, F. Using UAVs for map creation and updating. A case study in Rwanda. *Surv. Rev.* **2016**, *50*, 1–14. [CrossRef]
11. Ramadhani, S.A.; Bennett, R.M.; Nex, F.C. Exploring UAV in Indonesian cadastral boundary data acquisition. *Earth Sci. Informatics* **2018**, *11*, 129–146. [CrossRef]
12. Crommelinck, S.; Koeva, M.N.; Yang, M.Y.; Vosselman, G. Interactive cadastral boundary delineation from UAV data. In Proceedings of the ISPRS Annals of Photogrammetry, Remote Sensing and Spatial Information Sciences, Riva del Garda, Italy, 4–7 June 2018; Volume 4, pp. 4–7.
13. Yu, X.; Zhang, Y. Sense and avoid technologies with applications to unmanned aircraft systems: Review and prospects. *Prog. Aerosp. Sci.* **2015**, *74*, 152–166. [CrossRef]
14. Fetai, B.; Oštir, K.; Fras, M.K.; Lisec, A. Extraction of visible boundaries for cadastral mapping based on UAV imagery. *Remote Sens.* **2019**, *11*, 1510. [CrossRef]
15. Xia, X.; Persello, C.; Koeva, M. Deep fully convolutional networks for cadastral boundary detection from UAV images. *Remote Sens.* **2019**, *11*, 1725. [CrossRef]
16. International Standardization Organization (ISO). *ISO 19157: 2013 Geographic Information-Data Quality*; European Committee for Standardization: Brussels, Belgium, 2013.

17. Grant, D.; Enemark, S.; Zevenbergen, J.; Mitchell, D.; McCamley, G. The Cadastral triangular model. *Land Use Policy* **2020**, *97*, 104758. [CrossRef]
18. Förstner, W.; Gülch, E. A Fast Operator for Detection and Precise Location of Distinct Points, Corners and Centres of Circular Features. In Proceedings of the ISPRS Intercommission Conference on Fast Processing of Photogrammetric Data, Interlaken, Switzerland, 2–4 June 1987; pp. 281–305.
19. Lowe, G. SIFT—The Scale Invariant Feature Transform. *Int. J.* **2004**, *2*, 91–110.
20. Snavely, N.; Seitz, S.M.; Szeliski, R. Photo tourism: Exploring image collections in 3D. *Proc. SIGGRAPH 2006* **2006**, *1*, 835–846.
21. Remondino, F.; Spera, M.G.; Nocerino, E.; Menna, F.; Nex, F. State of the art in high density image matching. *Photogramm. Rec.* **2014**, *29*, 144–166. [CrossRef]
22. Gruen, A. Development and status of image matching in photogrammetry. *Photogramm. Rec.* **2012**, *27*, 36–57. [CrossRef]
23. Singh, K.K.; Frazier, A.E. A meta-analysis and review of unmanned aircraft system (UAS) imagery for terrestrial applications. *Int. J. Remote Sens.* **2018**, *39*, 5078–5098. [CrossRef]
24. Rehak, M.; Skaloud, J. Applicability of new approaches of sensor orientation to micro aerial vehicles. *ISPRS Ann. Photogramm. Remote Sens. Spat. Inf. Sci.* **2016**, *3*, 441–447. [CrossRef]
25. Pfeifer, N.; Glira, P.; Briese, C. Direct georeferencing with on board navigation components of light weight UAV platforms. *Int. Arch. Photogramm. Remote Sens. Spat. Inf. Sci.-ISPRS Arch.* **2012**, *39*, 487–492. [CrossRef]
26. Turner, D.; Lucieer, A.; Wallace, L. Direct georeferencing of ultrahigh-resolution UAV imagery. *IEEE Trans. Geosci. Remote Sens.* **2014**, *52*, 2738–2745. [CrossRef]
27. Haala, N.; Cramer, M.; Weimer, F.; Trittler, M. Performance test on UAV-based photogrammetric data collection. *ISPRS-Int. Arch. Photogramm. Remote Sens. Spat. Inf. Sci.* **2012**, *38*, 7–12. [CrossRef]
28. Eling, C.; Klingbeil, L.; Kuhlmann, H. Development of an RTK-GPS System for Precise Real-time Positioning of Lightweight UAVs. In *Ingenieurvermessung 14, Proceedings of the 17. Ingenieurvermessungskurs, Zürich, Switzerland, 14–17 January 2014*; Wieser, A., Ed.; S. Wichmann Verlag: Berlin, Germang, 2014; pp. 111–123.
29. Stöcker, C.; Nex, F.; Koeva, M.; Gerke, M. Quality assessment of combined IMU/GNSS data for direct georeferencing in the context of UAV-based mapping. In Proceedings of the International Archives of the Photogrammetry, Remote Sensing and Spatial Information Sciences—ISPRS Archives, Colone, Germany, 4–7 September 2017; Volume 42.
30. Gerke, M.; Przybilla, H.J. Accuracy analysis of photogrammetric UAV image blocks: Influence of onboard RTK-GNSS and cross flight patterns. *Photogramm.-Fernerkundung-Geoinf.* **2016**, *2016*, 17–30. [CrossRef]
31. Hugenholtz, C.; Brown, O.; Walker, J.; Barchyn, T.; Nesbit, P.; Kucharczyk, M.; Myshak, S. Spatial accuracy of UAV-derived orthoimagery and topography: Comparing photogrammetric models processed with direct georeferencing and ground control points. *Geomatica* **2016**, *70*, 21–30. [CrossRef]
32. Forlani, G.; Dall'Asta, E.; Diotri, F.; di Cella, U.M.; Roncella, R.; Santise, M. Quality assessment of DSMs produced from UAV flights georeferenced with onboard RTK positioning. *Remote Sens.* **2018**, *10*, 311. [CrossRef]
33. Ekaso, D.; Nex, F.; Kerle, N. Accuracy assessment of real-time kinematics (RTK) measurements on unmanned aerial vehicles (UAV) for direct georeferencing. *Geo-Spatial Inf. Sci.* **2020**, *23*, 165–181. [CrossRef]
34. James, M.R.; Robson, S.; Centre, L.E.; Engineering, G. Mitigating systematic error in topographic models derived from UAV and ground-based image networks. *Earth Surf. Process. Landf.* **2014**, *1420*, 1413–1420. [CrossRef]
35. Manfreda, S.; Dvorak, P.; Mullerova, J.; Herban, S.; Vuono, P.; Arranz Justel, J.; Perks, M. Assessing the accuracy of digital surface models derived from optical imagery acquired with unmanned aerial systems. *Drones* **2019**, *3*, 15. [CrossRef]
36. Mesas-Carrascosa, F.J.; Torres-Sánchez, J.; Clavero-Rumbao, I.; García-Ferrer, A.; Peña, J.M.; Borra-Serrano, I.; López-Granados, F. Assessing optimal flight parameters for generating accurate multispectral orthomosaicks by uav to support site-specific crop management. *Remote Sens.* **2015**, *7*, 12793–12814. [CrossRef]
37. Villanueva, J.K.S.; Blanco, A.C. Optimization of ground control point (GCP) configuration for unmanned aerial vehicle (UAV) survey using structure from motion (SFM). *Int. Arch. Photogramm. Remote Sens. Spat. Inf. Sci.-ISPRS Arch.* **2019**, *42*, 167–174. [CrossRef]

38. Tonkin, T.N.; Midgley, N.G. Ground-control networks for image based surface reconstruction: An investigation of optimum survey designs using UAV derived imagery and structure-from-motion photogrammetry. *Remote Sens.* **2016**, *8*, 786. [CrossRef]
39. Sanz-Ablanedo, E.; Chandler, J.H.; Rodríguez-Pérez, J.R.; Ordóñez, C. Accuracy of Unmanned Aerial Vehicle (UAV) and SfM photogrammetry survey as a function of the number and location of ground control points used. *Remote Sens.* **2018**, *10*, 1606. [CrossRef]
40. Agüera-Vega, F.; Carvajal-Ramírez, F.; Martínez-Carricondo, P. Assessment of photogrammetric mapping accuracy based on variation ground control points number using unmanned aerial vehicle. *Measurement* **2017**, *98*, 221–227. [CrossRef]
41. Oniga, V.E.; Breaban, A.I.; Pfeifer, N.; Chirila, C. Determining the suitable number of ground control points for UAS images georeferencing by varying number and spatial distribution. *Remote Sens.* **2020**, *12*, 876. [CrossRef]
42. James, M.R.; Robson, S.; d'Oleire-Oltmanns, S.; Niethammer, U. Optimising UAV topographic surveys processed with structure-from-motion: Ground control quality, quantity and bundle adjustment. *Geomorphology* **2017**, *280*, 51–66. [CrossRef]
43. Lowe, D.G. Object recognition from local scale-invariant features. *Proc. IEEE Int. Conf. Comput. Vis.* **1999**, *2*, 1150–1157.
44. Bay, H.; Ess, A.; Tuytelaars, T.; Van Gool, L. Speeded-up robust features (SURF). *Comput. Vis. Image Underst.* **2008**, *110*, 346–359. [CrossRef]
45. Alcantarilla, P.F.; Nuevo, J.; Bartoli, A. Fast explicit diffusion for accelerated features in nonlinear scale spaces. In Proceedings of the BMVC 2013-Electronic Proceedings of the British Machine Vision Conference, Bristol, UK, 9–13 September 2013.
46. Gonzales-Aguilera, D.; Ruiz de Ona, E.; Lopez-Fernandez, L.; Farella, E.M.; Stathopoulou, E.; Toschi, I.; Remondino, F.; Fusiello, A.; Nex, F. Photomatch: An open-source multi-view and multi-modal feature matching tool for photogrammetric applications. *ISPRS Ann. Photogramm. Remote Sens. Spat. Inf. Sci.* **2020**, *43*, 213–219. [CrossRef]
47. Lu, C.; Xu, L.; Jia, J. Contrast preserving decolorization. In Proceedings of the 2012 IEEE International Conference on Computational Photography (ICCP), Seattle, WA, USA, 28–29 April 2012.
48. Crommelinck, S. Delineation Tool. Available online: https://github.com/SCrommelinck/delineation-tool (accessed on 10 July 2020).
49. Pont-Tuset, J.; Arbelaez, P.; Barron, J.T.; Marques, F.; Malik, J. Multiscale combinatorial grouping for image segmentation and object proposal generation. *IEEE Trans. Pattern Anal. Mach. Intell.* **2017**, *39*, 128–140. [CrossRef]
50. Crommelinck, S.; Koeva, M.; Yang, M.Y.; Vosselman, G. Application of deep learning for delineation of visible cadastral boundaries from remote sensing imagery. *Remote Sens.* **2019**, *11*, 2505. [CrossRef]
51. Seifert, E.; Seifert, S.; Vogt, H.; Drew, D.; van Aardt, J.; Kunneke, A.; Seifert, T. Influence of drone altitude, image overlap, and optical sensor resolution on multi-view reconstruction of forest images. *Remote Sens.* **2019**, *11*, 1252. [CrossRef]

 remote sensing

MDPI

Article

Remote Sensing for Property Valuation: A Data Source Comparison in Support of Fair Land Taxation in Rwanda

Mila Koeva [1,*], Oscar Gasuku [2], Monica Lengoiboni [1], Kwabena Asiama [3], Rohan Mark Bennett [4,5], Jossam Potel [6] and Jaap Zevenbergen [1]

1 Faculty of Geo-Information Science and Earth Observation (ITC), University of Twente, Drienerlolaan 5, 7522 NB Enschede, The Netherlands; m.n.lengoiboni@utwente.nl (M.L.); j.a.zevenbergen@utwente.nl (J.Z.)
2 Director of Infrastructure, Osc Infrastructures Private Limited, Rubavu District, Gisenyi 173, Rwanda; oscdirector@rubavu.gov.rw
3 Geodetic Institute, Leibniz Universität Hannover, Nienburger Strasse 1, 30167 Hannover, Germany; asiama@gih.uni-hannover.de
4 Department of Business Technology and Entrepreneurship, School of Business, Law and Entepreneurship, Swinburne University of Technology, John St, Hawthorn, VIC 3122, Australia; rohanbennett@swin.edu.au
5 Kadaster Netherlands, The Dutch Cadastre, Land Registry, and National Mapping Agency, 7311 KZ Apeldoorn, The Netherlands
6 Department of Land Administration and Management, Faculty of Applied Fundamental Sciences, INES, Musanze, NM Street, Ruhengeri 155, Rwanda; jossam2003@ines.ac rw
* Correspondence: m.n.koeva@utwente.nl; Tel.: +31-534-874-410

Abstract: Remotely sensed data is increasingly applied across many domains, including fit-for-purpose land administration (FFPLA), where the focus is on fast, affordable, and accurate property information collection. Property valuation, as one of the main functions of land administration systems, is influenced by locational, physical, legal, and economic factors. Despite the importance of property valuation to economic development, there are often no standardized rules or strict data requirements for property valuation for taxation in developing contexts, such as Rwanda. This study aims at assessing different remote sensing data in support of developing a new approach for property valuation for taxation in Rwanda; one that aligns with the FFPLA philosophy. Three different remote sensing technologies, (i) aerial images acquired with a digital camera, (ii) WorldView2 satellite images, and (iii) unmanned aerial vehicle (UAV) images obtained with a DJI Phantom 2 Vision Plus quadcopter, are compared and analyzed in terms of their fitness to fulfil the requirements for valuation for taxation purposes. Quantitative and qualitative methods are applied for the comparative analysis. Prior to the field visit, the fundamental concepts of property valuation for taxation and remote sensing were reviewed. In the field, reference data using high precision GNSS (Leica) was collected and used for quantitative assessment. Primary data was further collected via semi-structured interviews and focus group discussions. The results show that UAVs have the highest potential for collecting data to support property valuation for taxation. The main reasons are the prime need for accurate-enough and up-to-date information. The comparison of the different remote sensing techniques and the provided new approach can support land valuers and professionals in the field in bottom-up activities following the FFPLA principles and maintaining the temporal quality of data needed for fair taxation.

Keywords: property valuation; property taxation; remote sensing; land; UAV

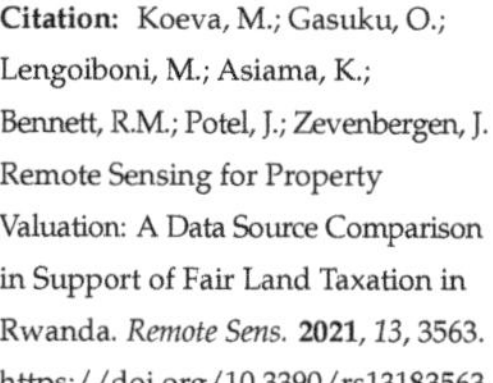

Citation: Koeva, M.; Gasuku, O.; Lengoiboni, M.; Asiama, K.; Bennett, R.M.; Potel, J.; Zevenbergen, J. Remote Sensing for Property Valuation: A Data Source Comparison in Support of Fair Land Taxation in Rwanda. *Remote Sens.* **2021**, *13*, 3563. https://doi.org/10.3390/rs13183563

Academic Editor: Yuji Murayama

Received: 21 July 2021
Accepted: 31 August 2021
Published: 8 September 2021

1. Introduction

Remote sensing (RS) data has been shown to be efficient in obtaining precise spatial information for a variety of applications, which is crucial to achieving sustainable development goals (SDGs). Property ownership, value, and rights are included as sub-goal 1.4 of the SDGs. Nowadays, being part of the fourth industrial revolution, remotely sensed images are ubiquitous in many socio-economic endeavors. Therefore, the use of remotely sensed data is highly advised for use in fit-for-purpose land administration (hereafter

FFPLA) [1]. Moreover, contemporary developments in photogrammetry and computer vision, coupled with high-resolution remote sensing data, has led many researchers to explore the use of machine learning to extract information automatically from images for cadastral applications [2].

Satellite images in various spatial and temporal resolutions are globally available, making them very useful for monitoring daily dynamics. Dabrowski and Latos [3] investigated the applicability of remote sensing images for land-related applications focusing on the effect of the different spatial, radiometric, temporal, and spectral resolution. Haeusler, Gomez, and Enßle [4] and Ali and Deininger [5] showed that remote sensing data, especially high-resolution satellite imagery (HRSI), can be used to extract or measure the height of buildings, which is useful for urban planning, assessment of property taxes, estimation of floor area, and so on. Jain [6] acquired socioeconomic attributes, roof material, shape, structure of buildings, and the age of construction from high-resolution imagery using object-based classification for the purpose of property taxation.

However, for deriving precise property characteristics needed for valuation for taxation purposes, higher spatial resolution is preferable. Recently, unmanned aerial vehicles (UAVs) have proven to be promising for many applications such as agriculture [7], mapping [8], surveying and cadaster applications [9–12], architecture and archeology [13], cultural heritage [14], among others. However, to the best of the authors' knowledge, the applicability of UAV images for property valuation has not been examined empirically [15]. Therefore, the current study aims to assess and compare different remote sensing data for property valuation for taxation, focusing on Rwanda as a case study.

Property valuation is done for different purposes, including taxation, sales, and insurance, amongst others. Valuation is a process of estimating the amount for which a property will be exchanged or the amount of taxes that should be paid for it on a particular date [16,17]. Being an art and a science, the complexity of the valuation process covers, among others, transparency of the market, diverse purposes, and stakeholders involved [16,18]. It is further strongly influenced by the background and experience of the valuer, as well as the global trend of economic development and investment interests [19,20].

There are two types of valuation procedures. As stated by Wallace and Williamson [21], the first is the valuation of a single property, and the second is so-called mass valuation, which refers to an area combining many properties. Valuations for taxation are usually carried out through a mass valuation. Property taxes are defined as the amount of money levied on a person, natural or unnatural, by a government, for the holding of real estate within a particular jurisdiction. Different approaches have been developed worldwide for the valuation of property for taxation, including artificial neural networks (ANNs), hedonic pricing methods, spatial analysis methods, and others [22].

Property tax collection approaches differ from country to country. However, the two main approaches are area-based taxation and value-based [23]. Typically, area-based taxations are used for determining the assessed value of the property where the property markets are not mature enough to support a value-based system [24]. For area-based tax, the taxes build on the unit, and therefore the unit assessment must be accounted for in the rate [25]. The rate is levied per m^2 of land area, building area, or a combination of the two. The unit's value reflects several factors such as the property's location, accessibility, land use/zoning, laws and regulations, socioeconomics, condition, age, and neighborhood development. Value-based taxation is based on market value (capital or rental).

There are many factors that affect property value such as locational, physical, legal, and economic factors [18,26]. They can be split into two groups, internal and external factors (Figure 1). This research is focused on the use of remotely sensed data for extraction of physical attributes or geospatial factors (e.g., parcel area, built-up area/gross floor area) and locational factors (e.g., accessibility, neighborhood development, and environment).

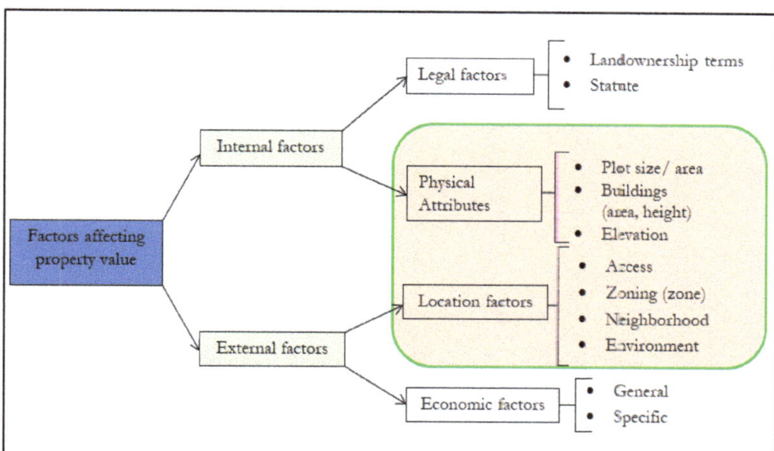

Figure 1. Factors affecting property value (Source: adapted based on [26]).

Rwanda has both property tax and a land lease fee, both of which vary depending on the tax base. The property tax is determined based on the open market value. In Rwanda, professional valuers prepare the valuation report of the property (parcel, buildings, and improvement on it). The tax amount is 0.1% of the total value of the property for industrial buildings, 1% for residential properties, and 0.5% for commercial buildings [27]. However, the land lease fee is based on the m^2 rate determined by the district [28,29]. The property tax relies on the percentage of the open market value of the property [30]. Both single property and mass valuation are used to assess the open market value of the property [21].

In 2008 and 2009, a traditional aerial survey with a 3000 m flying height was executed over the territory of Rwanda with a digital photogrammetric camera Vexcel UltraCamX. Post-processing procedures were completed by a Dutch photogrammetric company [31]. A digital elevation model and an orthophoto with a spatial resolution of 22 cm were further produced for the entire country [31]. The Rwandan land cadastre was built based on the orthophotos from these aerial and satellite images. Local citizens employed and mainly trained as 'para-surveyors' delineated the parcel limits on the imagery printouts that were scanned, geo-referenced, and digitized. Currently, property valuation relies on the data provided by the Rwanda land use management authority (RLMUA), especially the parcel area, the location of the property, and the land use; all this information can be obtained from an inspection of the land title and also from a field visit. For instance, the Rwanda National land-use masterplan, as well as the subsequent master plan for Kigali city, were developed based on the generated orthophotos from aerial images. These are currently sources of property valuation data [32]. However, substantial changes since 2009 have seen the database missing information such as buildings, improvements, accessibility, conditions, and zoning [33]. Therefore, methods for regular updating of the cadastral and valuation information are of high importance.

In summary, the usage of RS data for property valuation for taxation is still exploring innovations in the land administration domain, particularly as higher resolution data becomes cheaper and easier to capture more frequently. Therefore, the current research aims to assess and compare three different remote sensing technologies for property valuation for taxation, focusing on Rwanda as a case study, and ultimately to propose a new UAV-based approach. In alignment with the FFPLA principles, the study aims to glean lessons for that specific country context, but also for RS application in the domain more generally.

In the next section, the materials and methods used are presented, framed as a comparative analysis of different datasets over the case context of Rwanda. Results are presented first with regard to the data requirements for property valuation in Rwanda, and then an assessment of the different datasets against those requirements. The discussion and conclusion focus on discerning the immediacy of application of the results in Rwanda, and particularly the identification of the most relevant data source for property valuation for taxation purposes in Rwanda. Attention is also given to broader applications beyond Rwanda and further research requirements.

2. Materials and Methods

2.1. Study Area

The study area of Nyarutarama cell-Remera sector Gasabo districts (Figure 2) was selected for the current research. This area witnessed many changes over the years, which are reflected in differences in the values of the properties.

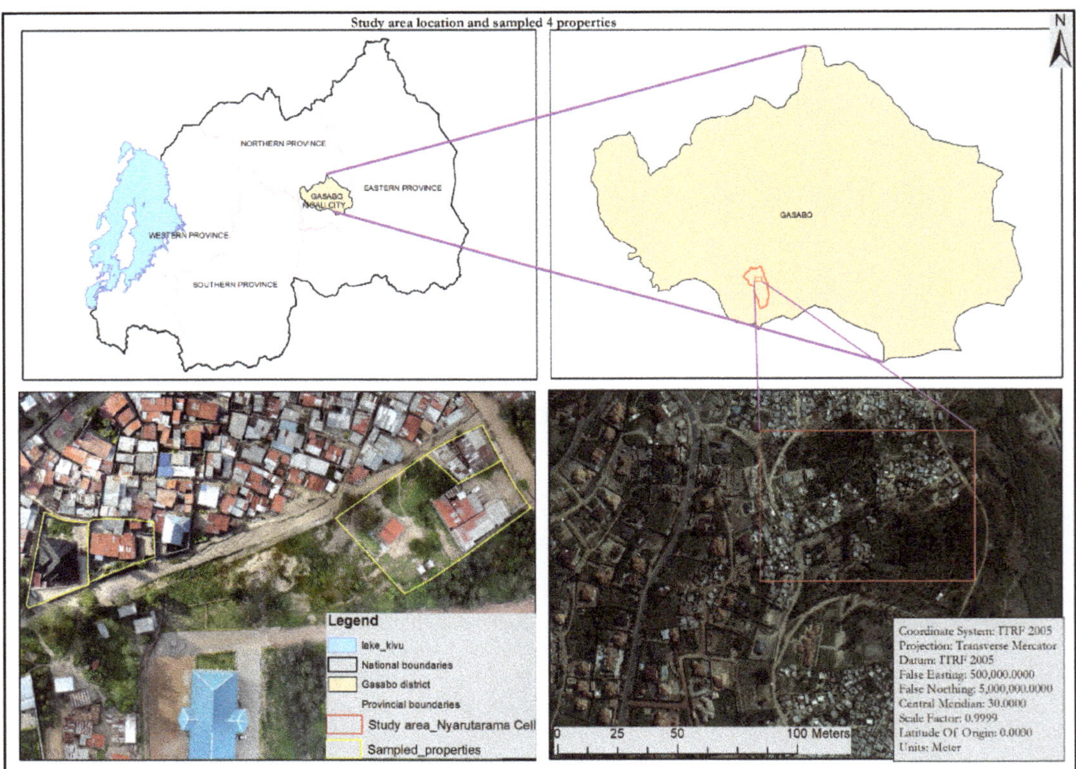

Figure 2. Location of the study area.

2.2. Data Sources

As shown in Table 1 and Figure 3 below, the orthophoto based on aerial images acquired in 2008/2009 with a digital camera on board an airplane was provided by the RLMU, and the satellite image was obtained from the Image repository of the University of Twente (ITC). The UAV images used for this research were collected in 2015 with a DJI Phantom 2 Vision Plus quadcopter at a flying height of 50 m. For the study, a total of 1172 geotagged nadir images were captured, and only 954 were obtained. However, an average of 85% forward and 75% side-overlaps was achieved. An orthophoto covering 950 m^2 with a spatial resolution of 3.3 cm and a radiometric resolution of 8 bits was produced with Pix4DCapture software. The final orthophoto with a positional accuracy of

6.0 cm was produced based on 13 premeasured ground control points with Leica GNSS with an accuracy of 2 cm [8].

Table 1. Used datasets and their sources.

Dataset	Source	Acquisition Date	Spatial Resolution	Radiometric Resolution	Spectral Resolution
Orthophoto from airplane aerial images	RLMU	2008/2009	22 cm	8-bit	3 bands
Satellite Worldview2 image	ITC image repository	2013	50 cm	16-bit	4 bands
Orthophoto from aerial UAV images	ITC image repository	2015	3.3 cm	8-bit	4 bands

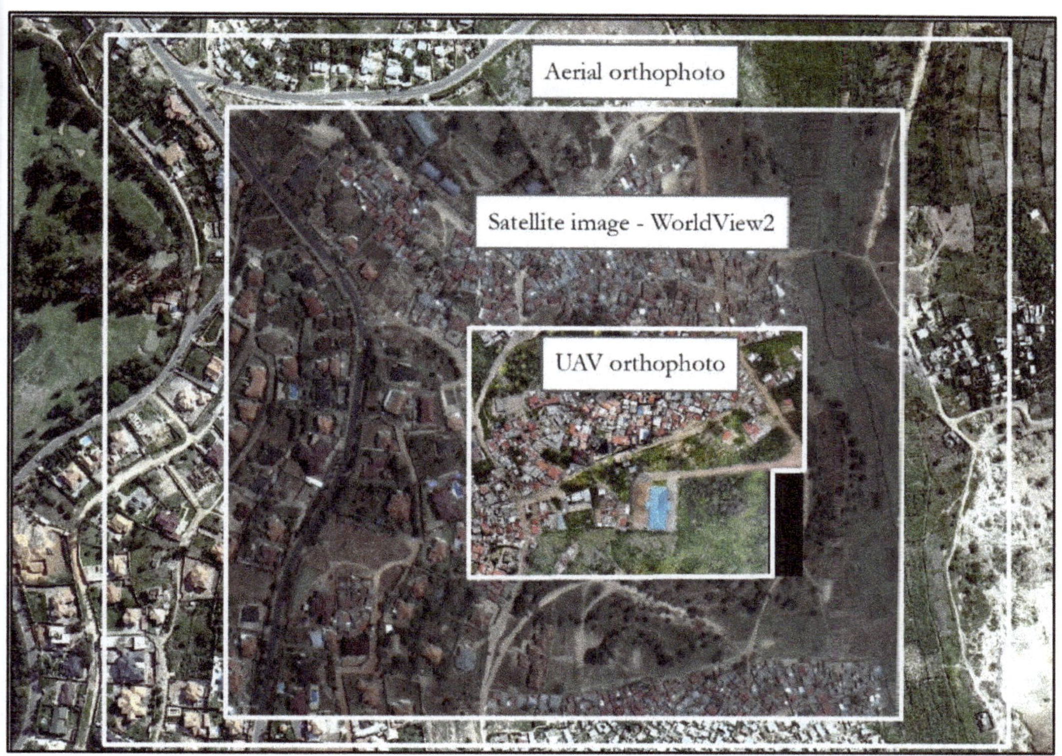

Figure 3. Coverage of the remote sensing data in Nyarutarama cell.

In this research, a mixed-methods approach combining quantitative and qualitative methods were used. The qualitative and quantitative data were given equal weighting and considered to be captured in parallel. Prior to the field visit, the concepts of property valuation for taxation and remote sensing were reviewed (Figure 4).

First, in terms of quantitative data collection, using high precision GNSS (Leica), reference data during the field survey was collected to be used for quantitative assessment of the extracted spatial features and property boundaries based on the images.

Second, in terms of qualitative data, primary data was collected via semi-structured interviews [34], focus group discussions, and field surveys aiming to assess the different RS data for valuation purposes. This interview technique was used to gather social perceptions and included the presentation of oral–verbal stimuli and replies [35]. In this research, face to face personal interviews were also completed. The experts sampling method [36] was used for the selection of the respondents based on their organization and functions, as shown in Table 2.

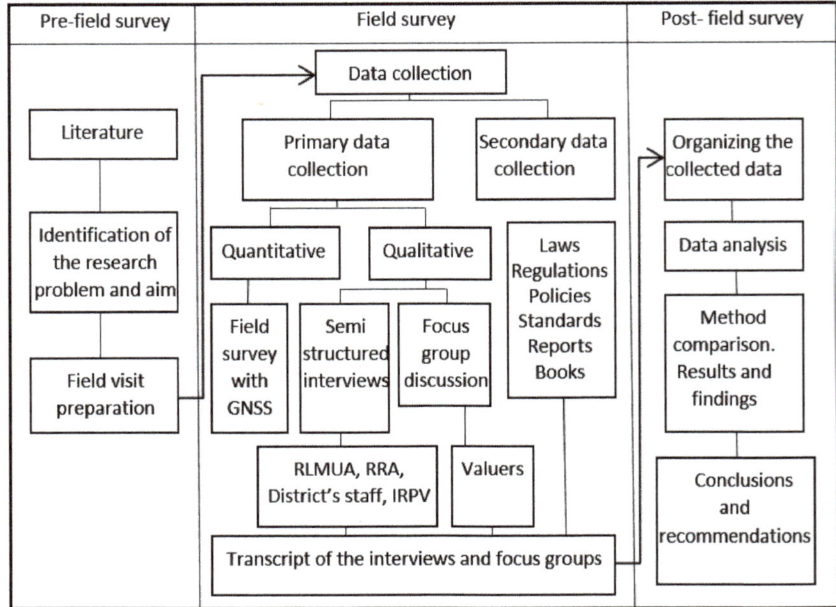

Figure 4. Research workflow.

Table 2. Number of respondents by institution.

Institutions	Category	Number of People
RLMUA former RNRA		2
RRA	Central government	3
Ministry of Infrastructure		1
District staff	Local government	3
RCMRD	Regional	1
IRPV	Private institution (private valuers)	3
	Total	13

Further data was collected using focus group discussions, involving thirteen people with a background and interest in valuation [37]. The topic of the discussion was "Comparison of remote sensing techniques for valuation for taxation in Rwanda". The aim of the interviews and the focus group discussion was to collect individual and group information on perceptions of the current property valuation for taxation purposes and to assess the advantages of the different remote sensing approaches for property valuation for taxation Open-ended questions were used [38] as a quantitative approach for collecting numerical data. The qualitative method was used to understand how the current property valuation for taxation system works in Rwanda and who are the key players and their specific roles. The perceptions of the stakeholders, especially government institutions, on using the remote sensing data for property valuation for taxation purposes was also investigated.

2.3. Methods for Data Processing and Analysis

In terms of analysis, for the RS dataset comparison, that is, the assessment of the usefulness of the different remote sensing techniques, four properties in the study area were selected based on the criteria, which are shown in Table 3, inspired by FFPLA elements. Moreover, as the acquisition years of the three datasets are different (2008, 2013, and 2015), the properties have been selected where there is no change during these years in terms of their area. This selection was especially needed for the accuracy assessment.

Table 3. Criteria of sampled and surveyed properties.

Sampled Property	Criteria
Property 1	Accessibility Clearly visible boundary Developed land (Building improvements) Visibility of the building footprints Comparison based on all RS data
Property 2	Accessibility
Property 3	Comparison based on all RS data
Property 4	Developed land (Building improvements) Visibility of the building footprints Comparison based on all RS data

An analysis of the qualitative data guided by FFPLA elements' [1] text-based analysis is used, dividing the factors into the following themes: (1) factors affecting or influencing the property value; (2) characteristics of generated orthophotos from remote sensing data; (3) time/availability of the platform; and (4) cost of acquiring the data, including the cost of hiring the platform. ArcGIS software was used for the quantitative assessment and the comparison of the parcel areas, location visualization of the subject property and neighborhood, and the coordinate comparison. The interviews and focus group discussions from participants were transcribed and analyzed using ATLAS.Ti 8.0 based on the developed themes. Thematic analysis based on the literature review on the factors was used to examine the collected data.

3. Results

In this section, the results are presented in separate sections following the idea of the legal, governmental, and technical framework introduced in FFPLA. In the first section, policies, laws, standards, and types of property taxes are described. This is followed by the data requirements and methods for data collection and analysis. Afterwards, the existing limitations for using RS data and considerations related to its application and comparison with the newly proposed UAV method are shown.

3.1. Policies, Laws, Standards and Types of Properties in Rwanda

The real property valuation profession in Rwanda is regulated by law N° 17/2010 [39]. It specifies the structure of the Institute of Real Property Valuers (IRPV) and defines their responsibilities. It describes the methods currently used for data analysis, but it does not specify any methods for data collection for property valuation [39]. Before 2010, property valuation was carried out mostly by civil engineers who had some training and experience in property valuation methods [40].

During the interviews with the IRPV officials, two of the three respondents stated that: "the valuation standard in Rwanda work as a practicing guide".

However, there is confusion in understanding and applying the standard and law at a local level. Therefore, international valuation standards are used. It is the responsibility of IRPV to develop the valuation standard, and it requires the approval of the property valuation regulatory council before using it.

The two forms of tenure in Rwanda are full ownership (freehold title) and long leasehold title. About 98% of Rwandan land is held under leasehold and 2% under freehold. Rwanda's land law requires landowners with a freehold title to pay property tax while emphyteutic leaseholders pay lease fees. This means that only 2% of Rwandan land is taxable. The Rwandan National Land Policy requires that "land transactions and land taxation be included in land administration as elements of land development" [41].

The current property tax (fixed asset tax) is amended by law N° 75/2018 of 7 September 2018, which determines the sources of revenue and property of decentralized entities. Additionally presidential order N° 25/01 of 9 July 2012 establishes the list of fees and other charges imposed by decentralized entities and determines their thresholds. The current laws related to property tax in Rwanda are categorized into three different taxes as specified by the taxation law and confirmed by the Rwanda Revenue Authority (RRA), which was confirmed by districts officials interviewed during fieldwork, including fixed asset tax, land lease fees, and rental income [29]. During the interviews, respondents highlighted that "fixed asset tax or property tax requires a valuation report and its taxes are based on the market value whereby the tax value is equal to 1/1000 of the open market value of the property".

Ministerial order N° 005/12/10/TC of 22 June 2012 determines the modalities for the implementation of law N° 59/2011 of 31 December 2011. Ministerial order N° 005/12/10/TC of 22 June 2012 determines the modalities for the implementation of taxation law. It specifies the process and suggests the steps to be followed by the taxpayer for all types of taxes. The steps and key actors involved in property tax (fixed asset tax) are visualized in Figure 5 below. These include RRA, districts, and taxpayers, Rwanda Land Management and Use Authority (RLMUA), valuers, Institute of Real Property Valuers (IRPV), and Rwanda Development Board (RDB).

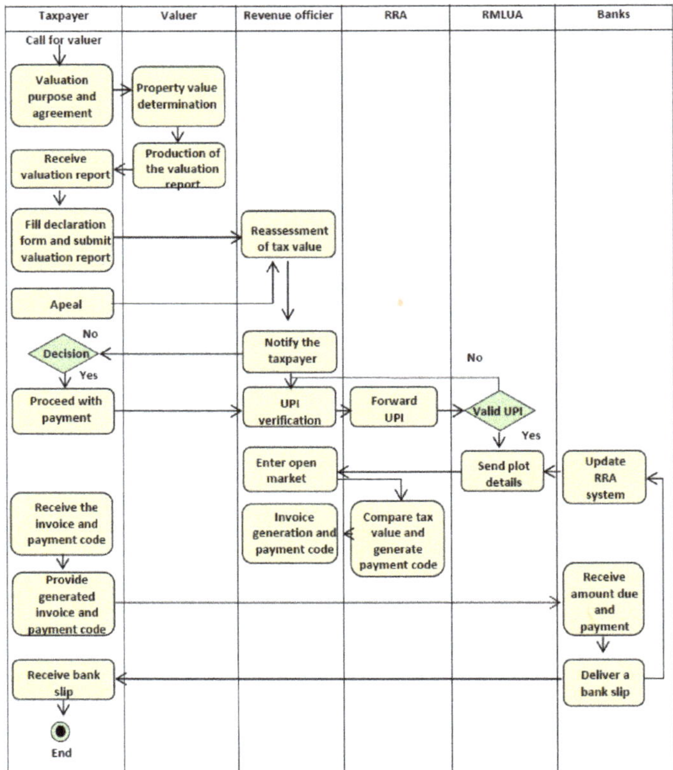

Figure 5. Property tax system in Rwanda.

Three types of property taxes are levied on property in Rwanda. They are lease fees, rental income taxes, and fixed asset taxes (derived from the interview with district and RRA officials). The difference between these three types of taxes is based on the land tenure type. Land lease fees are the tax levied from leaseholders; freeholders pay a fixed asset tax, and rental income tax can be levied on both land tenure regimes. All these taxes are levied on an annual basis. The only type of tax that requires a valuation report is fixed asset tax as it is based on open market value.

3.1.1. Land Lease Fees

Land lease fees are paid annually. The method used to determine the tax value is based on the rate per square meter or hectare. The rate is determined by the district council depending on the infrastructure and development in an area. The presidential order determining the list of fees and other charges levied by decentralized entities and determining the thresholds in article 9 states that: "Any person owning land and holding a land lease certificate issued by the competent organization shall pay an annual land lease fee based on the square meter or hectare". The same article declares that it is the responsibility of the district council to determine the fees to be paid annually based on the available infrastructure in the area where the land is located and its use. The thresholds of land lease fee rates are arranged from 30 Rwf (around 0.03 euro) to 80 Rwf (0.08 euro) per square meter and (4000 Rwf) per hectare levied from agriculture and livestock land with more than two hectares.

3.1.2. Rental Income Taxes

Rental income tax is a tax imposed on individuals who earn income from the rented immovable property. Taxes are paid annually based on rental fees the landowners collect from their tenants, and this tax is also paid in terms of percentages of generated income from the property after deduction of expenses. Taxation law article 50 states that the expenses should be 50% of the generated income per year. The tax rate ranges from 0–30%, depending on the generated income. The more income generated, the higher the rate applied.

3.1.3. Fixed Asset Taxes

Fixed asset tax, also called "property tax" is a tax imposed on immovable property with a freehold title. Through interviews from district and RRA officials, they highlighted that: "property tax requires a valuation report as it is based on the open market value that is why they require an expert in valuation to determine the open market value". Fixed asset tax value is normally based on the value of the property. The rate that is applied to the value is fixed at a thousandth (1/1000) of the taxable value per year. During fieldwork, interviewees underlined that: "they are still facing a problem of taxpayers who declare the property and hide some information related to their property for instance when they have a number of buildings within a few parcels, because the purpose is for taxation they undermine the value and report only one building". However, a mechanism of monitoring changes on the ground is needed so that taxpayers do not hide information that can be captured easily.

Fixed asset tax value can be updated once every four years as stipulated by taxation law article 15. However, if improvements or changes are made to the property, the taxpayer must file an updated valuation report (new self-assessed tax) and fill in a new tax declaration or assessment notice [29]. Interviewees highlighted that: "if a property is residentially used, the fixed asset tax value should be determined after deducting the amount equal to three million (≈3000 EUR) on its market value". This is specified in article 18, point number 8 of the taxation law, on tax exempted properties.

3.2. *Data Requirements for Property Valuation for Taxation in Rwanda*

The required data for property valuation, based on the current approaches for taxation purposes, depends on the type of property and its use. The most important requirement is the land ownership titles (lease or freeholds). A land title is the one that shows the basis upon which the land is held (whether it is under a leasehold or freehold title). Taxpayers with freehold titles require valuation reports as tax basis calculations. The necessary information required to value the property for taxation purposes, as said by participants, includes: "land certificate ownerships, built-up area, technical conditions of the property, construction materials, infrastructure attached to it, land use".

Most of the information is collected from the field, and others are retrieved from the land title (LAIS database) (Table 4).

Table 4. Required data and source availability.

Required Data	Fieldwork	Land Title LAIS	Masterplan	Google Earth	Valuers	Estate Agencies
Parcel area	✔	✔				
Current land use (on title)	✔	✔				
Built-up area	✔					
Construction materials	✔					
External works	✔			✔		
Status of the property	✔					
Infrastructure attached				✔		
Planned land use		✔	✔			
Sales comparable	✔	✔			✔	✔
Location	✔	✔	✔	✔		

The taxes for leaseholders do not require a valuation report. All the required data for tax calculation is based on the Unique Parcel Identifier (UPI), which is obtained from the land lease title and district council resolution. The tax value is based on a rate per m^2 rather than on the open market value. Thus, fixed asset tax (property tax) is based on the open market value and requires: "sales contract value or certificate of valuation by the certified valuer to fix the open market value".

The most highly trusted sources of the required data, as discovered during the fieldwork, include RLMUA, estate agencies, and valuers. However, RLMUA data were found to be incomplete, inaccurate, and outdated in some cases. Focus group participants confirmed this, stating that: "the land use on the title differs from land use on the ground, the size of the parcel on the ground differs from those in the system or title, while the number of houses within a parcel is missing from the Rwanda land cadastre". It is still a challenge in Rwanda to get all the required data for the valuers to support their value assessment. During the focus group discussion, the valuers highlighted that: "valuation depends on the available data, purposes of valuation, and use of the property".

For instance, for the income-generating properties, valuers need to look at the book account and see the income the property is generating. The property value is calculated by capitalizing of the future income from that property. Thus, many valuers depend on the property used to assess its value. This was also confirmed by participants from the focus group discussion, who claimed that: "there is a lack of data to use the appropriate methods to evaluate the property according to its type, which results in the replacement cost method being used more compared to other methods of property valuation".

3.3. Data Collection and Analysis Methods for Property Valuation for Taxation in Rwanda

The current data collection methods consist of field visit (inspection of the property), inspection of Google Earth/Maps, and consideration of the masterplans. During a field visit and inspection of the site, the valuer has to use different methods and tools for data collection for property value determination such as: "Tape measurement, digital camera, and handheld GPS and laser distance meter". Whilst the tape measurement is usually used for buildings and improvements, the parcel areas are obtained from the land title. Throughout focus group discussions, participants underlined that the tape measurements are used to: "measure the size of the houses, gross floor area, improvement, and other external works such as drainage system, parking, gardening" to complete the data from land titles (LAIS database). A digital camera is used to take pictures to: "ensure that the property exists at the date of inspection and to have a better visualization of their physical appearances, the materials and technical conditions of the subject property to be valued". This was highlighted by the valuers during the focus group discussion. Acquired images are used in property valuation as a source of information and prove their existence at the date of valuation. Currently, Rwandan valuers use handheld GPS to verify the location and measure the size of the property to be valued. During the interview and focus group discussions, the respondents said that: "even if handheld GPS is being used in data collection, its accuracy of 3 m, is not good as it should be".

A laser distance meter is used to measure the internal and external parts of the buildings. According to the users, this tool is more precise, and measurements can be done faster than with tape. Google Earth is used for location, verification, and visualization of the existence of the property at that parcel and confirms their existence on the ground at the date of valuation. In addition, the masterplan is one of the districts' planning outputs, and it is prepared for a period of 10 to 20 years. It is used mainly in urban areas to help valuers to know the allowed land use and zoning of a particular area.

The current technical steps involved in carrying out property valuation for taxation purposes in Rwanda are shown in Figure 6. The process starts with a call from the client (taxpayer) or tender bid advertisement of institutions (Private, Public, or NGOs). During this phase, an agreement for a meeting is scheduled (oral or written). After the agreement is done, the next step is to conduct a field inspection for fieldwork. It has to be conducted by a competent valuer in the presence of the landowner. If the valuation is for developed land, measurements of the buildings and notes on improvements within the compound of the parcel are required, while for undeveloped land the information provided on the land title is used, unless there are unreported changes on the land itself. During the fieldwork, the valuer has to collect the required data for valuation, and some pictures are taken for further analysis. The office work consists of analyzing the collected data and preparing the report. As a result, the validated valuation report is provided to the taxpayer. Updating of the valuation report also follows the same steps, unless there are no changes made to the property. However, the valuer should be well informed that there are no changes that happened in the last four years.

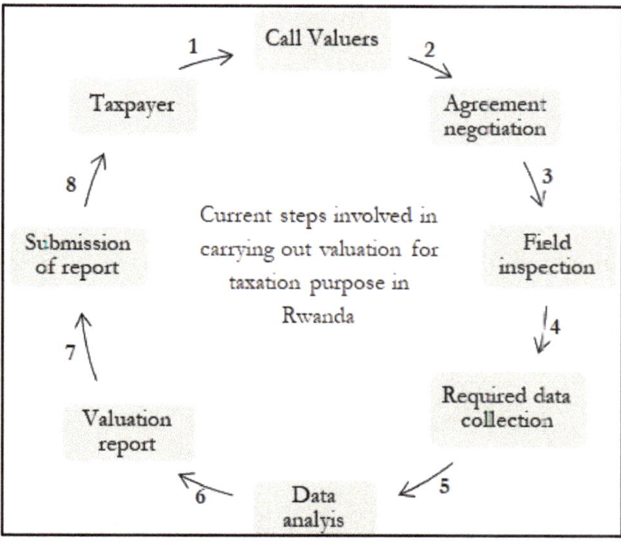

Figure 6. Steps of the current valuation system in Rwanda.

3.4. Existing Use of RS Data in Rwanda

Remote sensing data has been used in Rwanda since 2008. During the interviews, we found that in public institutions, such as RLMUA, RS data were used mainly in the creation of the digital cadastral maps. The interviewees highlighted that: "the Rwanda land cadastre was built based on the high-resolution orthophotos captured in 2008 using aircraft and satellite images. Satellite images in particular were used in the north-western part of the country. Due to the topography of that area, it was not possible to cover that area using aircraft". The generated orthophotos were very useful in identifying entire properties. Given that the images of 2008 are becoming outdated, the valuers often overlay them on top of the latest available satellite images. RLMUA, as a government institution, highlighted the importance of high spatial resolution. Currently, the government of Rwanda has signed the "Memorandum of Understanding with the government of Gabon under which both governments will be sharing spatial information and expertise in land registration". The government of Gabon will be providing satellite images as they have satellite centers, while the government of Rwanda will be providing expertise in land registration using a fit-for-purpose approach. The interviewee shared that "generally, the usability of remote sensing data in property valuation for taxation purposes is very low, for instance, the most used images are those captured during fieldwork and those images that can be downloaded from Google Earth, which often have a low spatial resolution".

3.5. *Identified Limitations for Using RS Data for Property Valuation for Taxation in Rwanda*

The general findings from interviews and focus group discussions are that the biggest challenges are a lack of data due to a lack of funds and skilled professionals to use them. In Rwanda, there "are no specific laws or regulations governing the use of remote sensing tools". This was highlighted during the interview session. The use of remote sensing tools is governed by the law regulating Civil Aviation in Rwanda. The availability of the platform is another challenge emphasized by the interviewees because they have to be ordered outside of the country. During the interview discussion with the Ministry of Infrastructure officials, they highlighted that "only one company has been registered and has the right to fly UAVs, but, for other remote sensing techniques, they need to hire international companies to carry out the photogrammetric acquisition and processing".

During the interviews, the applicability of different remote sensing images and the possible challenges associated with them were discussed, with examples shown below in Figures 7 and 8. Figure 7 shows the property visualization of sampled property number four in the different remote sensing datasets at the same scale (1:700). The results show that the features on the UAVs-orthophoto in Figure 7a are better visible than the satellite and aerial orthophoto, as shown in Figure 7b,c. The construction materials such as roof cover and pavement or external structures are also much clearer in the UAV image.

Figure 7. Property visualization from three used datasets: (**a**) from UAV, (**b**) from aerial orthophoto, and (**c**) from Satellite-Worldview 2.

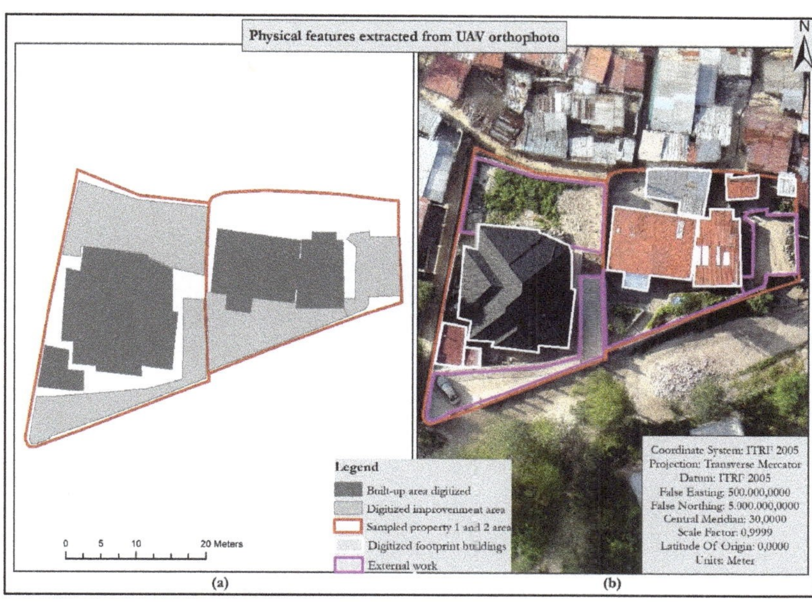

Figure 8. Features extracted from UAV orthophoto (**a**). Overlaid features on the UAV image (**b**).

Therefore, among these data sources, UAVs definitely outperform the others. UAV images seem suitable for determining the physical and locational characteristics of the property and its improvements. Based on the orientation of the camera (inclined or oblique) during image acquisition, information on the technical condition of the property, such as facades of the buildings and construction materials, can be obtained. The visualization of such information enables valuers to remotely evaluate the physical obsolescence of the property with high precision. Apart from information about the taxable property, the quantitative data such as parcel area, built-up area, perimeter, and distance to public facilities can be determined from the generated orthophoto (Table 5 and Figure 8).

Table 5. Physical characteristics of property identifiable on UAV-orthophoto for sampled property 1 and 2.

	Physical Factors					Location Factors				
Property (Land or Buildings)	**Parcel Area (m^2)**	**Built-Up Area (m^2)**	**Improvement (m^2)**	**Shape**	**Type**	**Neighborhood**	**Accessibility**	**Land Use**	**Environment**	**Utilities**
Property 1	708	275	282	regular	built	yes	yes	-	yes	yes
Property 2	557	247	175	regular	built	yes	yes	-	yes	yes

3.6. Consideration of the Potential for RS Data Application to Property Valuation for Taxation

The assessment of the different information that can be retrieved from remote sensing data is done based on the factors derived from the literature, visualized in Figure 1 and adapted for this study. These factors are classified into two main categories: internal and external factors. Internal factors include parcel size, built-up area; improvement; shape, and type of subject property, whereas external factors include accessibility, neighborhood, land use, environment, and utilities.

3.6.1. Internal Factors

The **parcel size** is used to determine the land value of either the developed or undeveloped parcels. This serves as the tax base for both leaseholders and freeholders. Moreover, for the land lease fee to be paid by the taxpayer, parcel size plays a very important role because the bigger the parcel, the more lease fee is paid if the location of the parcels are in the same area whereby the rate is the same. Parcel area serves as a land lease fee basis as the method of determining the lease fee is the rate per square meter. Additionally, for the property tax where the parcel is not developed, the open market value is determined based on the rate per square meter. However, the applied rate for lease fees and market value differs. The last one must be determined by the private valuers on the basis of recent sales, location, and infrastructure within the area and public facilities surrounding the area.

The valuers have the right only to report if the size on the title does not reflect the actual size on the ground. The authority who is intending to use the report has to decide on the above-reported errors if corrections should be made. The valuers must prove the mismatch of the areas. This can be completed using remotely sensed data. As shown in Figure 9, the parcel area for Property 2 manually digitized from UAV-orthophoto (b) is closer to the area measured with GNSS (ground truth) compared to manually digitized from the satellite image (c) and existing area in LAIS database, which is based on the orthophoto from aerial images from 2008 (a).

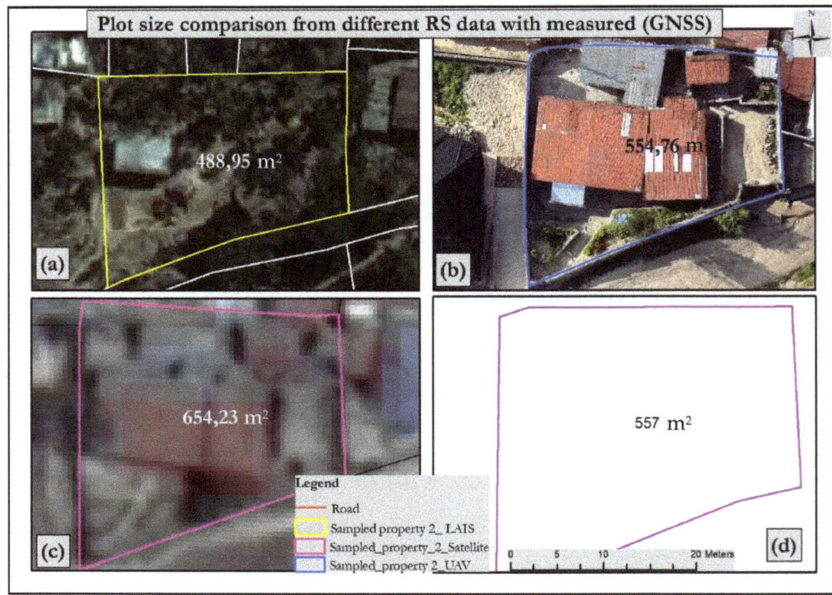

Figure 9. Plot size comparison of sampled property 2 from different remote sensing data: (**a**) existing parcel area in the LAIS database (based on orthophoto from aerial images 2008), (**b**) manually digitized from UAV-orthophoto, (**c**) manually digitized from satellite and (**d**) measured with GNSS during the field visit.

The **built-up area** is one of the most relevant factors for the property tax system. Property tax is imposed based on market value, and this includes the value of the land, the building, and any improvements. Therefore, precise area calculations are needed. Moreover, the size in the vertical dimension also should not be forgotten. In the built-up-areas, the numbers of floors can be visible if there are oblique images (Table 6). The participants from the focus group underlined that "due to the lack of data, especially related to buildings and improvements, the fieldwork measurement and image acquisitions are compulsory as the current property tax regime is based on open market value, and it depends on the determined information from the field".

Table 6. Built-up and parcel area manually digitized from the UAV orthophoto of sample properties 3 and 4.

	Physical Factors					Location Factors				
Property (Land or Buildings)	Parcel Area (m²)	Built-Up Area (m²)	Improvement (m²)	Shape of Parcel	Type	Neighborhood	Accessibility	Land Use	Environment	Utilities
Property 3	399	263	-	regular	built	yes	yes	-	yes	yes
Property 4	2471	417	278	regular	built	yes	yes	-	yes	yes

The purpose of acquired images is to show the existence of the property and to visualize technical conditions and the physical appearance of the property (Figure 10).

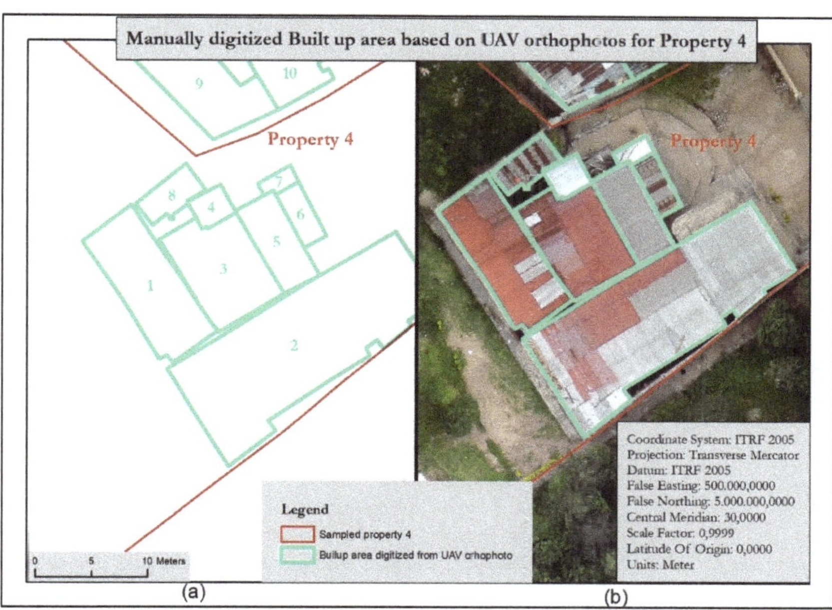

Figure 10. Digitized building footprints from UAV-orthophoto of property 4 (**a**). Footprint outlines overlaid on the UAV-orthophoto (**b**).

Improvement refers to the level of external work done to increase the value of the property; this includes property maintenance, concrete parking, gardening, sewage drainage system, water tank, and others. These features are also part of the market value of the property to be taxed and need to be assessed and considered during data collection for property valuation for taxation purposes. Therefore, the size of these external structures is required. However, they are not recorded in the current cadastre, but, can be extracted from images as shown in Figure 11. Moreover, if terrestrial images can be obtained, they can be used as supplementary materials, and in combination with UAV images, quite detailed information can be provided for valuation purposes.

Shape refers to the spatial form of a parcel, whether it is regular (perpendicular lines) or irregular (polygons with curves). The shape does not affect the value of the land and property directly, whereas it affects the improvement and design that can be put up in that parcel.

The **type** refers to whether the parcel of land is built or unbuilt. However, this goes hand in hand with land use, and development conditions that can be filled by specifying the number of buildings requires a construction coverage ratio and allows for a number of floors on an individual parcel.

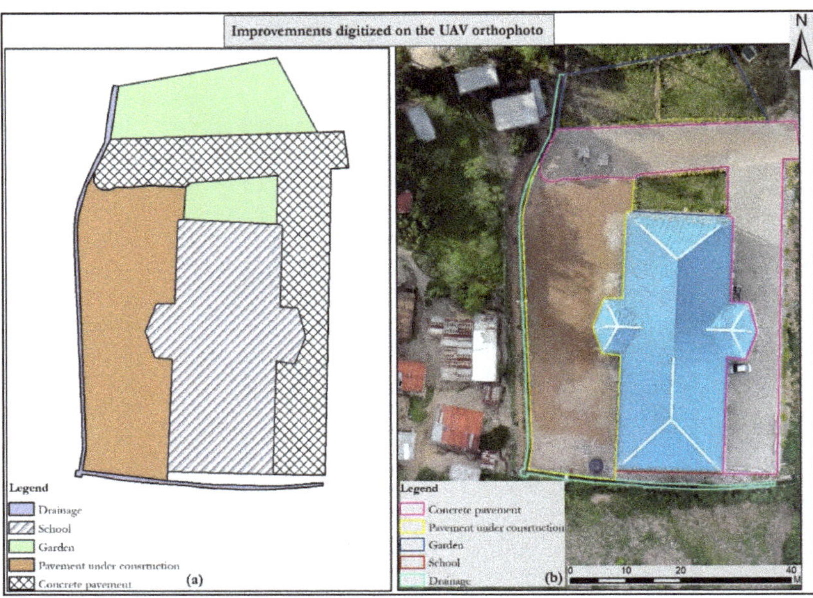

Figure 11. New features (improvements) digitized from UAV-orthophoto of property 4 (**a**). Outlines overlaid on the UAV-orthophoto (**b**).

3.6.2. External Factors—Location Factors

As shown in Figure 12, location factors such as neighborhood, accessibility (roads) of the property, public facilities, utilities, and infrastructures, all affect property value. The neighborhood is a crucial factor in the specific surroundings that affect the value of the property. The surrounding features of the property are determined by the buffer around that property in a particular area. Throughout fieldwork, the participants from the focus group discussions underscored that "location is the most important factor to be considered compared to the physical factors". They ranked the neighborhood factor as the first factor influencing the property value in a specific area. The levels of surrounding development to the property affect its value. These developments include transport facilities, retail outlets, service outlets, and public facilities, such as schools, markets, health centers, parks, churches, hospitals, manufacturers, and others. For instance, a property that stands out as being too different from the others will also differ in price, even if it is in the same neighborhood (Figure 12).

Accessibility is measured by how accessible the property is. During the discussion with valuers, they highlighted that "accessibility is more related to the road infrastructure and defines whether the property is accessible by primary roads or districts roads, water pipes, fiber optic internet, and electricity". For instance, if the property is directly attached to tarmac roads, as shown in Figure 13 for properties 3 and 4, it is more costly than a property located on a marram (**clayey/sand**) road in the same area. Accessibility was ranked as the third factor influencing the value of a specific property after neighborhood and construction materials.

Land-use zoning can either be the current use of the land or intended use as specified by the master plan. In Rwanda, the land use master plan for the entire country was developed in 2011. The district development plan was developed by referring to the developed national land-use masterplan. Property valuation considers the highest or best use of the land as it is being used as intended or planned to be used depending on the surrounding developments in those areas. Results from the focus group discussion with valuers in Rwanda underlined that "the highest and best use of the land is tangible potential, officially permitted, most economically feasible, and outstandingly profitable".

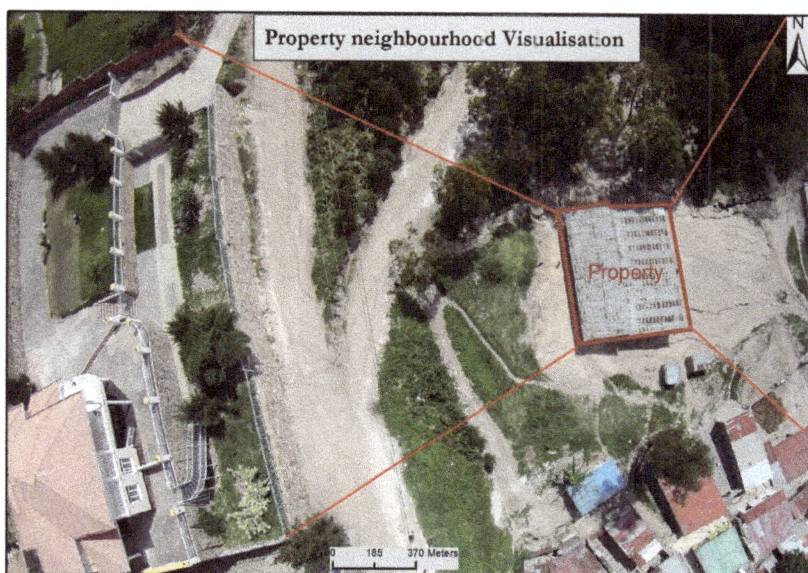

Figure 12. Neighborhood visualization of the property based on the UAV orthophoto.

Figure 13. Accessibility visualization of sampled properties 3 and 4 from UAV-orthophoto.

For valuation purposes, it is important to examine and put into consideration the issues of zoning and its changes. Additionally, land use planning laws need to be considered. For example, the development project must fulfill the requirements of the intended land use of the area by constructing the proposed structures on the specific parcel. Throughout our discussion, the valuers thought that "the masterplan of all districts should be online and open to the public as the Kigali master plan 2013 is user friendly and is used as a method of land use checking and to locate subject properties".

Figure 14 shows the planned land use of sample property 1. All permitted and prohibited constructions are specified in the zoning guidelines, which can be downloaded from the master plan (accessible online). The current master plan was developed based on the orthophotos from aerial images taken in 2008, and this is being used in the property valuation profession due to the fact that valuers use it as a credible source of information.

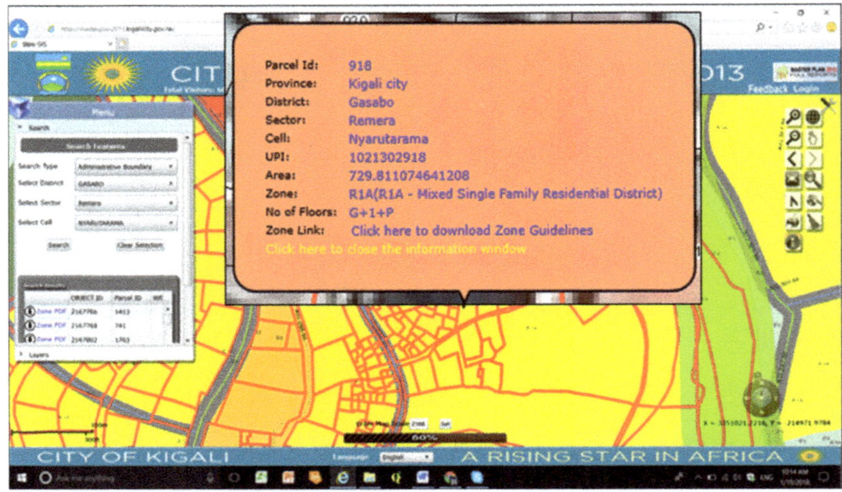

Figure 14. Land use of sample property 4 via the Kigali masterplan 2013 (Source: [42]).

The **environment** in this research refers to a geographical condition, a specific area where the property is located. The interviewee told us that "the area disposed to the effects of natural phenomena, such as flooding, high winds, and earthquakes, among others, are poor choices when buying property". Property that is located near wetlands (informal settlement areas) is worth less than that located in an approachable environment. Remote sensing data can be useful to analyze and visualize the prone area compared to the location of the property.

The **utilities** in this research refer to the services or features that are connected to the properties, such as water pipes, electricity, gas, sewerage drainage, and other facilities. With remote sensing data, utility features can be distinguished from other features. With UAV-orthophotos in particular, more accurate and precise information on the ground can be extracted compared to satellite images and orthophotos from aerial images.

3.7. Comparison of the Proposed New UAV Based Method for Property Valuation for Taxation

Based on the abovementioned factors, UAV images outperform the other remote sensing images. Therefore, during the interviews and focus group discussion, we assessed their applicability as a newly proposed method to be used for valuation for taxation in Rwanda. The elements for the assessment that we used were accuracy, completeness, up-to-datedness, cost, and availability of the platform.

Evaluation of the spatial accuracy was also done through a comparison of the area of property 1, 2, 3, and 4 digitized over the three remote sensing images and compared with the reference data measured with GNSS. As shown in Figure 15, since the resolution of the UAV image is much higher than the other images, the digitized area is much closer to the reference data. It should be mentioned that due to the coarse resolution, property 4 was not clearly visible for digitization on the satellite image. Therefore, it is missing in the figure below.

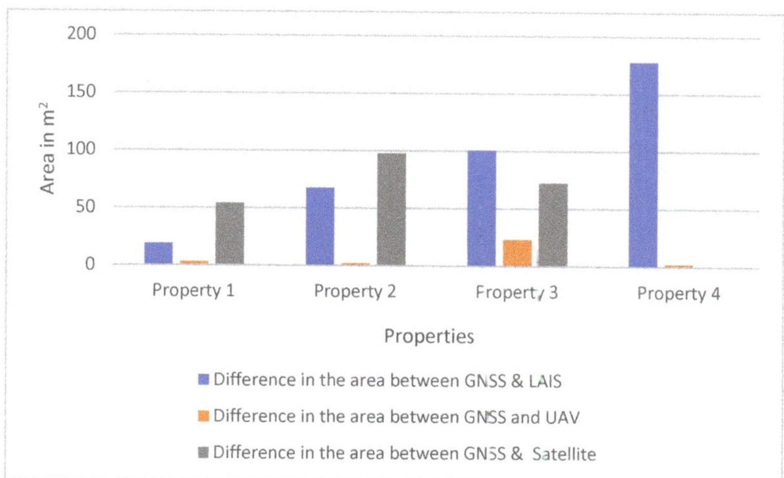

Figure 15. Area comparison between the measure of GNSS of four properties and polygons from the existing LAIS database based on digitization over an aerial image taken in 2008 (blue), UAV orthophotos (orange), and satellite images (grey).

Evaluation of **completeness** was based on the answers from both semi-structured interviews and focus group discussions with IRPV officials, valuers, and RRA officials. It was done by showing the features extracted from UAV-orthophoto and the other RS data, compared with the required data for property valuation. Findings from focus group discussions revealed that "the interior part of the property cannot be captured using remote sensing data and construction materials can be obtained in ascertained condition". Regarding the completeness of data, RLMUA interviewee and focus group discussion participants highlighted that "current information provided for property tax (fixed asset tax) is not enough (incomplete) and the information related to the building improvements within the compound of the parcel are not recorded in the LAIS database". Therefore, remotely sensed data, especially high-resolution ones such as UAV images, can be used to compensate for this gap. The only missing element that the participants shared is that RS data cannot provide information about the inside conditions of the properties.

The degree to which remote sensing data provides **updated** information was assessed via literature review, interviews, and focus group discussions data from RLMUA, IRPV, RRA, districts, and valuers. After the aerial data acquisition in 2008, the cadastral data has not been updated. Rwanda authorities have planned to update the existing spatial data in five years for urban areas and ten years for rural areas; however, satellite images from Google Earth and sporadic field measurements are still currently used where needed. During the interviews, valuers said that RLMUA, as an institution in charge of the spatial information related to land management and use, should find the approach for updating spatial information so that the data from the database reflects the reality on the ground. In relation to property taxation, the results from interviews with RRA and district officials concluded that the "classical approach of property to property self-declaration from landholders is being used for updating their data". This traditional approach is time-consuming and costly, and interviewees stated that "remote sensing techniques can be more useful to monitor and keep the spatial information regarding the changes on the ground up-to-date, which can save time compared to the traditional method".

The **cost** assessment is based on the previous research done in Rwanda and abroad on the usability of UAVs compared to the valuation fees in Rwanda. IRPV [43] has set valuation fees. They classify the properties into different categories. The categories include factors such as type (land or building); use (residential, commercial, and industrial), and location (urban, peri-urban, and rural area). What costs need to be considered for the tasks related to the current procedure and the one based on UAVs are shown in Figure 16.

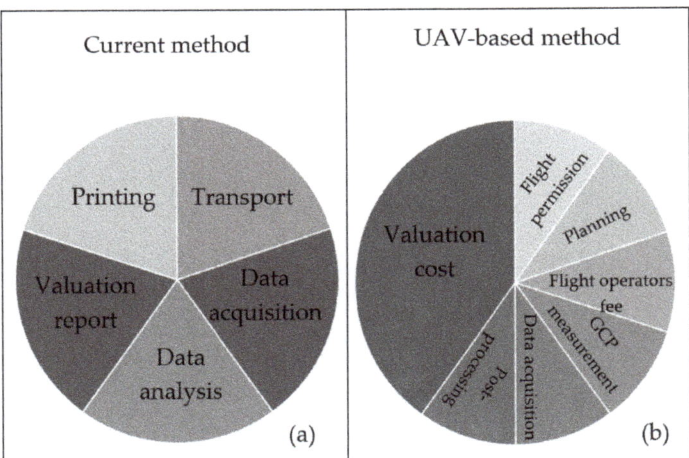

Figure 16. Cost assessment of the current property valuation method (**a**) and the proposed UAV-based method (**b**).

As shown in Figure 16, for the current valuation procedure, costs should be planned for transport, data collection, data analysis, valuation report preparation, and printing. According to [43], the average price of the charges for property valuation in an urban area is around 1050 Rwf per m^2. For the UAV-based method, the costs were calculated based on information provided by the Charis Company, the only company registered and allowed to fly in Rwandan territory. Based on their experience, a UAV approach should consider costs for obtaining flight permission, pre-planning activities, operator fees, ground control points (GCP) measurement, data acquisition, and post-processing. According to these data, the estimated price is $1100 per five hectares, which is equivalent to 18.95 Rwf per m^2. The price was converted into Rwf francs for better comparison. The online exchange was used as the exchange rate at the time was 861.42 Rwf /dollar. After the data acquisition, costs for valuation report preparation, analysis, and printing need to be included.

In terms of **availability of the platform**, during the interview discussion an official from the Ministry of Infrastructure shared that currently "only one company is registered and has the right to fly UAVs".

4. Discussion

This section is organized as follows. First, the overall potential of remotely sensed data in land valuation for taxation in Rwands is discussed. Second, a qualitative cost/benefit analysis of the proposed UAV methods for property valuation for taxation. Third, the comparison of, and discussion on, the current procedures and the proposed new methods are shown.

First, on overall potential for RS utilization in land valuation for taxation, the usability of remote sensing data in land administration in Rwanda is not new. The cadaster was built based on the orthophoto from aerial images acquired using aircraft and satellites in 2008. However, due to the high costs of such projects on a national level, these approaches are difficult to replicate for regular ad-hoc updates. Many changes happen on a daily basis, so up-to-date spatial and non-spatial information is of great importance. The socio-technical assessment completed via this research proves that remotely sensed data has great potential for all land management activities.

Second, the findings of the research show that the current application of remote sensing data and particularly high-resolution UAV images in property valuation for taxation is very limited: traditional methods are still applied. However, as reviewed in ministerial regulation N° 01/MOS/Trans/016, relating to UAVs, the usability of remote aircraft is allowed, and it provides the procedures of how the permit should be issued [44]. Therefore, they can be considered for the tasks for valuation for taxation.

From the interviews, it was confirmed that measuring the property and improvements on the proposed UAV image can reduce fieldwork time significantly. Moreover, from the comparison of the areas delineated over the three imagery data sources, compared to the reference data, the UAV images have the highest spatial resolution. Regarding spatial completeness and up-to-dateness, they have also been proven as very suitable input data. From the interviews, it was revealed that UAVs, as a new data acquisition technology, could be adopted at national and subnational government levels: the national government is the main geospatial data provider in Rwanda. Being low-cost, UAVs can be implemented by local stakeholders to support small-scale mapping with frequent flights. Such actions can contribute to land data gathering in a decentralized way. Similar findings were also found in [45]. Rwanda is one of the most advanced African countries in terms of usage of geospatial data. As such, it seems it would be easier to integrate UAV-based approaches for valuation and land administration. It will be easier to integrate such data into the existing spatial data infrastructure [45].

Although there are many works that outline the benefits of UAV images compared to the other RS techniques, including examples such as - higher spatial and temporal accuracy, and completeness. - the existence of regulations; the need for registered flight companies; license pilots, and so on, present challenges. As was argued in [45], the four main challenges for implementing UAVs as a land tenure data acquisition tool for Rwanda include: (1) Mixed terrain, which means that not all types of UAVs are suitable for flying; (2) Limitations of the current UAV regulations such as flying only in visual sight; (3) Ground truth data collection, especially in an urban environments (existence and reliability of national CORS data); and (4) Software and hardware requirements for data processing. Even cloud-based solutions are being used in many African countries, including Rwanda. However, this approach is often challenging due to the instability of the internet connection. In addition, from the interviews, the lack of human resources and funds for education is another main challenge, especially with regard to data processing and analysis skills.

Many researchers have investigated the optimal photogrammetric workflow configuration to minimize computational costs and reference data collection [46–48]. It is known that obtaining a high-precision GPS and IMU system onboard a UAV can minimize the collection of GCPs, which is usually quite challenging in African countries. Even though UAVs can provide very high-resolution images, the accuracy of captured coordinates may differ from centimeters to several meters, depending on the flight conditions and specific configuration. It has been proven in the research that the number of tie-points has a very significant impact on the correct image orientation process [49]. Based on six experiments in Europe and Africa, the authors can now identify the different flight configurations that will lead to more reliable results for UAV data acquisition. Following the guidelines can assist in achieving the best image outputs that can provide a reliable base for land administration tasks. In addition, based on remotely sensed data, innovative geospatial methods for automatic feature extraction for cadasters have been widely explored by the scientific photogrammetric and computer vision community. For this application, especially for urban areas, UAV data, with its high resolution capability, can outperform other remotely sensed data [50–52].

As mentioned, using remotely sensed data in Rwanda is not new. In terms using it for land valuation for taxation, the existing property valuation law [39] focuses on establishing IRPV and is administratively focused on: methods for valuation, the required data, or the need for standardization. Currently, most of the information is collected from on-ground field visits. The incompleteness and outdated spatial information from the cadaster in Rwanda have been noticed and reported by other researchers [33,53]. Currently, valuers have to create their own data collection method to execute their duties. There are no guidance documents or rules being followed on how that data can be collected, or which tools could be used, or which remote sensing data can be beneficial to the process. The issue of a lack of data on market transactions results in variations and inconsistencies [33]. In addition, the respondents from RLMUA indicated that the value of land, buildings, and improvements are not separated. Instead, they are recorded as a land value in the LAIS database.

Concerning property tax (fixed asset tax), the taxation system has been developed, but it is still hosted at the national level. According to the law, the property tax should be collected by the districts [29]. Districts have not invested in establishing appropriate revenue collection systems. This results in contracting the Rwanda Revenue Authority as a system holder to collect the tax on behalf of the districts. Currently, self-declaration and ad-hoc systems are being used in Rwanda as the taxation system. In Rwanda, it is the responsibility of the taxpayer to assess their properties and come up with the property value to fill the declaration form. Therefore, in Rwanda, the taxpayers are obliged to register and report their tax obligations to the tax collector [33]. A challenge to the current system is that not all taxpayers comply with this system of self-declaration (self-registration). With the self-assessment system, taxpayers are likely to undervalue their property as the tax is based on the open market value, and the fixed rate is applied. To avoid such results, the proposed remote sensing methods, incorporating UAVs as data acquisition techniques, can assure transparency, higher accuracy, and reliability. Further, to improve precision, more detailed information from terrestrial images or indoor models can also be added. Moreover, it is recommended that in the future this method be combined with additional information such as construction permit data, and indoor information linked to the vertical dimension. This will further improve the accuracy of current valuation.

5. Conclusions

RS has been used in Rwanda for the land administration since 2008, and the results of this research show RS continues to grow as an opportunity for land administraiton tasks in that country. In terms of national imagery datasets, there has been no update of the underlying imagery sources since 2008. The current spatial data from RLMUA is outdated and the current methods for spatial data capture are cumbersome and rely on physical inspection using tape measurements, handheld GPS, digital cameras, and so on. Therefore, identifying the best method to update imagery is a pressing concern. Many countries face this challenge.

As such, this research explored the differences between three different remote sensing data acquisition techniques and utimately suggests the usage of the UAV-based approach (not widely available in 2008) for property valuation for taxation purposes, at least for ad-hoc assessments, to be in alignment with the FFPLA principles. That said all RS sources can support assessment of internal factors (parcel size, built-up area; improvement; shape, and type of the subject property) and external factors (accessibility, neighborhood, land use, environment, and utilities). However, the UAV captured RS data was superior in many regards.

In terms of policy, legal, standards, RS approaches, whilst used in land valuation for taxation, could be more systematically applied in the country, especially when it recognised that these land valuation processes, by decree, should be decentralised to the district level (i.e. smaller areas to cover). That said, the lack of finances and trained staff, and only one company being registered to perform UAV flights in the country, present serious inhibutors to the use of UAVs. Many innovative approaches being adopted and applied in practice need initial and substantial governmental support to gain traction, build awareness, and create technical capacity.

Whilst the technical feasibility of the UAV-based approach is clear, workflow and procedures aligned with existing laws and integrated with other land administration processes requires more attention. Moreover, the transferability of the developed approach to other contexts also necessarily requires further investigation. It is suggested that the results of the work here could be the source material for, or at least provide guidance to the ongoing developments of, ISO 19152 LADM, 2nd edition, particularly as that standard will have a dedicated link to property valuation data and processes [54].

This comparison of different remote sensing sources, for land valuation for taxation purposes, can help practitioners, government, and involved institutions to assess the value of remotely sensed data, and the most appropriate sources. Moreover, the work illustrates the benefits of using data with higher spatial and temporal resolution for delivering transparency and a fair land valuation. The use of UAVs can fill the gap between land administration and land management authorities, and eventually strengthen multi-sectoral collaboration.

Author Contributions: Conceptualization, M.K., O.G., M.L. and K.A.; writing—original draft, M.K., O.G., M.L. and K.A.; introduction, M.K., R.M.B., O.G., M.L. and K.A.; materials and methods, M.K., O.G., R.M.B., M.L. and K.A.; results, M.K., O.G., R.M.B., M.L. and K.A.; discussion and conclusions, M.K., O.G., R.M.B., M.L., J.Z., K.A. and J.P.; fieldwork, O.G.; review and editing, M.K., R.M.B., O.G., M.L., J.Z., K.A. and J.P. All authors have read and agreed to the published version of the manuscript.

Funding: This research received no external funding.

Institutional Review Board Statement: The study was conducted according to the guidelines of the Declaration of Helsinki, and approved by the Institutional Review Board.

Informed Consent Statement: Informed consent was obtained from all subjects involved in the study.

Acknowledgments: The authors acknowledge cadastral experts, including professionals from the Rwanda Land Management Authority, the Rwanda Natural Resources Authority, the Ministry of Infrastructure, district staff, RCMRD, IRPV, RRA, the Organization of Surveyors in Rwanda, and its4land for supporting this study. We acknowledge the land registration department of Rwanda land management authority for providing legal boundaries. We are grateful to ITC for providing image data. We also thank Pix4D for providing us with the research license of Pix4dmapper.

Conflicts of Interest: The authors declare no conflict of interest.

References

1. Enemark, S.; Bell, C.K.; Lemmen, C.; Mclaren, R. *Fit-For-Purpose Land Administration*; International Federation of Surveyors (FIG): Copenhagen, Denmark, 2014; ISBN 9788792853103.
2. Crommelinck, S.; Koeva, M.; Yang, M.Y.; Vosselman, G. Application of Deep Learning for Delineation of Visible Cadastral Boundaries from Remote Sensing Imagery. *Remote Sens.* **2019**, *11*, 2505. [CrossRef]
3. Dąbrowski, R.; Latos, D. Possibilities of the Practical Application of Remote Sensing Data in Real Property Appraisal. *Real Estate Manag. Valuat.* **2015**, *23*, 68–76. [CrossRef]
4. Haeusler, T.; Gomez, S.; Enßle, F. Using satellite data for improved urban development. In Proceedings of the 2016 World Bank Conference on Land and Poverty, Washington, DC, USA, 20–24 March 2016; Word Bank: Washington, DC, USA, 2017; p. 16.
5. Ayalew, A.D.; Deininger, K. Property Tax in Kigali: Using Satellite Imagery to Assess Collection Potential. In Proceedings of the 2016 World Bank Conference on Land and Poverty, Washington, DC, USA, 20–24 March 2016; Word Bank: Washington, DC, USA, 2017.
6. Jain, S. Remote sensing application for property tax evaluation. *Int. J. Appl. Earth Obs. Geoinf.* **2008**, *10*, 109–121. [CrossRef]
7. Grenzdörffer, G.J.; Niemeyer, F. Uav Based Brdf-Measurements of Agricultural Surfaces with Pfiffikus. *ISPRS—Int. Arch. Photogramm. Remote Sens. Spat. Inf. Sci.* **2012**, *38*, 229–234. [CrossRef]
8. Koeva, M.; Muneza, M.; Gevaert, C.; Gerke, M.; Nex, F. Using UAVs for map creation and updating. A case study in Rwanda. *Surv. Rev.* **2016**, *50*, 312–325. [CrossRef]
9. Bennett, R.; Oosterom, P.; Lemmen, C.; Koeva, M. Remote Sensing for Land Administration. *Remote Sens.* **2020**, *12*, 2497. [CrossRef]
10. Koeva, M.; Stöcker, C.; Crommelinck, S.; Ho, S.; Chipofya, M.; Sahib, J.; Bennett, R.; Zevenbergen, J.; Vosselman, G.; Lemmen, C.; et al. Innovative Remote Sensing Methodologies for Kenyan Land Tenure Mapping. *Remote Sens.* **2020**, *12*, 273. [CrossRef]

11. Koeva, M.; Bennett, R.; Gerke, M.; Crommelinck, S.; Stöcker, C.; Crompvoets, J.; Ho, S.; Schwering, A.; Chipofya, M.; Schultz, C.; et al. Towards Innovative Geospatial Tools for Fit-For-Purpose Land Rights Mapping. *ISPRS—Int. Arch. Photogramm. Remote Sens. Spat. Inf. Sci.* **2017**, *42*, 37–43. [CrossRef]
12. Koeva, M.; Stöcker, C.; Crommelinck, S.; Chipofya, M.; Kundert, K.; Schwering, A.; Sahib, J.; Zein, T.; Timm, C.; Humayun, M.; et al. Innovative Geospatial Solutions for Land Tenure Mapping. *Rwanda J. Eng. Sci. Technol. Environ.* **2020**, 3. [CrossRef]
13. Chiabrando, F.; Nex, F.; Piatti, D.; Rinaudo, F. UAV and RPV systems for photogrammetric surveys in archaelogical areas: Two tests in the Piedmont region (Italy). *J. Archaeol. Sci.* **2011**, *38*, 697–710. [CrossRef]
14. Rinaudo, F.; Chiabrando, F.; Lingua, A.; Spanò, A. Archaeological Site Monitoring: Uav Photogrammetry Can Be an Answer. *ISPRS—Int. Arch. Photogramm. Remote Sens. Spat. Inf. Sci.* **2012**, *39*, 583–588. [CrossRef]
15. Ali, F. Fit-for-Purpose Boundary Mapping and Valuation of Agricultural Land Using UAVs: The Case of A1 Farms in Zimbabwe. Master's Thesis, University of Twente, Enschede, The Netherlands, 2017.
16. RICS. *Valuation—Global Standards*; RICS: London, UK, 2017.
17. Wyatt, P. *Property Valuation*, 2nd ed.; John Wiley & Sons, Ltd.: Chichester, UK, 2013; ISBN 9781405130455.
18. Sayce, S.; Cooper, R.; Smith, J.; Venmore-Rowland, P. *Real Estate Appraisal: From Value to Worth*, 1st ed.; Blackwell Publishing Ltd.: Oxford, UK, 2006; ISBN 140510001X.
19. Źróbek, S.; Grzesik, C. Modern Challenges Facing the Valuation Profession and Allied University Education in Poland. *Real Estate Manag. Valuat.* **2013**, *21*, 14–18. [CrossRef]
20. Ndungu, K.; Makathimo, M.; Kaaria, M. *The Challenges in Globalisation of Valuation Profession-Lessons from Nairobi, Kenya*; XXII International Congress: Washington, DC, USA, 2002.
21. Wallace, J.; Williamson, I. Building land markets. *Land Use Policy* **2006**, *23*, 123–135. [CrossRef]
22. Pagourtzi, E.; Assimakopoulos, V.; Hatzichristos, T.; French, N. Practice Briefing Real estate appraisal: A review of valuation methods. *J. Prop. Invest. Financ.* **2003**, *21*, 383–401. [CrossRef]
23. Mangioni, V.; Kauko, V. Valuing Land for Land Tax Purposes in Highly Urbanized Cities. In Proceedings of the 25th FIG Congress 2014 Engaging the Challenges-Enhancing the Relevance, Kuala Lumpur, Malaysia, 16–21 June 2014; pp. 1–22.
24. Mccluskey, W.; Plimmer, F. *The Potential for the Property Tax in the 2004 Accession Countries of Central and Eastern Europe*; Stephen, B., Amy, R., Eds.; RICS: London, UK, 2007; Volume 7, ISBN 9781842193716.
25. William, J.M.; Franzsen, R.C.D. Property tax reform in Africa: Challenges and potential. In Proceedings of the 2016 World Bank Conference on Land and Poverty, Washington, DC, USA, 14–18 March 2016; Word Bank: Washington, DC, USA, 2016; p. 18.
26. Wyatt, P.J. The development of a GIS-based property information system for real estate valuation. *Int. J. Geogr. Inf. Sci.* **1997**, *11*, 435–450. [CrossRef]
27. Immovable Tax Property: Welcome to RRA—Rwanda Revenue Authority. Available online: https://www.rra.gov.rw/index.php?id=64#:~{}:text= (accessed on 22 December 2020).
28. Government of the Republic of Rwanda. Presidential Order No 25/01 of 09/07/2012 Establishing the List of Fees and other Charges Levied by Decentralized Entities and Determining Their Thresholds, Pub. L. No. 25/01/2012/. Rwanda. Ministry of Natural Resources. Available online: http://www.minirena.gov.rw/fileadmin/Land_Subsector/Laws__Policies_and_Programmes/Laws/New_Fees_Presidential_Order_Official_Gazette_no_Special_of_27_07_2012.pdf (accessed on 22 December 2020).
29. Government of the Republic of Rwanda. Law Establishing the Sources of Revenue and Property of Decentralized Entities and Governing their Management; Ministry of Finance, Pub. L. No. N° 59/2011 Kigali, Rwanda, Law Establishing the Sources of Revenue and Property of Decentralized Entities and Governing Their Management, Pub. L. No. N° 59/2011. Available online: http://www.rra.gov.rw/fileadmin/user_upload/law_establishing_management_2012.pdf (accessed on 22 December 2020).
30. Kauko, V.; Heidi, F.; Vicent, M.; Juhopekka, V.; Hyyppa, H. Property taxation of 3D type properties. In Proceedings of the 2016 World Bank Conference on Land and Poverty, Washington, DC, USA, 14–18 March 2016; Word Bank: Washington, DC, USA, 2016.
31. Maurice, M.J.; Koeva, M.N.; Gerke, M.; Nex, F.; Gevaert, C. A photogrammetric approach for map updating using UAV in Rwanda. In Proceedings of the International Conference on Geospatial Technologies for Sustainable Urban and Rural Development GeoTechRwanda, Kigali, Rwanda, 18–20 November 2015; pp. 1–8.
32. Sagashya, G.D. National Land Use and Development Master Plan (Presentation to National Forum on Sustainable Urbanisation in support of EDPRS2 2014, RNRA. Available online: https://www.theigc.org/wp-content/uploads/2014/08/Panel-7-Sagashya-0.pdf (accessed on 22 December 2020).
33. Uwihoreye, M.J.B. Investigating the Contribution of Land Records on Property Taxation: A Case Study of Huye District, Rwanda. Master's Thesis, University of Twente, Enschede, The Netherlands, 2016.
34. Silverman, D. *Doing Qualitative Research: A Practical Handbook*, 2nd ed.; Sage Publication Ltd.: London, UK, 2005.
35. Kothari, C.R. *Research Methodology: Methods & Techniques*, 2nd ed.; New Age International (P) Ltd.: Jaipur, India, 2004.
36. Kumar, R. *Selecting a Sample*, 3rd ed.; Sage Publication Ltd.: London, UK, 2011; ISBN 978-1-84920-300-5.
37. Data Collection Tools—Enhanced Reader. Available online: moz-extension://340320db-84df-40dd-9492-dfe7ac0b9434/enhanced-reader.html?openApp&pdf=https%3A%2F%2Fwww.ndcompass.org%2Fhealth%2FGFMCHC%2FRevised%2520Data%2520Collection%2520Tools%25203-1-12.pdf (accessed on 3 September 2021).
38. Bryman, A.; Teevan, J.J. *Social Research Methods*; Oxford University Press: North York, ON, Canada, 2012.

39. Government of the Republic of Rwanda. *Law Establishing and Organising the Real Property Valuation Profession in Rwanda*; Ministry of Natural Resources: Kigali, Rwanda, 2010. Available online: http://www.minicom.gov.rw/fileadmin/minicom_publications/law_and_regurations/Law_relating_to_electronic_messages_electronic_signatures_and_electronic_transactions.pdf (accessed on 22 December 2020).
40. Thierry, H.N. An Assesment of Land Valuation for taxation in Rwanda. *Int. Real Estate Constr. Stud.* **2014**, *4*, 62–72.
41. Government of the Republic of Rwanda. Ministry of Lands National Land Policy, Kigali. Mnistry of Land, Environment, Forest, Water and Mines. Available online: http://www.ektaparishad.com/Portals/0/Documents/National_land_policy_Rwanda.pdf (accessed on 22 December 2020).
42. Masterplan Kigali. Available online: http://www.masterplan2013.kigalicity.gov.rw/ (accessed on 3 September 2021).
43. Valuation Fees Structure. Available online: https://www.irpv.rw/index.php?option=com_content&view=article&id=60:vps-protection&catid=1:latest-news&Itemid=114 (accessed on 3 September 2021).
44. Global Drone Regulations Database. Available online: https://www.droneregulations.info/Rwanda/RW.html (accessed on 3 September 2021).
45. Stöcker, C.; Ho, S.; Nkerabigwi, P.; Schmidt, C.; Koeva, M.; Bennett, R.; Zevenbergen, J. Unmanned Aerial System Imagery, Land Data and User Needs: A Socio-Technical Assessment in Rwanda. *Remote Sens.* **2019**, *11*, 1035 [CrossRef]
46. Lowe, G. Sift-the scale invariant feature transform. *Int. J.* **2004**, *2*, 91–110.
47. Snavely, N.; Seitz, S.M.; Szeliski, R. Photo tourism: Exploring photo collections in 3D. *ACM Trans. Graph.* **2006**, *25*, 835–846. [CrossRef]
48. Remondino, F.; Spera, M.G.; Nocerino, E.; Menna, F.; Nex, F.C. State of the art in high density image matching. *Photogramm. Rec.* **2014**, *29*, 144–166. [CrossRef]
49. Stöcker, C.; Nex, F.; Koeva, M.; Gerke, M. High-Quality UAV-Based Orthophotos for Cadastral Mapping: Guidance for Optimal Flight Configurations. *Remote Sens.* **2020**, *12*, 3625. [CrossRef]
50. Xia, X.; Persello, C.; Koeva, M. Deep Fully Convolutional Networks for Cadastral Boundary Detection from UAV Images. *Remote Sens.* **2019**, *11*, 1725. [CrossRef]
51. Crommelinck, S.; Bennett, R.; Gerke, M.; Yang, M.Y.; Vosselman, G. Contour Detection for UAV-Based Cadastral Mapping. *Remote Sens.* **2017**, *9*, 171. [CrossRef]
52. Crommelinck, S.; Bennett, R.; Gerke, M.; Koeva, M.N.; Yang, M.Y.; Vosselman, G. SLIC superpixels for object delineation UAV data. In Proceedings of the ISPRS Annals of the Photogrammetry, Remote Sensing and Spatial Information Sciences: International Conference on Unmanned Aerial Vehicles in Geomatics (UAV-G 2017), Bonn, Germany, 4–7 September 2017; Volume 4.
53. Zevenbergen, J.; De Vries, W.; Bennett, R. *Advances in Responsible Land Administration*; CRC Press: Boca Raton, FL, USA, 2015; ISBN 1498719619.
54. Kalogianni, E.; Janečka, K.; Kalantari, M.; Dimopoulou, E.; Bydłosz, J.; Radulović, A.; Vučić, N.; Sladić, D.; Govedarica, M.; Lemmen, C.; et al. Methodology for the development of LADM country profiles. *Land Use Policy* **2021**, *105*, 105380. [CrossRef]

remote sensing

MDPI

Article

Testing and Validating the Suitability of Geospatially Informed Proxies on Land Tenure in North Korea for Korean (Re-)Unification

Cheonjae Lee * and Walter Timo de Vries

Chair of Land Management, TUM School of Engineering & Design, Technical University of Munich (TUM), Arcisstrasse 21, 80333 Munich, Germany; wt.de-vries@tum.de
* Correspondence: Cheonjae.lee@tum.de; Tel.: +49-(0)89-289-25790

Abstract: The role of remote sensing data in detecting, estimating, and monitoring socioeconomic status (SES) such as quality of life dimensions and sustainable development prospects has received increased attention. Geospatial data has emerged as powerful source of information for enabling both socio-technical assessment and socio-legal analysis in land administration domain. In the context of Korean (re-)unification, there is a notable paucity of evidence how to identify unknowns in North Korea. The main challenge is the lack of complete and adequate information when it comes to clarifying unknown land tenure relations and land governance arrangements. Deriving informative land tenure relations from geospatial data in line with socio-economic land attributes is currently the most innovative approach. In-close and in-depth investigations of validating the suitability of a set of geospatially informed proxies combining multiple values were taken into consideration, as were the forms of knowledge co-production. Thus, the primary aim is to provide empirical evidence of whether proposed proxies are scientifically valid, policy-relevant, and socially robust. We revealed differences in the distributions of agreements relating to land ownership and land transfer rights identification among scientists, bureaucrats, and stakeholders. Moreover, we were able to measure intrinsic, contextual, representational, and accessibility attributes of information quality regarding the associations between earth observation (EO) data and land tenure relations in North Korea from a number of different viewpoints. This paper offers valuable insights into new techniques for validating suitability of EO data proxies in the land administration domain off the reliance on conventional practices formed and customized to the specific artefacts and guidelines of the remote sensing community.

Keywords: remote sensing; land tenure; land administration; geospatially informed analysis; knowledge co-production

Citation: Lee, C.; de Vries, W.T. Testing and Validating the Suitability of Geospatially Informed Proxies on Land Tenure in North Korea for Korean (Re-)Unification. *Remote Sens.* 2021, *13*, 1301. https://doi.org/10.3390/rs13071301

Academic Editor: Mila Koeva

Received: 18 February 2021
Accepted: 26 March 2021
Published: 29 March 2021

Publisher's Note: MDPI stays neutral with regard to jurisdictional claims in published maps and institutional affiliations.

1. Introduction

The role of remote sensing data in detecting, estimating, and monitoring socioeconomic status (SES) such as quality of life dimensions and sustainable development prospects has received increased attention across number of disciplines in recent years [1–5]. Geospatial thinking and technology have provided an important opportunity to advance the understanding of the specific questions which drives SES perspectives that include humanitarian health [6,7], rural household poverty [4,8], neighborhood deprivation [9], valuation of land and property [10], and urban dynamics [3]. This argument has given rise to much debate on not only making an association of remotely sensed spatial data and socio-economic parameters but also predicting or interpreting them.

Land administration deals with the people-to-land relationship, by describing, analyzing, designing, and measuring their relations that include social, economic, spatial, legal, and engineering perspectives [11]. Along with the growth of remote sensing applications in SES, there has been growing recognition and evidence of the vital links between remote sensing and land administration [12–19]. Geospatial data has emerged as powerful source of information for enabling both socio-technical assessment [20] and socio-legal analysis [21] in land administration sphere. Moreover, recent advancements have led to a renewed interest in identifying socio-spatial footprints of land boundaries and associated land rights through integration of different sets of geospatial and socio-economic data and informative interpretation [12,21]. Thus, a novel approach to land administration sheds a contemporary light on grounded theories and practices of land tenure, use, value, development, and governance.

However, the challenge now is how to make better use of geospatial technologies to generate evidence and the resulting data as robust evidence in spatial decision-making processes. Although, the effectiveness of the earth observation (EO) data for public decision-making in the land sector has been well-exemplified by [22,23], it was also claimed that the existing accounts fail to fully resolve the government demands for better land policy-making and planning [22]. This argument is line with a longstanding quarrel of evidence-informed policy-making [24–27] that aims to ensure transparent use of sound evidence and appropriate consultation processes in policy making [24]. It is therefore that such evidence must meet certain criteria on technical quality, a relevant source of information and effective communication [28]. The term "geospatially informed analysis (GIA)" is used here to refer to evidence generation and provision of salient and legitimate evidence in spatial decision-making, supported by geospatial technologies and geospatial data. With regard to land management, it is important to bear in mind that spatial decision making is conducted on various spatial scales, ranging from local to national, and thus incorporates multiple spatial cognitions and perspectives of both state and non-state agencies.

It is possible that GIA might not be applicable to the contexts in which the limited access to data exists, quality of data matters and reliability of data sources have been raised. North Korea belongs to this category. Studies on North Korea [29–34] and governing North Korean land tenure [35–38] have attracted considerable attention, both scholarly and popular. The majority of existing research accounts for land (tenure) reform in North Korea and land governance arrangements in unified Koreas. To date, however, there is a notable paucity of evidence-based literature describing and investigating how to identify unknown land tenure relations in North Korea due to the obvious difficulties in obtaining and analyzing empirical data. Several ways of overcoming these barriers to capturing the relationship between land tenure and governance and Korean (re-)unification process have recently been suggested that involve understanding and suggesting methods and solutions to problems [21,39,40]. Drawing upon both land administrative and geospatial engineering approaches, these enabled to provide reasonably consistent evidence and knowledge-base of an association between land tenure/land governance and (re-)unification of which relatively little is known. Whereas previous approaches suggested here were based on documented spatial knowledge and reasoning relating to land tenure and land governance, this study aims to supplement and extend these insights by incorporating and reflecting on local spatial knowledge and expertise.

Although a number of different studies have been conducted on the subject of North Korea and its refugees, there are still insufficient insights into the fundamental differences between North Korean and South Korean perceptions, beliefs, and experiences [41]. Such an information gap may hamper a smooth transition towards unification. It is therefore of fundamental importance to gain further insights directly from North Korean refugees. Currently, the only possible way of doing this is to interview and engage with North Korean refugees (that refers to new settlers or defectors; *saeteomin* or *bukhanitaljumin* in Korean terms). Urban studies in North Korea tend to involve on-site work, social contact, and face-to-face interaction with the population of interest rather than relying on existing literature and available data. However, most of the work carried out to date has not been able to provide robust evidence on the basis of persistent observation and in-depth analysis [42].

In general, gathering information regarding land tenure from multiple non-human resources such as legislation, policy documents and case studies is possible. However, the main challenge faced by many decision-makers is the incompleteness and inadequacy (inadequate proof or evidence) of information when it comes to clarifying how land tenure relations and land governance arrangements are really constructed and maintained in North Korea. In the light of these unknowns, additional data from multiple stakeholders is valuable for obtaining a more detailed insight into the broad spectrum of personal experience, views, and judgements [43,44]. In this respect, North Korean refugees can also act as human capital, not only as a supplement to publications on North Korea but also as a way of conducting an empirical analysis of primary data on North Korea [45].

When it comes to abnormal circumstances for gathering empirical evidence in this study, it is necessary to explain the main challenges and most common practices. More recent attention in urban planning and land management of North Korea has focused on using focus group interview (FGI) methods with North Korean refugees and spatial analysis with Google Earth images (EO data). Notwithstanding the fact that these methods seem to be most feasible and effective, the current research on land tenure and land governance in North Korea has been impeded by the lack of empirical data, rigorous methodologies, and reliability and validity of information for the in-close and in-depth investigation. Therefore, in addition to methods commonly used, such a content analysis on internal documents in North Korea and joint expert consultation processes make up for a dearth of evidence-based knowledge base [42,46].

Adopting a similar position aforementioned, findings from a previous study demonstrate how a process of socializing the pixel (combining with land administrative and geospatial engineering approaches) can take place through (re-)interpreted semantic land tenure relations [21]. As a result, GIA has been proposed based on a mixed-methods design and an information fusion approach, to construct a strong and consistent association between land tenure and EO data. However, a further investigation into the validation of elaborated meaning and interpreted information throughout extensive consultations with outside experts and multiple stakeholders needs to be undertaken before the association between land tenure and EO data can be more clearly understood in line with algorithmic approaches. Thus, the primary aim of this paper is to provide empirical evidence for the claim that it is possible to standardize the identification and categorization of certain objects, environments, and semantics visible in EO data that can be used to (re-)interpret land tenure relations.

Given our awareness of critical consequences in misidentifying geospatially informed proxies due to the lack of appropriate and rigorous validation of suitability, a threefold approach to empirical knowledge elicitation for GIA is taken, comprising (1) extracting scientific knowledge with a high-level of expertise based on topical (i.e., land tenure and land administration), methodological (i.e., remote sensing and earth observation), and contextual (i.e., Korean (re-)unification) interests; and (2) identifying bureaucratic knowledge (i.e., government officials); and (3) deriving knowledge of local communities in geographic areas of interest from people (i.e., North Korean refugees) who have the most accurate understanding (through familiarity or personal experience) of land tenure relations, land governance and land use practices. This study seeks to answer the following specific research questions:

- To what extent does scientific, bureaucratic and stakeholder knowledge agree or disagree with a set of identified pixel-based proxies related to land tenure in North Korea?
- How does a knowledge co-production process help to validate suitability of geospatially informed proxies and become legitimate land tenure knowledge?

This paper comprises four sections. The first deals with the conceptual and methodological accounts of research and analysis. The following section brings together the key findings relating to proxy identification and the measurement of information quality. The remaining sections of the paper comprises a summary and discussion of the findings and further implications for future research, respectively.

2. Validating the Suitability of Geospatially Informed Proxies

Validating the suitability of the remote sensing data and products against the social context of the location is critical. However, despite the explosive growth in the use of remote sensing in a wide range of applications in many different fields, there is increasing concern about the lack of rigorous social and contextual validation methods and techniques, which, in turn, may result in the misidentification or misinterpretation of proxies. Therefore, to ensure better-informed geospatial analysis, it is necessary to consider not only the procedure and legitimacy of validating the suitability of relevant proxies and combining multiple values through knowledge co-production processes.

2.1. On the Need of Tailored Approaches to EO Data Validation

"Validation" is a term frequently used in the remote sensing literature, yet it is used in different disciplines to mean different things. To avoid, terminological confusion, it is important to bear in mind that the term validation throughout this paper has come to be used in its broadest sense to refer to validating suitability or usability of proxies using EO data for GIA. A validation is a fundamental requirement when using EO data in any mapping project. It provides a basis for identifying classification errors and enables the overall accuracy and uncertainty of mapping outcomes to be estimated with sample data [47]. Congalton [48] lists three reasons why validation has become so important. It enables (1) the identification and correction of usage errors in images, (2) a robust quantitative comparison of methods, and (3) the provision of more reliable information to enable better informed decision making. Much of the available literature on remote sensing deals with the question of accuracy; however, Campbell and Wynne [49] critically warn that validation is a much more complex process, as considered by many, and it displays obvious difficulties in convincingly addressing whether the outcomes are correct. The design of sampling, response, and analysis processes are an important component of accuracy assessment and play a key role in yielding rigorously defensible validation in remote sensing science [47]. As conventional methods, there are several possible validation techniques available to examine the accuracy or error of EO data, such as visual inspection, non-site-specific analysis, difference image creation, error budgeting and quantitative accuracy assessment [48,49]. Indeed, the renewed interest in various image classification methods, such as geographic object-based image analysis (GEOBIA), necessitates a range of different validation efforts to meet the respective characteristics [50].

Despite the cutting-edge advancements in EO data validation, it should be noted that remote sensors on satellites and aircraft cannot directly detect and record a particular social, political, economic, or historical context of landscapes and their internal dynamics [51]. Unlike the remote sensing community, some studies confirmed that including participatory techniques with professionals in land administration domain helps to validate quality and usability [52,53]. Moreover, a qualitative approach in conjunction with visual interpretation and quantitative analysis to measure the scalability of the semi-automated cadastral boundary feature extraction from remote sensing data has been applied to the validation [54]. In the same vein, using EO data to derive proxies for identifying and interpreting unknown land tenure relations requires a rigorous interpretation of various contextual information and a more nuanced insight into the socio-legal-spatial properties. Therefore, there are limits to how far conventional and solid validation procedures, which have already been formed and customized for specific artefacts and validation objectives in the remote sensing community [55], can be taken in land administration science. In other words, a tailored validation protocol would help to establish a higher accuracy and feasibility based on the results obtained from the EO data interpretation.

2.2. Knowledge Co-Production: Scientific, Bureaucratic and Stakeholder Knowledge

The notion of "knowledge co-production" is commonly used to refer to the collaborative and interactive process of synthesizing different sources and types of knowledge [56–58]. Co-produced knowledge blurs the boundaries between science and practice [59]. Therefore, not only experts, scientists and professionals now play a pivotal role in the decision-making process, but the committed knowledge of non-scientific stakeholders should also be taken into account.

According to Freedman [60], if scientifically valid trials of a useful or interesting hypothesis are conducted or provide reliable information on the hypothesis being tested, the values and validity are recognized as scientific. To enable this, scientific validation needs to consider the inclusion of expert knowledge at higher levels of education and professionalism in order to test for scientific acceptability such as transparency and replicability of results [61]. Bureaucratic knowledge is now considered essential, as bureaucrats and civil servants possess advanced knowledge from governance processes that include top-down political representation, bottom-up citizen participation and informal knowledge-sharing networks [62]. In administrative and government practice, "bureaucratic knowledge" serves to navigate complex decision making. It associates the political and strategic use of knowledge rather than the intrinsic (non-instrumental) value of knowledge, such as the norms of ethos and ethics, with bureaucratic works [57,62–64]. Thus, bureaucratic validation is such that it synthesizes knowledge from both internal and external resources and creates new forms of knowledge from perceptions of political feasibility and institutional arrangement [65]. Unlike scientific validation, which is based on replication logic, bureaucratic knowledge relies on either pragmatic plausibility or feasibility logic, similar to political logic.

However, critical questions have been raised about the uncertainty of decision making in resolving multifaceted local and societal problems on the sole basis of scientific and expert knowledge [66]. One of the most significant current discussions in this argument concerns the incorporation of varying stakeholder knowledge that reduces rigidity, represents multiple perspectives, and promotes adaptability in decision making [67]. There is also a growing body of literature that recognizes the importance of stakeholder or lay knowledge as a key informant that emphasizes intense contextual and localized knowledge of people in their local environments [68]. Therefore, the potential advantage of using stakeholder validation is the increased precision both in terms of context and localization in validating the suitability of proxies that cannot be verified through disciplinary expert assessment or administrative capacity. In view of all that has been mentioned so far, one may suppose, as argued by Edelenbos et al., [57] that only coproduced knowledge fully assesses pre-identified proxies for GIA that considers scientific validity, policy relevance and social robustness.

2.3. Geospatially Informed Analysis (GIA)

In recent decades, describing, analyzing, and understanding people-to-land relations using geospatial technology has given rise to effective legal, social, and spatial solutions to multifaceted problems relating to land. Recently, more advanced geospatial intelligence has not only offered administratively straightforward, technically feasible, and financially affordable approaches [69], but it has also provided a rich set of data and information that conventional analytical techniques would have been unable to identify or access. We also view land management as a combination of interventions in governance, based on questions of how and under what conditions such land interventions are responsible and how these can be supported by technologies. It is possible, therefore, that GIA supports both smart and responsible land management [70,71], especially of difficult-to-access regions where unknown or unsupported land governance exists [21,39,40].

On the one hand, geospatial intelligence is not currently sufficiently embedded in decision-making processes, while on the other, decision makers do not sufficiently rely on geospatial intelligence, even though it is available. Even geospatial intelligence is too product-oriented and insufficiently process-oriented. Based on this line of argument, we note that, despite the above claims of GIA, it still needs to be clarified how, where, and when it can be used to enrich both scientific and bureaucratic knowledge. Building on the critical insights from GIA, it enables us to address proxy development in a smart and responsible manner, where significant uncertainty exists regarding data access, data integration and data reliability.

GIA is fast becoming the ultimate driver of spatial decision making in land management and sheds new light on recent insights into societies, the environment, the earth, resilience, and sustainable development. However, scientifically framed knowledge and technical expertise in remote sensing and earth observation tends to greatly exaggerate the excellence of laboratory experiments conducted under highly controlled conditions and technocratic approaches to dealing with land/spatial problems. At the same time, the dominance of bureaucratic knowledge in land policymaking devalues other forms of knowledge and undermines the local context, the political representation of citizens and the social processes of land governance and land use. Using geospatial tools and instruments, the citizen (stakeholder) is now able not only to consume and produce geospatial information but also to contribute grounded knowledge more effectively to spatial decision-making processes. Synchronization, complementing and contradiction with views, judgements and experiences in the knowledge coproduction process affect the GIA utilized in spatial decision making and thus determine exactly how, where and when different forms of knowledge can legitimize scientific standards and conformity as bureaucratic and social norms.

3. A Case Study: Geospatially Informed Analysis of North Korea

The method was designed for which proxies are considered relevant and useful by scientists, government professionals and stakeholders and to elicit their evaluation of the information quality. This necessitated very careful investigation. A survey was conducted with 77 sample respondents recruited from scientific, bureaucratic and stakeholder groups. Data for this study were collected using a web-based questionnaire, and the analysis used both the Chi-square test and the one-way ANOVA test. The following subsections describe in greater detail what was investigated and who was involved, how the survey was conducted and how the data was analyzed.

3.1. Identification of Proxies and Quality of Information

The first important stage of the analysis was to identify proxies by which to derive unknown land tenure relations in North Korea in conjunction with EO data. A preliminary investigation proposed a set of candidate proxies relating to the key questions based on the elements of image interpretation used in remote sensing, divided into the four categories of land ownership, land use, land transfer, and land access [21]. Within these categories, a total of 66 proxies were derived from 32 groups of objects, environments and semantics that were visible in the EO data and that could be (re-)interpreted to discern unknown land tenure relations in North Korea (see Figure 1). These proxies generally consisted of combinations of shape patterns, colors, textures relating to physical structures, types of buildings, infrastructures, types of land use, and proximity of comparable features. The line of reasoning attached to the proxies is significantly associated with central concepts of tenure claims and interests such as collective ownership, land lease and use, occupation, transactions, and land access. Hence, in line with our approach to validating the suitability of remotely sensed proxies, we set out to test the hypothesis that determines "whether proposed proxies are (1) scientifically valid, (2) administratively relevant or useful, and (3) contextualized and localized." Hence, the null hypothesis (H_0) is "no difference between scientific, bureaucratic, and stakeholder distributions (of agreements) for identified proxies."

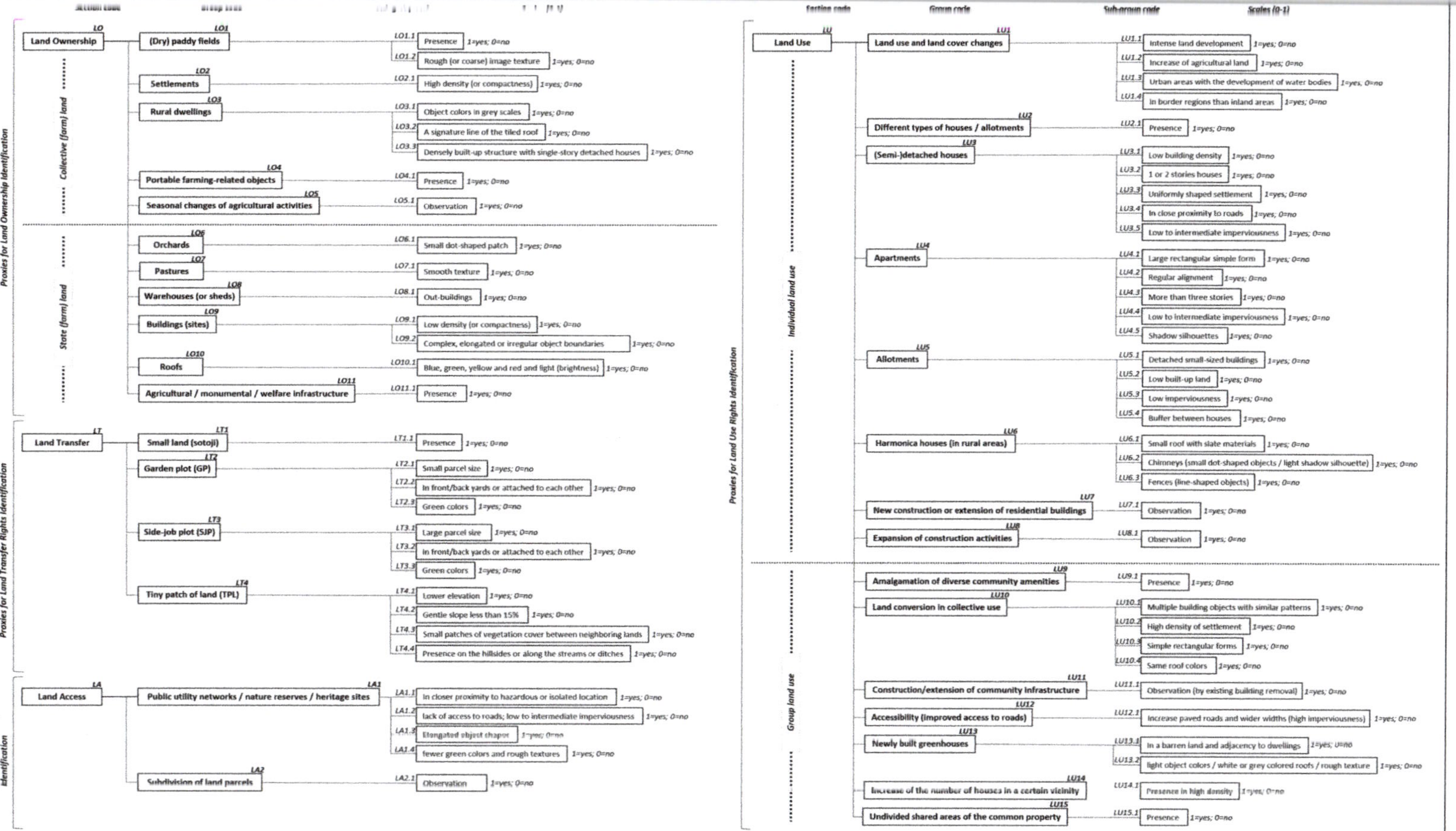

Figure 1. The organization of geospatially informed proxies on land ownership, land use, land transfer and land access rights (source: the authors, based on Lee and de Vries (2020) [21]).

One of the most influential accounts of a methodology for information quality assessment comes from [72], which sets out the theoretical dimensions of information quality (IQ) and comprehensively examines four key quality attributes from both academics' and practitioners' perspectives: (1) intrinsic, (2) contextual, (3) representational, and (4) accessible. Both intrinsic and contextual quality underline the informative factors, but intrinsic attributes are associated with accuracy, believability, reputation, and objectivity, whereas the latter considers tasks that require added value, relevance, completeness, timeliness, and an appropriate amount. On the other hand, both representational and accessibility dimensions stress the technical accounts of the system by which information must not only be interpretable, easy to understand and represented clearly and consistently, but also emphasize accessibility and security [72]. To test the hypothesis, two different approaches were taken in an attempt to account for the identification of proxies (Part I of the questionnaire) and the measurement of information (proxy) quality (Part II of the questionnaire).

3.2. Selection of Proxies and Measurement of Information Quality

3.2.1. Participants

The participants (see Table 1) were divided into three groups on the basis of their knowledge production methods: (1) scientific knowledge, focusing in particular on topical, methodological, and contextual interests (scientists); (2) bureaucratic knowledge that exists in the context of administrative and governmental practices; and (3) stakeholder knowledge of key informants with the emphasis on contextual and localized knowledge of North Korea. A random sample of participants (i.e., thus, it is unable to report the participation rate exactly.) with a different set of knowledge was identified from personal networks and connections with government agencies (from Korea Land and the Geospatial Informatix Corporation (LX) as well as local authorities under the Ministry of Land, Infrastructure and Transport (MOLIT)), (non-)governmental organizations (the Korea Hana Foundation, the Together Foundation, and the Saeil Academy), academic and research institutions (from universities, the Korea Research Institute for Human Settlements (KRIHS), the Land & Housing Institute (LHI), and the Spatial Information Research Institute (SIRI)) in South Korea. Therefore, group A and B were recruited based on South Korean's expertise, while group C consisted of North Korean refugees living in South Korea.

A total of 77 participants took part in the study. Of the total cohort of 77 participants, 29 were members of scientific knowledge groups (38%) while 30 and 18 respondents were from bureaucratic (39%) and stakeholder knowledge groups (23%), respectively (see Table 1). The Participants A group comprised scientists representing a broad range of expertise and domains in the fields of land management, land administration, land governance, land tenure, and cadastral surveying. This group also included eligible specialists with substantial knowledge and skills in remote sensing and earth observation technologies. In addition, participants were recruited from independent entities who share knowledge and a deeper understanding of Korean (re-)unification. The Participant B group represented bureaucratic knowledge and policy usefulness, with the following parameters: government professionals and officials who demonstrated a set of professional skills and had gained relevant work experience in public sectors involving land tenure/administration/management, land/cadastral surveying, and geospatial information. To create our stakeholder sample, we considered people with a declared or conceivable interest or stake in land tenure relations, land governance arrangements and land-use practices in North Korea. Thus, the Participants C group involves judgements of stakeholders who have the most direct and accurate understanding of land systems in North Korea by virtue of their life experience; in our context, this refers to North Korean refugees.

To begin this process, each participant group was invited via multiple contact points to participate in the study, with a link to the online questionnaire included. The participants were asked to complete two tasks relating to the identification of proxies and the measurement of information quality. The invitations included a clear explanation of the purpose of the research, along with an introductory statement and instructions attached to the questionnaire. Originally, the questionnaire was compiled in English. However, it was subsequently translated into Korean to gain a better understanding of the possibilities of identifying proxies and measuring information quality. The participants were asked to complete two parts of the anonymized questionnaire within two weeks (29 June–15 July 2020).

Table 1. Characteristics of participants.

	Total	Knowledge Groups		
		Scientific (A)	Bureaucratic (B)	Stakeholder (C)
N	77	29	30	18
Gender (% female)	32%	28%	27%	50%
Age				
30 years or younger	23%	17%	23%	33%
31–50 years	64%	69%	63%	56%
51 years or older	13%	14%	14%	11%
Completed educational level				
Middle-level applied: Middle & high school	8%	3%	3%	22%
Higher vocational: Bachelor's degree	35%	10%	40%	67%
Higher academic: Master's degree	29%	35%	34%	11%
Postgraduate academic: PhD	28%	52%	23%	0%
Work experience *				
0–5 years	47%	48%	37%	61%
6–10 years	15%	14%	10%	28%
10 or more years	38%	38%	53%	11%

Note. * For scientific and bureaucratic groups had land administration, management, remote sensing, or unification-related work experiences; on the other hand, it did not require relevant professional knowledge and experiences for stakeholder groups.

3.2.2. Questionnaire

The questionnaire was developed in consultation and discussion with international and local scientific communities (e.g., universities and research institutions), (1) by sharing cutting-edge scientific knowledge on smart and responsible land management; (2) by comparing how local contexts influence land tenure relations, especially in developing countries; and (3) by underpinning a new conceptual and methodological account of a geospatially informed analysis in a remote sensing community.

Due to budget constraints, time-limitations, and travel restrictions (owing to COVID-19), the data was collected using a web-based questionnaire based on the freely available Google Forms questionnaire. In order to identify the most transferrable and applicable proxies, the participants were asked whether they agreed or disagreed by choosing one of two possible values on a binary scale. One advantage of the binary scale is that it avoids the problem of nuanced and neutral answers from respondents. By forcing respondents' options, we obtained precise data with which to clarify and confirm the proxies identified beforehand (See Figure 1). In addition, after each proxy group category, participants were able to add their comments or suggestions for additional candidate proxies in a supplementary space (see further details of a questionnaire with a following link: https://forms.gle/8SzK323vYWBhRfhF8; it is available only in Korean.). To provide a good and full understanding of the questionnaire's survey content, the questionnaire was fully described with the background, purpose, and key terms of study, and provided sample satellite images in each section with detailed descriptions. We also amended the question format into a more respondents-friendly form with the help of a communication expert as well as many land administration specialists in South Korea to make it easier for respondents to answer the questions.

Unlike the binary scale format, Likert items allow more finely tuned responses and enable respondents to indicate the extent of their agreement, including a neutral response to the questions. For the attitude questions measuring information quality, a 5-point Likert scale was used to ask respondents whether they agreed or disagreed, with the following possible variations: excellent, good, fair, poor, very poor. Questions on measurement were in part adopted from AIMQ (a methodology for information quality assessment) methodology [67] and referred to such as aspects as believability, completeness, consistent representation, interpretability, objectivity, relevancy, timeliness, and understandability. Finally, the participants were asked to leave an e-mail address if they wished to know the results of the study.

3.2.3. Data Analysis

To formally compare the views and judgements of different group samples in identifying geospatially informed proxies, the Chi-square test was selected to test whether there were any (significant) differences in the distributions across scientific experts, bureaucrats, and stakeholders. We adopted the Chi-square test since this test is also suitable for more than two nominal variables or arbitrary dimension (R $\times$ C rather than 2 $\times$ 2) [73,74]. Proxies representing only a few statistical differences ($p < 0.05$) were considered to be in agreement. The experiment was conducted with two possible outcomes (agree or disagree) with the results of the proxy identification (validation) being expressed as a proportion of the overall respondents from the scientific, bureaucratic and stakeholder groups, since our data were derived from random sampling.

To measure information quality, a one-way ANOVA test was conducted. The one-way ANOVA test is one of the most commonly-used techniques for determining whether or not there are any statistically significant differences between the means of two or more independent variables (e.g., between groups, within groups). In an experiment, the measurement variable is the independent variable; thus, scientific, bureaucratic, and stakeholders' standpoints, respectively, were determined. The nominal variable is the dependent variable and can take one of five values (very poor/poor/fair/good/excellent) relating to the quality of the information on proxy identification, based on a 5-point Likert scale. It is equal to 1 if respondents give the answer "very poor" with regard to believability, completeness, consistent representation, interpretability, objectivity, relevancy, timeliness, and understandability. On the other hand, it is equal to 5 when participants consider it to be "excellent". Prior to the one-way ANOVA test, we also conducted D'Agostino–Pearson normality and lognormality tests to determine whether the data set was well-modelled (that the given sample comes from a normally distributed population). We followed up the one-way ANOVA test with Tukey's multiple comparison test (Tukey–Kramer test) to compare every variable with every other variable.

4. Results

4.1. *Identification of Proxies*

4.1.1. Land Ownership (LO)

To identify land ownership, 15 proxies were incorporated in the analysis (See Figure 1). Of these 15 proxies, first eight (LO. 1 to LO. 8) were associated with the identification of collective (farm)land and the remainder (LO. 9 to LO. 15) with state (farm)land. Null hypothesis (H_0) cannot be rejected, i.e., that there is no difference between scientific, bureaucratic and stakeholder distributions for nine proxies. For these nine proxies, the judgements elicited from scientific knowledge are in agreement with those observed in bureaucratic groups. On the other hand, we reject the null hypothesis that there is no difference between the three different knowledge groups for the following six proxies: rough/coarse image texture of (dry)paddy fields, high density/compactness of settlements, signature line of a slanting roof of rural dwellings, observation of seasonal changes in agricultural activities, small dot-shaped patch of orchards, smooth texture of pastures, and low density of building (sites).

It was found that the Table 2 compares the results obtained in the Chi-square test of validating the suitability of proxies for land ownership identification. In general, when a p-value is less than 0.05 for each proxy (No 2, 3, 5, 8 to 10, and 12), it means that the agreements elicited from the scientific, bureaucratic, and stakeholder groups are highly inconsistent, and thus hinder validating the suitability of a set of proxies. Although there is higher rates of disagreement among scientific group (ranging from 13.8 % to 48.3%) arising from the interpretations of the identified proxies, the possible proxies for land ownership identification derived from EO data (LOs. 1, 4, 6, 7, 11, and 13 to 15) achieved a better understanding among the bureaucratic (mean average: 46.2%) and stakeholder groups (mean average: 63.9%).

Table 2. Validating the suitability of proxies for land ownership identification and differences between knowledge groups.

LO.	Proxies for Land Ownership Identification	Chi-Square Test	Knowledge Groups (Agreement, %)		
		χ^2 (*p*-Value)	Scientific (A)	Bureaucratic (B)	Stakeholder (C)
1	Presence of (dry)paddy fields	5.732 (0.056)	24.1%	43.3%	66.7%
2	Rough/coarse image texture of (dry)paddy fields	12.950 (0.0001 **)	13.8%	33.3%	72.2%
3	High density/compactness of settlements	8.337 (0.015 *)	34.5%	50.0%	77.8%
4	Object colors in grey scales of rural dwellings	5.873 (0.053)	31.0%	50.0%	66.7%
5	A signature line of the slanting oof of rural dwellings	12.260 (0.002 **)	20.7%	40.0%	72.2%
6	Densely built-up structure with single-story detached houses	5.732 (0.056)	31.0%	43.3%	66.7%
7	Presence of portable farming-related objects	5.366 (0.068)	37.9%	46.7%	72.2%
8	Observation of seasonal changes of agricultural activities	16.140 (0.000 ***)	24.1%	26.7%	77.8%
9	Small dot-shaped patch of orchards	12.440 (0.002 **)	17.2%	30.0%	66.7%
10	Smooth texture of pastures	7.631 (0.022 *)	17.2%	30.0%	55.6%
11	Outbuildings of warehouses	4.186 (0.123)	31.0%	40.0%	61.1%
12	Low density of building (sites)	6.407 (0.040 *)	24.1%	40.0%	61.1%
13	Complex, elongated/irregular boundaries of buildings (sites)	5.155 (0.076)	20.7%	43.3%	50.0%
14	Blue, green, yellow, red, and light roof colors	2.465 (0.291)	37.9%	50.0%	61.1%
15	Presence of agricultural, monumental and welfare infrastructure	0.462 (0.462)	48.3%	53.3%	66.7%

Note. Agreement of knowledge groups means the percentage of yes in the survey. * p value $\leq$ 0.05; statistically significant between knowledge groups. ** p value $\leq$ 0.01; statistically very significant between knowledge groups. *** p value $\leq$ 0.001; statistically extremely significant between knowledge groups.

4.1.2. Land Use Rights (LU)

There were 35 proxies incorporated for the identification of LU from EO data. Proxies LU. 1 to LU. 24 reflect aspects of individual land use rights, while another explanation of group land use right is associated with proxies LU. 25 to LU. 35. Significant associations for the difference between scientific, bureaucratic and stakeholder's agreements were not found to be related throughout a Chi-square test for all the possible proxies of land use rights. In other words, we retain the null hypothesis (p value $\leq$ 0.05) that there is no difference between the knowledge groups. All participant groups agreed that much uncertainty (judging by the agreement ratio of $\leq$ 50%) still exists concerning the relationship between EO data and the identification of land use rights at some points. However, statistical difference does not fully account for difference in actual opinions. In other words, the χ^2 and p values only demonstrate a statistically significant difference, which is the result of a rational exercise with numbers but does not denote any practical significance in that there is no difference.

Table 3 shows the breakdown of χ^2 and p values along with the fraction of total agreement for validating the suitability of proxies for identifying land use rights. There is still no systematic understanding of how EO data contributes to land use rights identification (LUs. 11 to 15, and 24) among scientific, bureaucratic and stakeholder knowledge groups (less than 30% are in agreement); however, strong evidence was found in support of the validation of the six proxies in individual and seven proxies in group land use rights with agreement of at least 40 percent in two separate groups (LUs. 5, 6, 7, 16, 17, 22, 25 to 27, 29 to 31, and 34).

Table 3. Validating the suitability of proxies for land use rights identification and differences between knowledge groups.

LU	Proxies for Land Use Rights Identification	Chi-Square Test	Knowledge Groups (Agreement, %)		
		χ^2 (*p*-Value)	Scientific (A)	Bureaucratic (B)	Stakeholder
1	LULC changes with intense land development	3.237 (0.198)	31.0%	50.0%	27.8%
2	LULC changes with increase in agricultural land	1.149 (0.563)	27.6%	40.0%	38.9%
3	LULC changes in urban areas with the development of water bodies	0.449 (0.798)	31.0%	30.0%	38.9%
4	LULC changes in border regions than inland area	1.515 (0.468)	31.0%	46.7%	38.9%
5	Presence of different types of houses/allotments	0.119 (0.942)	55.2%	53.3%	50.0%
6	Low building density of (semi-)detached houses	0.026 (0.986)	37.9%	40.0%	38.9%
7	Half-stories in (semi-)detached houses	2.018 (0.364)	27.6%	43.3%	44.4%
8	Uniformly shaped settlement of (semi-)detached houses	1.637 (0.441)	27.6%	43.3%	33.3%
9	In close proximity to roads with (semi-)detached houses	1.637 (0.441)	27.6%	43.3%	33.3%
10	Low to intermediate imperviousness of (semi-)detached houses	1.527 (0.465)	31.0%	43.3%	27.8%
11	Large, simple rectangular form of apartments	1.795 (0.407)	17.2%	30.0%	16.7%
12	Regular alignment of apartments	0.761 (0.683)	17.2%	26.7%	22.2%
13	More than three stories of apartments	1.157 (0.560)	17.2%	23.3%	11.1%
14	Low to intermediate imperviousness of apartments	0.184 (0.912)	17.2%	20.0%	22.2%
15	Shadow silhouettes of apartments	0.590 (0.744)	10.3%	16.7%	16.7%
16	Detached small-size allotment buildings	2.128 (0.345)	55.2%	46.7%	33.3%
17	Low built-up allotment land	0.423 (0.809)	48.3%	46.7%	38.9%
18	Low imperviousness of allotments	0.967 (0.616)	41.4%	33.3%	27.8%
19	Buffer between allotment houses	0.043 (0.978)	31.0%	33.3%	33.3%
20	Small roofs with slate material of harmonica houses	0.281 (0.868)	27.6%	33.3%	27.8%
21	Chimneys (small dot-shaped objects/light shadow silhouette) of harmonica houses	4.481 (0.106)	10.3%	30.0%	33.3%
22	Fences (line-shaped objects) of harmonica houses	1.383 (0.500)	34.5%	46.7%	50.0%
23	Observation of new construction or extension of residential buildings	0.663 (0.717)	27.6%	33.3%	38.9%
24	Observation of expansion of construction activities	1.103 (0.576)	27.6%	30.0%	16.7%
25	Presence of amalgamation of various community amenities	0.539 (0.763)	41.4%	50.0%	50.0%
26	Multiple building objects with similar patterns for land conversion in collective use	0.835 (0.658)	48.3%	36.7%	44.4%

Table 3. *Cont.*

LU	Proxies for Land Use Rights Identification	Chi-Square Test	Knowledge Groups (Agreement, %)		
		χ^2 (*p*-Value)	Scientific (A)	Bureaucratic (B)	Stakeholder
27	High density of settlement for land conversion in collective use	0.483 (0.785)	37.9%	46.7%	44.4%
28	Simple rectangular forms for land conversion in collective use	2.537 (0.281)	24.1%	43.3%	38.9%
29	Same roof colors for land conversion in collective use	0.715 (0.699)	37.9%	40.0%	50.0%
30	Observation of construction/extension of community infrastructure	0.377 (0.828)	48.3%	46.7%	55.6%
31	Improved accessibility with increased paved roads and wider widths	2.491 (0.287)	27.6%	40.0%	50.0%
32	Newly built greenhouses on barren land adjacent to dwellings	0.490 (0.782)	34.5%	36.7%	44.4%
33	Light object colors/white or grey colored roofs/rough texture of newly built greenhouses	3.524 (0.171)	24.1%	30.0%	50.0%
34	Increase in the number of houses in a certain vicinity present in a high density	2.264 (0.322)	31.0%	50.0%	44.4%
35	Presence of undivided shared areas of common property	2.413 (0.299)	27.6%	36.7%	50.0%

Note. Agreement of knowledge groups means the percentage of yes in the survey.

4.1.3. Land Transfer Rights (LT)

The proxies used to identify LT had 11 responses to the questions of each knowledge group based on the following key components: small plots (*sotoji*) divided into a garden plot (GP), side-job plot (SJP), and a tiny patch of land (TPL) in North Korea. As with data obtained in the previous section on land use rights identification, we also found that there is no statistical difference in a set of given observations (p value ≥ 0.05). Therefore, we do not reject the null hypothesis for the difference in views, judgements, and experiences between scientific, bureaucratic, and stakeholder distributions on the proposed proxies for land transfer rights. There remain several aspects concerning small plots (*sotoji*) about which relatively little is known to scientific and bureaucratic knowledge groups in South Korea (only less than a third (30%) agreed on confirming land transfer rights). However, if we could turn for a moment to look at both Table 4, we can see that the stakeholder group with the most accurate understanding of land tenure relations, land governance and land use practices had a higher mean estimated percentage (32%) of agreement than the average ratio of other groups, with the validation of eleven proxies. This is especially the case with LT. 2 (38.9%), LT. 3 (44.4%), LT. 4 (38.9%), and LT. 6 (44.4%).

4.1.4. Land Access Rights (LA)

As mentioned in the previous study [21], assuming and identifying EO data proxies for LA in North Korea is one of the most challenging problems, as private land tenure is not recognized in North Korea, and thus there are no land use regulations arising through the restriction of private rights. With regard to restrictions of land access rights for public purpose only, five proxies were included in the analysis. On average, these proxies received the highest agreement among identified land tenure claims, ranging from 34.5% to 70% among scientific, bureaucratic, and stakeholder knowledge groups. As Table 5 shows, there is a significant difference between the bureaucratic and scientific/stakeholder groups in the proxy with fewer green colors and rough textures in public utility networks/nature reserves/heritage sites. Thus, the null hypothesis that there is no difference between scientific, bureaucratic, and stakeholder distributions for this proxy cannot be rejected.

What is interesting about the data here is that the bureaucratic knowledge group obtained the highest level of agreement on proxy identification (63.3%, 60.0%, 70.0%, 66.7%, and 50.0%, respectively in order).

Table 4. Validating the suitability of proxies for land transfer rights identification and differences between knowledge groups.

LT	Proxies for Land Transfer Rights Identification	Chi-Square Test	Knowledge Groups (Agreement, %)		
		χ^2 (*p*-value)	Scientific (A)	Bureaucratic (B)	Stakeholder (C)
1	Presence of small plots (sotoji)	2.167 (0.338)	38.0%	26.7%	33.3%
2	Small parcel size of garden plot (GP)	0.783 (0.675)	31.0%	26.7%	38.9%
3	GP in front/back yards or attached to each other	1.038 (0.592)	34.5%	30.0%	44.4%
4	GP with green colors	0.918 (0.631)	27.6%	26.7%	38.9%
5	Large parcel size of side-job plot (SJP)	1.034 (0.596)	17.2%	16.7%	27.8%
6	SJP in front/back yards or attached to each other	1.415 (0.492)	27.6%	33.3%	44.4%
7	SJP with green colors	0.258 (0.878)	24.1%	30.0%	27.8%
8	Lower elevation of tiny patch of land (TPL)	1.413 (0.493)	17.2%	30.0%	27.8%
9	Gentle slope less than 15% of TPL	1.413 (0.493)	17.2%	30.0%	27.8%
10	TPL with small patches of vegetation cover between neighboring lands	0.761 (0.683)	17.2%	26.7%	22.2%
11	Presence on the hillsides or along the streams or ditches of TPL	0.761 (0.683)	17.2%	26.7%	22.2%

Note. Agreement of knowledge groups means the percentage of yes in the survey.

Table 5. Validating the suitability of proxies for land access rights identification and differences between knowledge groups.

LA	Proxies for Land Transfer Rights Identification	Chi-Square Test	Knowledge Groups (Agreement, %)		
		χ^2 (*p*-Value)	Scientific (A)	Bureaucratic (B)	Stakeholder (C)
1	Public utility networks/nature reserves/heritage sites in close proximity to hazardous or isolated area	1.768 (0.413)	51.7%	63.3%	44.4%
2	Public utility networks/nature reserves/heritage sites with a lack of access to roads; low to intermediate imperviousness	2.083 (0.352)	48.3%	60.0%	38.9%
3	Elongated shapes of public utility networks/nature reserves/heritage site objects	4.115 (0.127)	44.8%	70.0%	50.0%
4	Fewer green colors and rough textures of public utility networks/nature reserves/heritage sites	6.909 (0.031 *)	34.5%	66.7%	38.9%
5	Observation of subdivision of land parcels	1.474 (0.478)	34.5%	50.0%	44.4%

Note. Agreement of knowledge groups means the percentage of yes in the survey. * p value ≤ 0.05; statistically significant between knowledge groups.

4.2. Measurement of Information Quality

The participants were asked to consider data, information, or knowledge with regard to whether an elaborated meaning or an interpreted information element is valid or not and then to complete an eight-question survey about information quality. Of the eight aspects, there was no statistically significant differences between group mean values (1–5) for believability, completeness, consistent representation, interpretability, objectivity, and timeliness as determined by one-way ANOVA. This indicates a high level of consensus on information quality among the different knowledge groups. On the other hand, there was a significant difference from those of variables between the means of three groups both on relevancy at the p value ≤ 0.05 level for the three conditions (between/within/total) and understandability of information (see Table 6). However, the one-way ANOVA test does not tell us where the difference exists, and which specific groups differed. The post hoc Tukey test indicated that the relevancy in scientific and bureaucratic groups (A–B) and bureaucratic and stakeholder (B–C) groups differed significantly at $p \leq 0.05$; regarding understandability, there was a statistically significant difference ($p = 0.018$) between the scientific and bureaucratic (A–B) groups (see Table 7).

Table 6. Differences in information quality between knowledge groups.

One-Way Anova Test		SUM of Squares	Df [(1)]	MEAN Square	F	*p*-Value
Believability	Between groups	8.429	2	4.215	2.801	0.067
	Within groups	111.400	74	1.505		
	Total	119.800	76			
Completeness	Between groups	9.419	2	4.710	3.074	0.052
	Within groups	113.400	74	1.532		
	Total	122.800	76			
Consistent representation	Between groups	3.283	2	1.642	1.105	0.336
	Within groups	109.900	74	1.486		
	Total	113.200	76			
Interpretability	Between groups	5.464	2	2.732	1.633	0.202
	Within groups	123.800	74	1.673		
	Total	129.200	76			
Objectivity	Between groups	9.193	2	4.597	2.650	0.077
	Within groups	128.300	74	1.734		
	Total	137.500	76			
Relevancy	Between groups	21.820	2	10.910	7.526	0.001 **
	Within groups	107.300	74	1.450		
	Total	129.100	76			
Timeliness	Between groups	8.902	2	4.451	2.750	0.070
	Within groups	119.800	74	1.619		
	Total	128.700	76			
Understandability	Between groups	11.740	2	5.870	3.895	0.024 *
	Within groups	111.500	74	1.507		
	Total	123.200	76			

Note. [(1)] Degrees of Freedom; * p value ≤ 0.05; statistically significant between knowledge groups. ** p value ≤ 0.01; statistically very significant between knowledge groups.

Table 7. Post-hoc test of differences in relevancy and understandability between knowledge groups.

Tukey's Multiple Comparisons Test	Difference of Levels	Mean Difference	Std. Error	95.00% CI of Diff.		*p*-Value
				Lower Bound	Upper Bound	
Relevancy	A–B	−0.7908	0.3135	−1.541	−0.04089	0.036 *
	A–C	0.5536	0.3613	−0.3104	1.418	0.281
	B–C	1.344	0.3590	0.4859	2.2	0.001 **
Understandability	A–B	−0.8897	0.3197	−1.654	−0.1251	0.018 *
	A–C	−0.5230	0.3683	−1.404	0.3580	0.336
	B–C	0.3667	0.3660	−0.5087	1.242	0.578

Note. * p value ≤ 0.05; statistically significant between knowledge groups; ** p value ≤ 0.01; statistically very significant between knowledge groups.

Figure 2 also displays an average of a range of values (1–5) for each level of information quality in different knowledge groups. What stands out is that the bureaucratic group has the highest median within the samples in all aspects of information quality, while scientific ($M = 2.5$) and stakeholder knowledge ($M = 2.7$) groups had a lower mean score compared to the bureaucratic groups, except in relevancy of information. The results, indicate that relevancy of information received relatively positive scores from bureaucratic ($M = 4.1$) and scientific groups ($M = 3.3$); on average, respondents from all groups reported lower levels of consistent representation ($M = 2.46602$) and interpretability ($M = 2.5$).

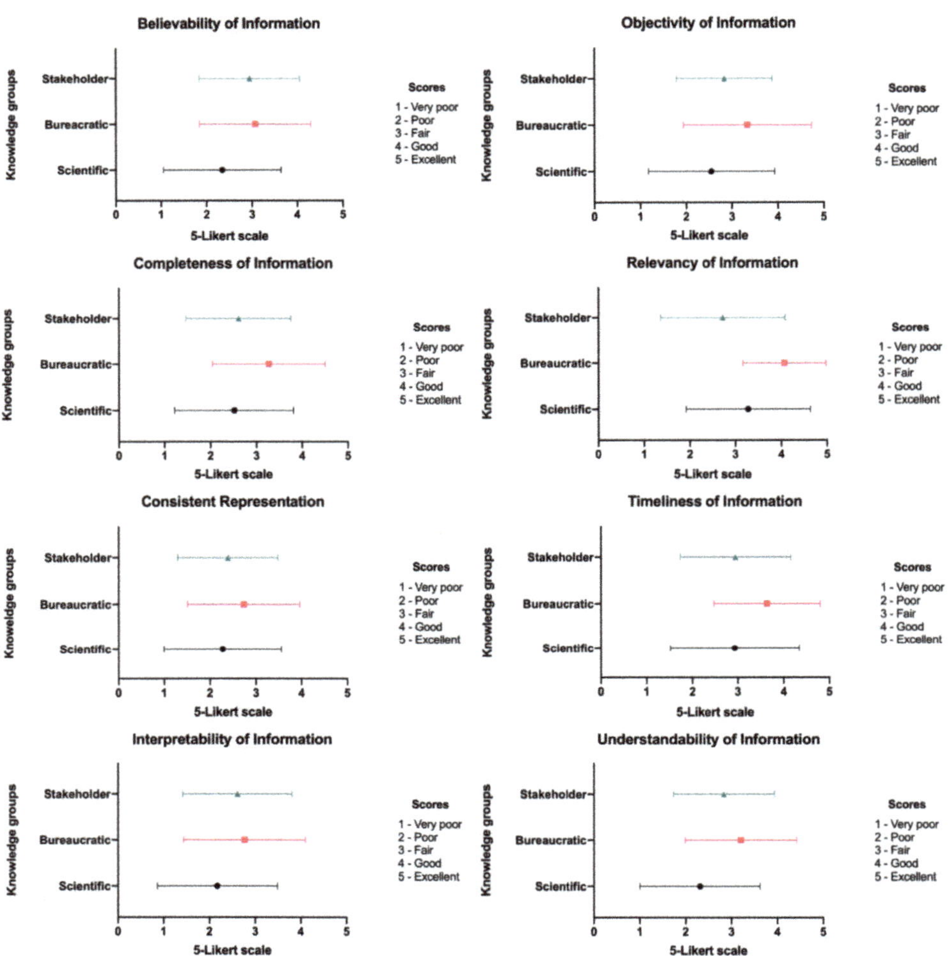

Figure 2. An arithmetic average of a range of values estimated. The horizontal bar plot shows minimum, mean, and maximum values of information quality within groups. The dependent variable consists of five values (very poor/poor/fair/good/excellent) on a 5-point Likert scale. The independent variables are scientific, bureaucratic, and stakeholders' agreements.

5. Summary and Discussion

The principal limitation of closed questions in a questionnaire restricted respondents' answers and expressiveness, enabling us to clarify and confirm the proxies we preidentified (in Figure 1). A small sample was chosen because of the expected difficulty in obtaining a high-level of expertise and an accurate understanding in the given context, based on the fact that significant uncertainty exists regarding geospatial and socio-economic data access, data integration, and data reliability.

Up to now, there have been no controlled studies which compare differences in findings. However, the experimental work presented here provides first investigation how pixel-based land tenure information become legitimate land tenure knowledge to some extent. Although differences of agreement still exist, the most obvious findings to emerge from the analysis is that there appears to be some agreement in judgements of proxies among scientific, bureaucratic, and stakeholder groups. They agree on some proxies but not on others. Weak associations of EO data and land ownership were identified for eight proxies, including such as coarse image texture of (dry)paddy fields, high density/compactness of settlements, a linear roof of rural dwellings, seasonal changes of agricultural activities, and small dot-shaped patch or orchards.. On the other hand, some proxies are associated with both collective and state land ownership and are strongly supported by bureaucratic and stakeholder groups that stress the political and strategic use of proxies and possess the most localized and contextualized knowledge.

With regard to proxies for land use rights identification, we can confirm that six proxies relating to individual use rights such as presence of different types of houses/allotments, detached small-size allotment buildings and low built-up allotment land and seven to group land use rights (e.g., amalgamation of community amenities and increase in the number of houses in a certain vicinity) were identified out of a total of 35 proxies. These were found to be particularly associated with houses, allotments, land conversions and improvements to the location. However, no significant agreement was found for all groups, especially in apartment-types of proxies (e.g., rectangular forms, regular alinements, imperviousness, or shadow silhouettes of apartments). These proxies could have been generated by misclassification bias or an erroneous assumption when identifying geospatially informed proxies. The reason for this is not clear but it may have something to do with the nature of apartments where multiple objects reside.

In addition, the findings indicate that elements of EO data interpretation such as color, shape, size, height, and site (e.g., large parcel size of SJP, SJP with green colors, lower elevation, and gentle slope of TPL) may not be associated with land transfer rights. However, there is a knowledge gap resulting from a lack of clear understanding of specific aspects and details of small plots (*sotoji*) in North Korea by South Korean scientists and government professionals. The stakeholder group that has the most accurate understanding of land tenure relations, land governance and land use practice showed significantly higher ratio of agreement. Regarding the identification of land access rights, whilst there was strong agreement in the validation of proxies between all knowledge groups and considered to be most negotiated knowledge that is scientifically most valid, policy-relevant, and socially robust among others in this study. As far as infrastructure elements are concerned, three groups have shown a higher level of agreement among other proxy selections. We may assume that a proxy identification in relation to infrastructure in North Korea could be more important than anything else. In other words, it is possible that these identified proxies could account for unknown aspects of land access rights in North Korea.

However, these results also need to be interpreted with caution. Firstly, we revealed a strong and consistent association between land ownership and EO data and the mean average of agreement in stakeholder groups—for those with the most localized knowledge of land tenure—were higher compared to those of other groups. If the debate is to be moved forward, a better understanding of different perceptions on land tenure among North Korean refugees needs to be developed. It can be relatively easier for North Korean refugees to distinguish collective and state (farm)land through EO data because they have empirically familiar with the socialist land tenure system. Another reason to support this claim may be that there was an obvious difficulty with defining the term which have accustomed with South Korean land management practices.

Secondly, the present results were significant in at least two major respects. The experimental data suggested that the three groups considered in this study were all in a higher degree of agreement on identifying land use rights, nothing in particular really stood out, but the agreement was distributed evenly at a relatively higher level than the average of those observed in other claims. However, some of those experts still argued that idea was not feasible to empirically derive changes in land use rights in North Korea with EO data. Despite the fact that there has been increased numbers of North Korean refugees (approximately 30,000 residents), one argued that it still remained challenging to understand the notion of "individual" land use rights, according to his/her own experiences of having worked and lived in North Korea.

Thirdly, all knowledge groups showed that the proxies for land transfer rights identification were appeared to the lowest in the level of agreement. However, the questions came up against the great problem of reliability of reported data and we are often not in a position to know whether it enabled participants to provide fairer, more objective, and more accurate and reliable assessments for validating the suitability of geospatially informed proxies. For instance, making a judgement on this, however, inevitably makes additional demands for expertise of either land management or remote sensing. Furthermore, these validations require in-depth local knowledge of the distinguishing feature in North Korea (i.e., *sotoji*) that is necessary to make association between EO data and land transfer rights. To further identify the proxies, it is necessary to rely on multiple techniques and methods, which combine both direct responses of individual, ranked, and stated choice responses of either individuals or groups and indirect collections of perceptions, beliefs, and social values. In addition to interviews and focus group discussions (FGD), one could add a number of other relevant tools and techniques, such as, Q methodology (which combines quantitative and qualitative data collections techniques with statistical and interpretative data analyses methods), the Delphi technique (which relies on consecutive perceptions and interpretations), and multi-criteria decision analysis (MCDA) (which combines and infers from multiple opinions and preferences).

In order to ensure whether proposed proxies for land access rights were valid, the participants were asked to select between the two, either agree or disagree. Of the 77 participants who completed the questionnaire, nearly half reported that land access rights could be identified in line with the proxies and the EO data. One could argue that this finding is largely biased by the selection of respondents who had similar epistemic backgrounds. The consequence of this bias could be that the understanding of what land rights constitute and what not, would reflect the acquaintance with private rights tenure regimes only and perhaps more limited awareness and experience with State-based tenure regimes. Hence, it is important to keep this possible bias in these responses in mind, especially from scientific and bureaucratic groups.

Nevertheless, regarding the measurement of information quality (see Figure 2), the result was that we expected. Although EO data proxy identification for land tenure relations in North Korea seems to be strongly relevant to respondents' research, policies, and social interests (i.e., relevancy of information; timeless of information), many participants did suffer from a lack of consistent representation and interpretability of information. In order to further investigate and confirm this finding, a provision of multi-disciplinary training will enhance both researchers' capacity to understand land tenure and land governance in question and policymakers' confidence in making spatial decisions in the context of Korean (re-)unification based on GIA. This also enables multiple engaged stakeholders to reveal the interconnection of geospatial science in land management practice. All these require affinity with multiple technical disciplines such as geoinformation and earth observation sciences, civil and environmental engineering as well as sensitivity of social and political processes including public administration, law, economics, and (human) geography. Secondly, one major drawback when implementing GIA was that non-remote sensing scientists, government professionals and North Korean refugees were suffered from scientific and technological literacy [75,76] (i.e., a high density of technical terms of remote sensing used in research). As this case very clearly demonstrated, it is important that reformulating the scientific language in a communicative style should be considered to facilitate active stakeholders' engagement in advancing GIA. Lastly, the existing and grounded knowledge of GIA needs to be translated into voluntary guidelines, policy briefs for scientists, policymakers, and other interest groups. In addition, fact-finding projects from around the globe where unknown land tenure and unsupported land governance exists needs to be implemented.

Different forms of evidence can be used to inform spatial decision making in land management, with data being gathered via statistical and administrative evidence (from government), analytical evidence (by scientific experts), evidence from citizens and stakeholders and evidence from evaluations [77]. This also accords with our approaches, which showed that how a knowledge co-production process helps to validate suitability of geospatially informed proxies and become legitimate land tenure knowledge. Given the fact that this study has to be conducted with the best of all qualities, it should confirm or reject our hypotheses as analytical evidence that may report a possible association between EO data and land tenure with a case study in North Korea. Then, by incorporating and reflecting on local spatial knowledge from multiple stakeholders (i.e., scientists, government professionals and North Korean refugees), it enabled us to tell policymakers what land tenure knowledge they consider legitimate (i.e., scientific validity, policy relevance and social robustness) and what counts as geospatially informed evidence. It is only after knowledge (evidence) co-production processes, the finding of this study supports the view that we can bridge and close the gap between technical aspects of the EO data evidence generation and operational contexts in spatial decision making in land administration and management. Much of the available literature so far on remote sensing for land administration is too product-oriented for skilled and trained technicians [12–19] and insufficiently process-oriented for policymakers and end-users, allowing them to make decisions in the most rational and informed way possible with EO data [22]. The point is not to go against the promising ideas on RS applications, techniques, products, and methods, but to really emphasize that it is an opportune time to undertake the most engaged and negotiated knowledge for both evidence generation and provision of salient and legitimate evidence in responsible and smart decision-making in land administration. This approach can be a way forward remote sensing for land administration 2.0.

6. Conclusions

The aim of the research question in this study was to determine the extent to which scientific, bureaucratic, and stakeholder knowledge coincides with a set of identified proxies that would enable us to conclude whether certain proposed proxies are scientifically valid, administratively relevant, contextualized, and localized. The findings from this study could then be used to standardize the identification and categorization of certain objects, environments, and semantics visible in EO data that can (re-)interpret land tenure relations in North Korea in preparation for Korean (re-)unification.

Of the four different land tenure claims, both Chi-square and one-way ANOVA analysis revealed that the distribution of agreements relating to land ownership and land transfer rights identification varied among scientific experts, bureaucrats, and stakeholders. Moreover, it was possible to measure intrinsic, contextual, representational, and accessibility attributes of comprehensive information to ascertain associations between EO data and land tenure relations in North Korea based on different viewpoints. From here, the step towards enhancing and developing the existing account is clearly supported by the current findings on information quality.

The findings of this investigation complement those of a previous study relating to a conceptual and methodological development of a geospatially informed analysis in the land administration domain [21]. These findings contribute in several ways to our understanding of how the pixel can be converted to legitimate land tenure knowledge. First, it can help us establish a tailored validation protocol with a higher accuracy and feasibility based on the identification and interpretation of unknown land tenure relations derived from EO data and various types of contextual information as well as a more nuanced view of socio-legal-spatial properties. Second, these findings, being based on knowledge co-production, are relevant to scientists, policy-makers, and practitioners involved in the decision making process relating to land tenure reform and land governance rearrangement on the basis of emerging geospatial technologies and datasets in the context of Korean (re-)unification. Furthermore, the methods used in this study can also be applied to other cases elsewhere in the world, in particular, difficult-to-access regions or fragile and conflict-affected areas. Lastly, the present study contributes additional evidence of geospatially better-informed analysis that emphasizes scientific validity, policy relevance, and social robustness within a responsible and smart land management framework. The geospatially better-informed analysis is not about how geospatial intelligence can directly detect information but how technology can smartly and responsibly support better information regarding land issues for the benefit of scientists, policymakers, and stakeholders.

Although the current study is based on a small sample of participants and used a focus group questionnaire, it offers valuable insights into new validation techniques of suitability for EO data in the land administration domain based on conventional practices that have been formed and customized to accommodate the specific artefacts and validation objectives used in the remote sensing community. The scope of this study was limited in terms of participants' knowledge, for example their level of expertise (scientific), administrative involvement (bureaucratic), and knowledge of locales in geographic areas of interest (stakeholder). However, the limited number of samples adds further caution regarding the generalizability of these findings. Thus, further investigation and experimentation to develop the internal and external validity of findings and GIA methodology would be of great help in understanding the associations between EO data and land tenure claims. Considerably more work will need to be done to identify intrinsic links between geospatial data and land tenure relations. It will then be necessary to concentrate on the development of EO data interpretation in line with artificial intelligence (AI) so as to be able to delve deeper into the future of land administration.

Author Contributions: Conceptualization, C.L.; methodology, C.L.; software, C.L.; validation, C.L. and W.T.d.V.; formal analysis, C.L.; investigation, C.L.; resources, C.L.; data curation, C.L.; writing—original draft preparation, C.L.; writing—review and editing, C.L. and W.T.d.V.; visualization, C.L.; supervision, W.T.d.V.; project administration, C.L. All authors have read and agreed to the published version of the manuscript.

Funding: This work was supported by the German Research Foundation (DFG) and the Technical University of Munich (TUM) in the framework of the Open Access Publishing Program.

Institutional Review Board Statement: Not applicable.

Informed Consent Statement: All prospective participants had been informed all the information (i.e., voluntary nature of study, risks, benefits, procedures to maintain confidentiality, etc.) and agreed to participate before proceeding the E-survey.

Acknowledgments: We thank three anonymous reviewers for their insightful comments in narrowing the gap between our claims and the actual content of the manuscript. We are also immensely grateful to colleagues in Korea who have helped and supported in conducting surveys, including survey creation and distribution. Last but not the least, I am also very thankful to the respondents who provided their insight, expertise and views, although any errors are our own and should not tarnish the reputations of these esteemed persons.

Conflicts of Interest: The authors declare no conflict of interest.

References

1. Avtar, R.; Komolafe, A.A.; Kouser, A.; Singh, D.; Yunus, A.P.; Dou, J.; Kumar, P.; Gupta, R.D.; Johnson, B.A.; Thu Minh, H.V.; et al. Assessing sustainable development prospects through remote sensing: A review. *Remote Sens. Appl. Soc. Environ.* **2020**, *20*, 100402. [CrossRef]
2. Sapena, M.; Wurm, M.; Taubenböck, H.; Tuia, D.; Ruiz, L.A. Estimating quality of life dimensions from urban spatial pattern metrics. *Comput. Environ. Urban Syst.* **2021**, *85*, 101549. [CrossRef]
3. Warth, G.; Braun, A.; Assmann, O.; Fleckenstein, K.; Hochschild, V. Prediction of Socio-Economic Indicators for Urban Planning Using VHR Satellite Imagery and Spatial Analysis. *Remote Sens.* **2020**, *12*, 1730. [CrossRef]
4. Watmough, G.R.; Marcinko, C.L.J.; Sullivan, C.; Tschirhart, K.; Mutuo, P.K.; Palm, C.A.; Svenning, J.-C. Socioecologically informed use of remote sensing data to predict rural household poverty. *Proc. Natl. Acad. Sci. USA* **2019**, *116*, 1213. [CrossRef]
5. You, Z.; Shi, H.; Feng, Z.; Yang, Y. Creation and validation of a socioeconomic development index: A case study on the countries in the Belt and Road Initiative. *J. Clean. Prod.* **2020**, *258*, 120634. [CrossRef]
6. Brown, M.E.; Grace, K.; Shively, G.; Johnson, K.B.; Carroll, M. Using satellite remote sensing and household survey data to assess human health and nutrition response to environmental change. *Popul. Environ.* **2014**, *36*, 48–72. [CrossRef]
7. Greenough, P.G.; Nelson, E.L. Beyond mapping: A case for geospatial analytics in humanitarian health. *Confl. Health* **2019**, *13*, 50. [CrossRef] [PubMed]

8. Jean, N.; Burke, M.; Xie, M.; Davis, W.M.; Lobell, D.B.; Ermon, S. Combining satellite imagery and machine learning to predict poverty. *Science* **2016**, *353*, 790–794. [CrossRef]
9. Kuffer, M.; Thomson, D.R.; Boo, G.; Mahabir, R.; Grippa, T.; Vanhuysse, S.; Engstrom, R.; Ndugwa, R.; Makau, J.; Darin, E.; et al. The Role of Earth Observation in an Integrated Deprived Area Mapping "System" for Low-to-Middle Income Countries. *Remote Sens.* **2020**, *12*, 982. [CrossRef]
10. Balaji, L.; Muthukannan, M. Investigation into valuation of land using remote sensing and GIS in Madurai, Tamilnadu, India. *Eur. J. Remote Sens.* **2020**, 1–9. [CrossRef]
11. Dale, P.; McLaughlin, J. *Land Administration*; Oxford University Press: Oxford, UK, 2000.
12. Bennett, R.; Oosterom, P.v.; Lemmen, C.; Koeva, M. Remote Sensing for Land Administration. *Remote Sens.* **2020**, *12*, 2497. [CrossRef]
13. Crommelinck, S.; Koeva, M.; Yang, M.Y.; Vosselman, G. Application of Deep Learning for Delineation of Visible Cadastral Boundaries from Remote Sensing Imagery. *Remote Sens.* **2019**, *11*, 2505. [CrossRef]
14. Fetai, B.; Oštir, K.; Kosmatin Fras, M.; Lisec, A. Extraction of Visible Boundaries for Cadastral Mapping Based on UAV Imagery. *Remote Sens.* **2019**, *11*, 1510. [CrossRef]
15. Koeva, M.; Nikoohemat, S.; Oude Elberink, S.; Morales, J.; Lemmen, C.; Zevenbergen, J. Towards 3D Indoor Cadastre Based on Change Detection from Point Clouds. *Remote Sens.* **2019**, *11*, 1972. [CrossRef]
16. Koeva, M.; Stöcker, C.; Crommelinck, S.; Ho, S.; Chipofya, M.; Sahib, J.; Bennett, R.; Zevenbergen, J.; Vosselman, G.; Lemmen, C.; et al. Innovative Remote Sensing Methodologies for Kenyan Land Tenure Mapping. *Remote Sens.* **2020**, *12*, 273. [CrossRef]
17. Park, S.; Song, A. Discrepancy Analysis for Detecting Candidate Parcels Requiring Update of Land Category in Cadastral Map Using Hyperspectral UAV Images: A Case Study in Jeonju, South Korea. *Remote Sens.* **2020**, *12*, 354. [CrossRef]
18. Xia, X.; Persello, C.; Koeva, M. Deep Fully Convolutional Networks for Cadastral Boundary Detection from UAV Images. *Remote Sens.* **2019**, *11*, 1725. [CrossRef]
19. Yan, J.; Jaw, S.W.; Soon, K.H.; Wieser, A.; Schrotter, G. Towards an Underground Utilities 3D Data Model for Land Administration. *Remote Sens.* **2019**, *11*, 1957. [CrossRef]
20. Stöcker, C.; Ho, S.; Nkerabigwi, P.; Schmidt, C.; Koeva, M.; Bennett, R.; Zevenbergen, J. Unmanned Aerial System Imagery, Land Data and User Needs: A Socio-Technical Assessment in Rwanda. *Remote Sens.* **2019**, *11*, 1035. [CrossRef]
21. Lee, C.; de Vries, W.T. Bridging the Semantic Gap between Land Tenure and EO Data: Conceptual and Methodological Underpinnings for a Geospatially Informed Analysis. *Remote Sens.* **2020**, *12*, 255. [CrossRef]
22. Bégué, A.; Leroux, L.; Soumaré, M.; Faure, J.-F.; Diouf, A.A.; Augusseau, X.; Touré, L.; Tonneau, J.-P. Remote Sensing Products and Services in Support of Agricultural Public Policies in Africa: Overview and Challenges. *Front. Sustain. Food Syst.* **2020**, *4*. [CrossRef]
23. Jewiss, J.L.; Brown, M.E.; Escobar, V.M. Satellite Remote Sensing Data for Decision Support in Emerging Agricultural Economies: How Satellite Data Can Transform Agricultural Decision Making [Perspectives]. *IEEE Geosci. Remote Sens. Mag.* **2020**, *8*, 117–133. [CrossRef]
24. Head, B.W. Toward More "Evidence-Informed" Policy Making? *Public Adm. Rev.* **2016**, *76*, 472–484. [CrossRef]
25. Parkhurst, J. *The Politics of Evidence: From Evidence-Based Policy to the Good Governance of Evidence*; Taylor & Francis: Abingdon, UK, 2017.
26. Nutley, S.M.; Nutley, S.; Walter, I.; Davies, H.T. *Using Evidence: How Research Can Inform Public Services*; Policy Press: Bristol, UK, 2007.
27. Oliver, K.; Lorenc, T.; Innvær, S. New directions in evidence-based policy research: A critical analysis of the literature. *Health Res. Policy Syst.* **2014**, *12*, 1–11. [CrossRef]
28. Shaxson, L.; Datta, A.; Tshangela, M.; Matomela, B. *Understanding the Organisational Context for Evidence-Informed Policy-Making*; Department of Environmental Affairs: Pretoria, South Africa, 2016.
29. Jung, E.-E. A Study on the Research Methodology of the North Korean Economy. In *KDI Review of the North Korean Economy*; Korea Development Institute: Sejong, Korea, 2019; pp. 63–65.
30. Kang, J.W. North Korean Studies and the Uses of Qualitative Methodology. *J. Asiat. Stud.* **2015**, *58*, 66–97.
31. Koh, Y.-H. A Study on the Research Trends of North Korean Studies after the Division of South and North Korea in 1945. *Unification Policy Stud.* **2015**, *24*, 29–54.
32. Koh, Y.-H. A Study on Trends and Issues of North Korean Studies. *J. Peace Unification Stud.* **2019**, *11*, 5–32.
33. Lee, H.K. The Present Status and Desirable Direction of North Korean Study. *J. Peace Stud.* **2010**, *11*, 83–104.
34. Ryu, K.; Kim, Y.H. A Jasmine Revolution in North Korea? Looking for Alternative Approaches to the Study of North Korean Regime Change. *North Korean Stud. Rev.* **2012**, *16*, 399–431.
35. Choe, S.C.; Lee, Y.S. A Study on Land Ownership and Use in North Korea. *J. Korean Reg. Sci. Assoc.* **1998**, *14*, 1–33.
36. Choi, M.J.; Kim, H.-S.; Kim, Y.S.; Park, E.-S. An Application of South Korean Land and Housing Legislations to Real Estate Asset Distribution for North Korean Residents After Unification. *J. Korea Plan. Assoc.* **2015**, *50*, 89–103. [CrossRef]
37. Kim, H.-S.; Seo, S.T.; Kim, D.-H.; Jeong, Y.W.; Choi, D.-S.; Cho, K.-H. Urban Planning of North Korea after Unification. *Urban Inf. Serv.* **2014**, *389*, 3–17.
38. Kim, S.Y. A Study on the Formation and Changes of Socialist Land System in North Korea and its Future Directions after Unification. In *Real Estate Focus*; Research Institute of Korea Appraisal Board: Seoul, Korea, 2012; pp. 64–78.
39. Lee, C.; de Vries, W.T. A divided nation: Rethinking and rescaling land tenure in the Korean (re-)unification. *Land Use Policy* **2018**, *75*, 127–136. [CrossRef]

40. Lee, C.; de Vries, W.T.; Chigbu, U.E. Land Governance Re-Arrangements: The One-Country One-System (OCOS) Versus One-Country Two-System (OCTS) Approach. *Adm. Sci.* **2019**, *9*, 21. [CrossRef]
41. Park, M.; Do, J. Research on North Korean Defectors' Values using the Focus Group Interview (FGI) Method: Its Objectives, Methodology, and Significance. *J. Humanit. Unification* **2019**, *79*, 5–35. [CrossRef]
42. Hong, M. *The Marketization and Social Mobility in North Korea: Spatial Structure·Urban Politics·Social Stratum*; Korea Institute for National Unification: Seoul, Korea, 2015.
43. Shadbolt, N.R.; Smart, P.R.; Wilson, J.; Sharples, S. Knowledge elicitation: Methods, Tools and Techniques. In *Evaluation of Human Work*; CRC Press: Boca Raton, FL, USA, 2015; pp. 163–200.
44. Mukherjee, N.; Zabala, A.; Huge, J.; Nyumba, T.O.; Adem Esmail, B.; Sutherland, W.J. Comparison of techniques for eliciting views and judgements in decision-making. *Methods Ecol. Evol.* **2018**, *9*, 54–63. [CrossRef]
45. Jeong, E.-M. Application and Trend of Researches on North Korean Refugees as a Method of North Korea Studies. *Rev. North Korean Stud.* **2005**, *8*, 139–176.
46. Park, S.; Kim, T.; Kim, S.; Song, J. *Urban Planning and Development Practices in North Korea: Urban Consequences of Informal Market*; Korea Research Institute for Human Settlements: Sejong, Korea, 2016.
47. Olofsson, P.; Foody, G.M.; Herold, M.; Stehman, S.V.; Woodcock, C.E.; Wulder, M.A. Good practices for estimating area and assessing accuracy of land change. *Remote Sens. Environ.* **2014**, *148*, 42–57. [CrossRef]
48. Congalton, R.G. Accuracy assessment and validation of remotely sensed and other spatial information. *Int. J. Wildland Fire* **2001**, *10*, 321–328. [CrossRef]
49. Campbell, J.B.; Wynne, R.H. *Introduction to Remote Sensing*, 5th ed.; Guilford Press: New York, NY, USA, 2011.
50. Radoux, J.; Bogaert, P. Good Practices for Object-Based Accuracy Assessment. *Remote Sens.* **2017**, *9*, 646. [CrossRef]
51. Kelly, A.B.; Kelly, N.M. Validating the remotely sensed geography of crime: A review of emerging issues. *Remote Sens.* **2014**, *6*, 12723–12751. [CrossRef]
52. Aditya, T.; Maria-Unger, E.; Bennett, R.; Saers, P.; Lukman Syahid, H.; Erwan, D.; Wits, T.; Widjajanti, N.; Budi Santosa, P.; Atunggal, D. Participatory Land Administration in Indonesia: Quality and Usability Assessment. *Land* **2020**, *9*, 79. [CrossRef]
53. Asiama, K.; Bennett, R.; Zevenbergen, J. Participatory land administration on customary lands: A practical VGI experiment in Nanton, Ghana. *ISPRS Int. J. Geo Inf.* **2017**, *6*, 186. [CrossRef]
54. Wassie, Y.A.; Koeva, M.N.; Bennett, R.M.; Lemmen, C.H.J. A procedure for semi-automated cadastral boundary feature extraction from high-resolution satellite imagery. *J. Spat. Sci.* **2018**, *63*, 75–92. [CrossRef]
55. Loew, A.; Bell, W.; Brocca, L.; Bulgin, C.E.; Burdanowitz, J.; Calbet, X.; Donner, R.V.; Ghent, D.; Gruber, A.; Kaminski, T.; et al. Validation practices for satellite-based Earth observation data across communities. *Rev. Geophys.* **2017**, *55*, 779–817. [CrossRef]
56. Armitage, D.; Berkes, F.; Dale, A.; Kocho-Schellenberg, E.; Patton, E. Co-management and the co-production of knowledge: Learning to adapt in Canada's Arctic. *Glob. Environ. Chang.* **2011**, *21*, 995–1004. [CrossRef]
57. Edelenbos, J.; van Buuren, A.; van Schie, N. Co-producing knowledge: Joint knowledge production between experts, bureaucrats and stakeholders in Dutch water management projects. *Environ. Sci. Policy* **2011**, *14*, 675–684. [CrossRef]
58. Pohl, C.; Rist, S.; Zimmermann, A.; Fry, P.; Gurung, G.S.; Schneider, F.; Speranza, C.I.; Kiteme, B.; Boillat, S.; Serrano, E.; et al. Researchers' roles in knowledge co-production: Experience from sustainability research in Kenya, Switzerland, Bolivia and Nepal. *Sci. Public Policy* **2010**, *37*, 267–281. [CrossRef]
59. Lee, C.; de Vries, W.T. Sustaining a Culture of Excellence: Massive Open Online Course (MOOC) on Land Management. *Sustainability* **2019**, *11*, 3280. [CrossRef]
60. Freedman, B. Scientific Value and Validity as Ethical Requirements for Research: A Proposed Explication. *IRB Ethics Hum. Res.* **1987**, *9*, 7–10. [CrossRef]
61. Schie, N.V.; Duijn, M.; Edelenbos, J. Co-Valuation: Exploring methods for expert and stakeholder valuation. *J. Environ. Assess. Policy Manag.* **2011**, *13*, 619–650. [CrossRef]
62. Eckhard, S. Bridging the citizen gap: Bureaucratic representation and knowledge linkage in (international) public administration. *Governance* **2020**. [CrossRef]
63. Kingdon, J.W.; Stano, E. *Agendas, Alternatives, and Public Policies*; Little, Brown: Boston, MA, USA, 1984; Volume 45.
64. McClean, S.; Shaw, A. From Schism to Continuum? The Problematic Relationship Between Expert and Lay Knowledge—An Exploratory Conceptual Synthesis of Two Qualitative Studies. *Qual. Health Res.* **2005**, *15*, 729–749. [CrossRef]
65. Hunt, J.; Shackley, S. Reconceiving Science and Policy: Academic, Fiducial and Bureaucratic Knowledge. *Minerva* **1999**, *37*, 141–164. [CrossRef]
66. Stilgoe, J. The (co-)production of public uncertainty: UK scientific advice on mobile phone health risks. *Public Underst. Sci.* **2007**, *16*, 45–61. [CrossRef]
67. Gray, S.; Chan, A.; Clark, D.; Jordan, R. Modeling the integration of stakeholder knowledge in social–ecological decision-making: Benefits and limitations to knowledge diversity. *Ecol. Model.* **2012**, *229*, 88–96. [CrossRef]
68. Petts, J.; Brooks, C. Expert Conceptualisations of the Role of Lay Knowledge in Environmental Decisionmaking: Challenges for Deliberative Democracy. *Environ. Plan. A Econ. Space* **2006**, *38*, 1045–1059. [CrossRef]
69. Enemark, S.; Bell, K.; Lemmen, C.; McLaren, R. *Fit-For-Purpose Land Administration*; FIG Publication. No. 60; International Federation of Surveyors: Copenhagen, Denmark, 2014.

70. de Vries, W.T.; Bugri, J.T.; Mandhu, F. *Responsible and Smart Land Management Interventions: An African Context*; CRC Press: Boca Raton, FL, USA, 2020.
71. de Vries, W.T.; Chigbu, U.E. Responsible land management-Concept and application in a territorial rural context. *Fub. Flächenmanag. Bodenordn.* **2017**, *79*, 65–73.
72. Lee, Y.W.; Strong, D.M.; Kahn, B.K.; Wang, R.Y. AIMQ: A methodology for information quality assessment. *Inf. Manag.* **2002**, *40*, 133–146. [CrossRef]
73. McDonald, J.H. *Handbook of Biological Statistics*; Sparky House Publishing: Baltimore, MD, USA, 2009; Volume 2.
74. Warner, P. Testing association with Fisher's Exact test. *J. Fam. Plan. Reprod. Health Care* **2013**, *39*, 281–284. [CrossRef]
75. Bubela, T.; Nisbet, M.C.; Borchelt, R.; Brunger, F.; Critchley, C.; Einsiedel, E.; Geller, G.; Gupta, A.; Hampel, J.; Hyde-Lay, R. Science communication reconsidered. *Nat. Biotechnol.* **2009**, *27*, 514–518. [CrossRef] [PubMed]
76. Fourez, G. Scientific and technological literacy as a social practice. *Soc. Stud. Sci.* **1997**, *27*, 903–936. [CrossRef]
77. Jones, H. Promoting evidence-based decision-making in development agencies. *ODI Backgr. Note* **2012**, *1*, 1–6.

MDPI

Article

Deep Learning for Detection of Visible Land Boundaries from UAV Imagery

Bujar Fetai *, Matej Račič and Anka Lisec

Faculty of Civil and Geodetic Engineering, University of Ljubljana, Jamova Cesta 2, 1000 Ljubljana, Slovenia; matej.racic@fgg.uni-lj.si (M.R.); anka.lisec@fgg.uni-lj.si (A.L.)
* Correspondence: bujar.fetai@fgg.uni-lj.si; Tel.: +386-1-4768629

Abstract: Current efforts aim to accelerate cadastral mapping through innovative and automated approaches and can be used to both create and update cadastral maps. This research aims to automate the detection of visible land boundaries from unmanned aerial vehicle (UAV) imagery using deep learning. In addition, we wanted to evaluate the advantages and disadvantages of programming-based deep learning compared to commercial software-based deep learning. For the first case, we used the convolutional neural network U-Net, implemented in Keras, written in Python using the TensorFlow library. For commercial software-based deep learning, we used ENVINet5. UAV imageries from different areas were used to train the U-Net model, which was performed in Google Collaboratory and tested in the study area in Odranci, Slovenia. The results were compared with the results of ENVINet5 using the same datasets. The results showed that both models achieved an overall accuracy of over 95%. The high accuracy is due to the problem of unbalanced classes, which is usually present in boundary detection tasks. U-Net provided a recall of 0.35 and a precision of 0.68 when the threshold was set to 0.5. A threshold can be viewed as a tool for filtering predicted boundary maps and balancing recall and precision. For equitable comparison with ENVINet5, the threshold was increased. U-Net provided more balanced results, a recall of 0.65 and a precision of 0.41, compared to ENVINet5 recall of 0.84 and a precision of 0.35. Programming-based deep learning provides a more flexible yet complex approach to boundary mapping than software-based, which is rigid and does not require programming. The predicted visible land boundaries can be used both to speed up the creation of cadastral maps and to automate the revision of existing cadastral maps and define areas where updates are needed. The predicted boundaries cannot be considered final at this stage but can be used as preliminary cadastral boundaries.

Keywords: land; cadastral mapping; visible boundary; UAV; deep learning

Citation: Fetai, B.; Račič, M.; Lisec, A. Deep Learning for Detection of Visible Land Boundaries from UAV Imagery. *Remote Sens.* **2021**, *13*, 2077. https://doi.org/10.3390/rs13112077

Academic Editors: Rohan Bennett, Claudio Persello and Mila Koeva

Received: 19 March 2021
Accepted: 24 May 2021
Published: 25 May 2021

Publisher's Note: MDPI stays neutral with regard to jurisdictional claims in published maps and institutional affiliations.

1. Introduction

Accelerating cadastral mapping to establish a complete cadastre and keeping it up-to-date is a contemporary challenge in the domain of land administration [1,2]. Cadastral mapping is considered the first step in establishing cadastral systems and serves as the basis for defining the boundaries of land units to which land rights refer [3]. Mapping the boundaries of land rights in a formal cadastral system helps to increase land tenure security [4]. More than 70% of land rights are unregistered globally and are not part of any formal cadastral system [1]. The challenge of accelerating the creation of cadastral maps is present mainly in developing regions with low cadastral coverage [5]. Cadastral maps are usually defined as spatial representations of cadastral records, showing the extent and ownership of land units [6]. An effective cadastral system should provide up-to-date land data [7]. In countries with complete cadastral coverage, this is considered one of the major challenges. To overcome the challenge of accelerating cadastral mapping while providing up-to-date land data, low-cost and rapid cadastral surveying and mapping techniques are required [5,8].

The proposed cadastral surveying techniques are indirect rather than direct surveying. Indirect cadastral surveying is based on the delineation of visible cadastral boundaries from high-resolution remote sensing imagery. In contrast, direct or ground-based surveying techniques are based on field survey and are often considered slow and expensive [1,5]. The application of image-based cadastral mapping is based on the recognition that many cadastral boundaries coincide with visible natural or man-made boundaries, such as hedgerows, land cover boundaries, building walls, roads, etc., and can be easily detected from remote sensing imagery [2,9]. The detection of such boundaries from data acquired with sensors on unmanned aerial vehicles (UAVs) has gained increasing popularity in cadastral applications [10–12].

In cadastral applications, UAVs have gained prominence as a powerful technology that can bridge the gap between slow but accurate field surveys and the fast approach of conventional aerial surveys [13]. Sensors on UAVs provide low-cost, efficient and flexible systems for high-resolution spatial data acquisition, enabling the production of orthoimages, digital surface models and point clouds [14]. Overall, UAVs have shown a high potential for detecting land boundaries in both rural and urban areas [8,15]. In addition, UAV-based orthoimages have been considered as base maps for the creation of cadastral maps and for updating or revising existing cadastral maps [10,12,16]. Besides the high visibility of cadastral boundaries on UAV imagery, manual delineations have been reported in many previous case studies [8]. The contemporary approach to cadastral mapping aims to simplify and speed up image-based cadastral mapping by automating the detection of visible cadastral boundaries from images acquired with high-resolution optical sensors [15,17,18].

1.1. Deep Learning for Cadastral Mapping

Only a limited number of studies have investigated the automatic approach to detect visible cadastral boundaries from UAV imagery. Mainly, tailored workflows using image segmentation and edge detection algorithms have been applied to automate cadastral mapping and thus provide more efficient approaches [8,15]. Multi-resolution segmentation (MRS) and globalized probability of boundary (gPb) are among the most popular segmentation and edge detection algorithms used in the cadastral mapping [15]. Early algorithms, such as Canny edge detection, extract edges by computing gradients of local brightness, which are then combined to form boundaries. However, the approach is characterized by the detection of irrelevant edges in textured regions [19]. Furthermore, gPb provides more accurate results compared to other approaches on edge detection (e.g., Canny detector and Prewitt, Sobel, Roberts operator) [20]. MRS, gPb and Canny are unsupervised techniques. Unsupervised techniques include methods that require segmentation parameters to be defined. The challenge is to define appropriate segmentation parameters for features that vary in size, shape, scale and spatial location. Then, the image is automatically segmented according to these parameters [19]. With respect to modern methods for automatic boundary detection in cadastral mapping, deep learning is becoming increasingly important—as a supervised technique [21]. However, the deeper understanding is challenging, so the abstraction of the process offers a solution.

Deep learning methods such as convolutional neural networks (CNNs) are very effective in extracting higher-level representations needed for classification or detection from raw input [22,23]. Moreover, recent studies indicate that deep learning ensures higher accuracy in delineating visible land boundaries than some object-based methods [15,17,24]. In the study by Crommelinck et al. [17], it was reported that CNNs, namely the VGG19 architecture, provide a more automated and accurate approach for detecting visible boundaries from UAV imagery than the machine learning approach random forest (RF). Furthermore, the study highlighted that the model based on VGG19 architecture provides more promising loss and accuracy metrics compared to other CNN architectures such as ResNet, Inception, Xception, MobileNet and DenseNet. The study conducted by Xia et al. [15] investigated the potential of fully CNNs for cadastral boundary detection in urban and semiurban areas. The results showed that fully CNNs outperformed other state-of-the-art machine learning techniques, including MRS and gPb. The results indicated 0.37 in recall, 0.79 in precision and 0.50 in *F*1 score. The study by Park and Song [25] aims to identify the inconsistencies between the existing land use information from existing cadastral maps and the current land use in the field. The proposed method involves updating the existing land cover attributes of cadastral maps using UAV hyperspectral imagery classified with CNNs and then creating a discrepancy map showing the differences in land use. CNNs bring innovative capabilities to cadastral mapping that can facilitate and accelerate the delineation of visible cadastral boundaries. In line with these studies, improving the accuracy of automatic visible boundary detection remains a challenge in contemporary image-based cadastral mapping [15].

One CNN architecture that has not been satisfactorily investigated for visible boundary detection in cadastral applications is U-Net. U-Net was originally developed for biomedical image segmentation and is considered a revolutionary architecture for semantic segmentation tasks [26–30]. Generally, it is claimed that the main challenge in CNNs is a large amount of training data preparation and computational requirements [26]. Thus, providing thousands of UAV training data can be considered as a limitation for visible land boundary detection with CNNs, especially when a model is trained from scratch. However, the U-Net architecture is designed to work with fewer training images preprocessed by an intensive data augmentation procedure and still provide precise segmentation [26]. In addition, a software-based module, ENVI deep learning, has recently been developed to simplify and perform deep learning procedures with geospatial data. The number of studies that have tested its potential is very small [31]; in particular, it has not been sufficiently explored for the detection of visible cadastral boundaries from UAV imagery.

1.2. Objective of the Study

The main objective of this study is to investigate the potential of CNN architecture, namely U-Net, based on UAV imagery training samples, as a deep learning-based detector for visible land boundaries. In addition, we wanted to evaluate the advantages and disadvantages of programming-based, e.g., custom, deep learning compared to a commercial software-based solution. Here, we compared the results of U-Net with those of the recently released software-based ENVI deep learning by focusing on the boundary mapping approaches and their conformity in the land administration domain.

2. Materials and Methods

2.1. UAV Data

It is argued that the number of visible cadastral boundaries is higher in rural areas than in dense urban areas (an example of a visible cadastral boundary in Figure 1b). A rural area in Odranci, Slovenia, was selected for this study. UAV images were acquired at a flight altitude of 90 m, resulting in 997 images to cover the study area. The images were acquired in September 2020, at midday, under clear skies. The UAV images were indirectly georeferenced using a uniform distribution of 18 ground control points (GCPs). The GCPs were surveyed with real-time kinematic (RTK) using the global navigation satellite system (GNSS) receiver Leica GS18. In addition, the GCPs were also surveyed with RTK, using a multifrequency low-cost GNSS instrument (base and rover), namely ZED-F9P receiver with u-blox ANN-MB-00 antenna—as a cheaper alternative to geodetic GNSS receivers (Figure 1b). The differences were insignificant for 2D cadastral mapping ($RMSE_{x,y}$ = 0.019 m). The obtained ground sampling distance (GSD) from the UAV orthoimage was 0.02 m. The study site had an area of 63.9 ha and was divided into areas for training and testing the CNNs (Figure 1a).

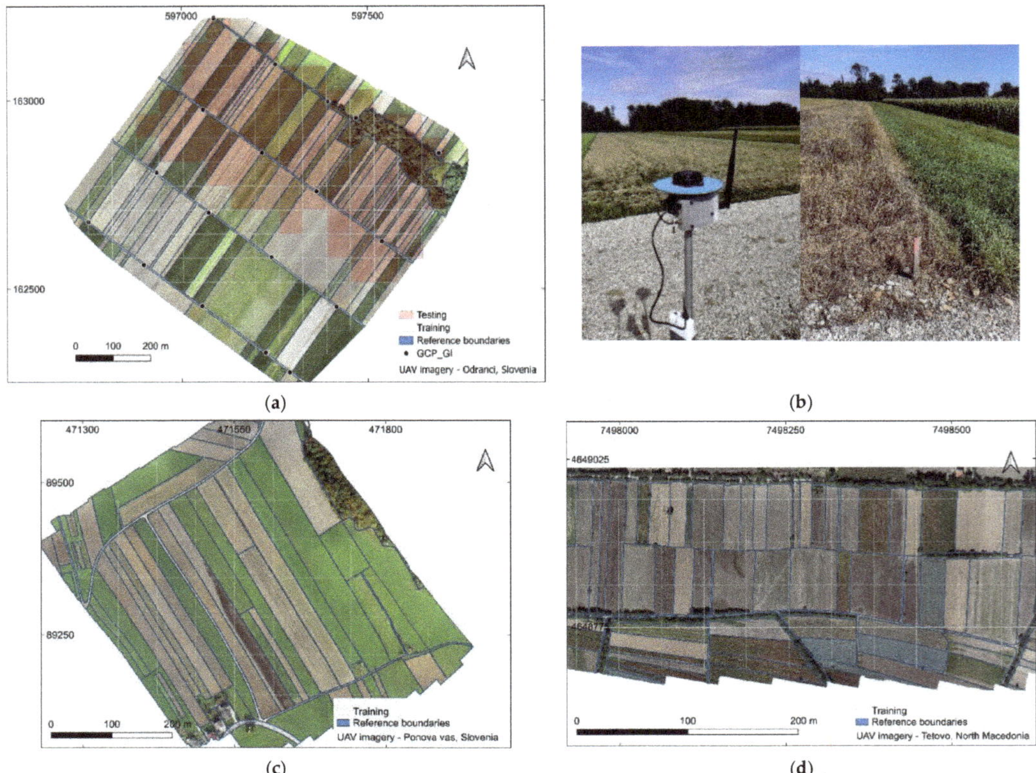

Figure 1. (**a**) UAV imagery of 0.25 ground sample distance (GSD) for Odranci–Slovenia, divided into areas for training and testing; (**b**) low-cost instrument ZED-F9P and example of visible cadastral boundaries. (**c**) UAV imagery of 0.25 (GSD) for Ponova vas—Slovenia, used for training; (**d**) UAV imagery of 0.25 (GSD) for Tetovo—North Macedonia, used for training.

With the aim of increasing the number and diversity of training data, additional UAV images with a rural scene from Ponova vas (Slovenia) and Tetovo (North Macedonia) were used (Figure 1c,d). The UAV data in Ponova vas was acquired at an altitude of 80 m and had a GSD of 0.02 m. The UAV data in Tetovo have a GSD of 0.03 m and were acquired at an altitude of 110 m. Figure 1a,c,d shows the UAV orthoimages of the study areas.

The selected areas contain agricultural fields, roads, fences, hedges and tree groups, which are assumed to represent cadastral boundaries [8]. The cadastral reference boundaries were derived from the UAV orthoimages by manual land delineation on-screen in all three study areas. All UAV images were acquired using a rotary-wing UAV, namely the DJI Phantom 4 Pro. Table 1 shows the specifications of the data acquisition.

Table 1. Specification of unmanned aerial vehicle (UAV) dataset for the selected study areas.

Location	UAV Model	Camera/Focal Length (mm)	Overlap Forward/Sideward	Flight Altitude	GSD (cm)	Coverage Area (ha)	Purpose
Odranci, Slovenia	DJI Phantom 4 Pro	1″ CMOS/24 mm	80/70	90 m	2.35	63.9	Training and Testing
Ponova vas, Slovenia				80 m	2.01	25.0	Training
Tetovo, North Macedonia				110 m	2.85	24.3	Training

2.2. Detection of Visible Land Boundaries

In general, the workflow of this study consists of three main parts, namely data preparation, visible land boundary detection and accuracy assessment. The specific steps for both the U-Net and ENVI deep learning boundary mapping approaches are described in the following subsections.

2.2.1. U-Net

In deep learning, CNNs can be trained in two approaches, from scratch or via transfer learning [17,32]. In our case, the U-Net model was trained from scratch based on UAV images.

The UAV orthoimages of the selected study areas (Figure 1a–c) were randomly tiled in 256 pixels × 256 pixels. To increase the field of view for each tile, the original spatial resolution of the UAV orthoimages had to be converted to a larger GSD, from 2–3 to 25 cm. The results were 219 original tiles, namely 144 tiles for training and 75 tiles for testing (Figure 1a,c,d). In addition, corresponding label images (also called ground truth images) were created for each UAV image. The label images, with a size of 256 × 256 × 1, were created from the manually digitized reference boundaries, which were initially in the vector format. The reference boundaries were buffered to 50 cm and later rasterized using GRASS GIS tools [33]. Additionally, the UAV tiles were then rotated, flipped and scaled to improve generalization and increase the number of training samples. This technique is known in deep learning as data augmentation and is used to supplement original training data. Once the data preparation and augmentation were completed, the next step was to train the U-Net model.

The CNN based on U-Net is symmetric and contains encoding and decoding parts, which gives it the U-shaped form. U-Net is described in detail in [26]. The left part, the encoding path, is a typical convolutional network that contains repetitive usage of 3 × 3 convolutions, each followed by a rectified linear unit (ReLU) and a max-pooling operation, i.e., 2 × 2 convolutions. During the encoding path, the contextual information (depth) of the images was increased while the resolution of the images was reduced. The right part, the decoding path, merged the contextual and resolution information of the images through a sequence of 2 × 2 up-convolutions. The goal of the decoding path is to provide precise localization using the contextual information from the encoding path. During the decoding path, the resolution of the image was upconverted to its original size. The U-Net architecture implemented in this study is shown in Figure 2.

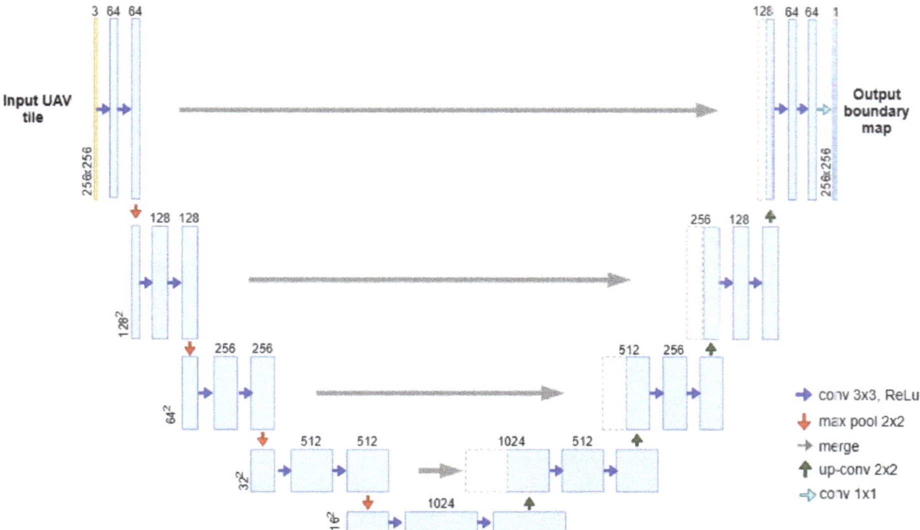

Figure 2. The implemented U-Net architecture (adapted from [26]).

Overall, training a CNN model requires a powerful graphics processing unit (GPU), lots of memory and efficient computations. To overcome this requirement while providing a cost-effective and fast approach for visible boundary detection and hence cadastral mapping, the training of U-Net was performed by Google Collaboratory [34]. U-Net was implemented in the high-level neural network API Keras [35]. The process was written in Python in combination with the TensorFlow library [36]. The implementation of the model in Keras was done by modifying and referencing to [37], which is an implementation for grayscale biomedical images. In this study, the U-Net model was adapted to work with three-band images, namely RGB UAV images, as input and produce a single band boundary map as output with the same image size as the input. However, the predicted boundary maps were not georeferenced.

Considering that georeferencing is the key component in cadastral mapping, further improvements were made. In this study, we considered two additional steps, namely georeferencing the predicted boundaries and merging the georeferenced tiles to obtain the boundary map for the entire extent of the test area. The processing and analysis were done using open-source modules, including Rasterio [38], GDAL [39] and Numpy [40]. The workflow and boundary mapping approach used in this study are shown in Figure 3.

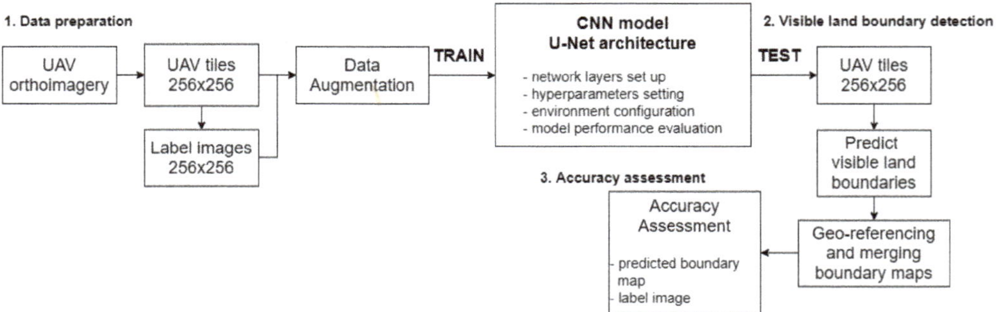

Figure 3. Workflow for the detection of visible land boundaries based on the U-Net model.

2.2.2. ENVI Deep Learning

ENVI deep learning [41] can be categorized as software-based deep learning technology that offers its own U-Net-like model. The model is called ENVINet5 and is described in detail in [42]. In this study, the ENVINet5 model was used to compare it with the U-Net model—both the results and the land boundary mapping approach.

The training approach is patch-based, i.e., the entire extent of the training UAV data can be used as input, and the model can learn based on the pixels specified in the patch. Considering this, a patch size of 256 pixels × 256 pixels was used for training and validating the ENVINet5 as a single-class model. Moreover, the training of the ENVINet5 model is based on a labelled raster that should be created within the software. Generally, there are two approaches: by on-screen manual digitizing or by directly uploading features in vector format. In our case, we uploaded the shapefile (.shp) of reference cadastral boundaries (buffered to 50 cm), defined as the region of interest (ROI), from which the label raster was created. We used the recently released version of ENVI deep learning, i.e., version 1.1.2, which has an option for data augmentation, unlike the previous version where data augmentation was not possible. Data augmentation is performed by rotating and scaling the original UAV training data.

The training of the ENVINet5 model was done using the toolbox deep learning guide map. Before starting the training, it was necessary to initialize a TensorFlow model, which defines the structure of the model, including the architecture (ENVINet5 for a single class), the patch size (256 × 256), and the number of the bands that are used for training (3 bands, RGB). After the model was initialized, the training data was uploaded. In the following, the values for the training parameters are required, such as the number of epochs, the number of patches per epoch, the number of patches per batch, class weight, etc. For the number of patches per epoch and per batch, it is suggested to leave them blank so that ENVI automatically determines the most appropriate values. For saving the model and the trained weights (output model), ENVI uses the HDF5 (.h5) format. The generated land boundary maps were georeferenced, and no post-processing step was required. The boundary mapping approach and workflow used in this study are shown in Figure 4.

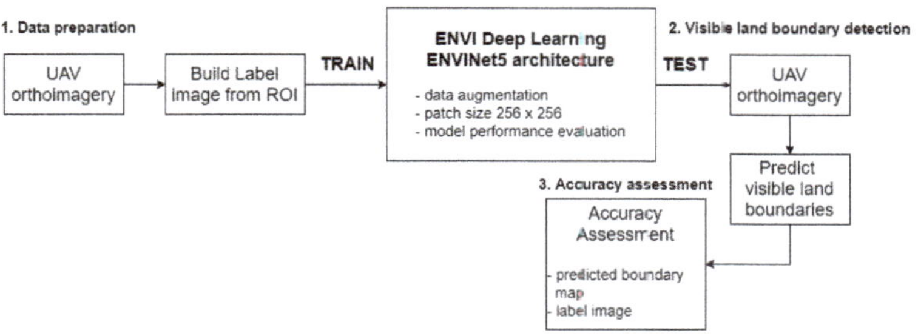

Figure 4. Workflow for the detection of visible land boundaries based on the ENVINet5 model.

However, there were some hardware and software requirements, such as NVIDIA GPU driver version 410.x or higher and NVIDIA graphics card with CUDA compute capability 3.5–7.5. Additionally, it is recommended to have at least 8 GB GPU memory to perform the training of the models with the GPU. If this requirement is not met, the training will be performed with the central processing unit (CPU), which is too slow for a large number of images.

2.3. Accuracy Assessment

The accuracy assessment in this study investigates two aspects—the evaluation of the two models U-Net and ENVINet5 and the evaluation of the detection quality of the visible land boundaries for the test UAV data (Figure 1a).

Both CNN models, U-Net and ENVINet5, were monitored with loss and accuracy during the training process. Loss is defined as the sum of errors for each sample in training between labels and predictions. To maximize the efficiency of the model, loss should be minimized. For this purpose, we used the cross-entropy loss expressed by the following formula:

$$cross - entropy\ loss = -(y_i \log(\hat{y}_i) + (1 - y_i) \log(1 - \hat{y}_i)) \quad (1)$$

where:

y_i—actual label value,

$\hat{y}_i$—predicted value.

To assess the performance of the models, overall accuracy was used as the evaluation metric. The overall accuracy was calculated by summing the percentages of pixels correctly identified as land boundaries by the model compared to the labelled reference boundaries and dividing by all boundaries. Overall accuracy is expressed with the following equation:

$$overall\ accuracy = \frac{TP + TN}{TP + FP + FN + TN} \quad (2)$$

where true positive (*TP*), true negative (*TN*), false positive (*FP*) and false negative (*FN*) are shown in Table 2, which is the confusion matrix used to evaluate the detection quality of the visible land boundaries.

Table 2. Confusion matrix.

		Ground Truth	
		Boundary	**No Boundary**
Prediction	Boundary	*TP*	*FP*
	No boundary	*FN*	*TN*

The detection quality of the visible land boundaries was evaluated by computing the F1 score derived from the confusion matrix. *F*1 score was calculated for test UAV data (not seen by the model during training) and represented the harmonic mean between recall and precision (Equations (3) and (4)). Larger values indicate higher accuracy.

$$recall = \frac{TP}{TP + FN} \tag{3}$$

$$precision = \frac{TP}{TP + FP} \tag{4}$$

The recall is the ratio of correctly predicted visible boundaries to all reference cadastral boundaries. The precision is the ratio of correctly predicted visible boundaries to all predicted positive visible boundaries. The *F*1 score combines precision and recall and is expressed with the following equation:

$$F1\ score = 2 * \frac{recall * precision}{recall + precision} \tag{5}$$

3. Results

3.1. CNN Architecture

In our study, the labelled images and RGB UAV images were used to train the deep CNN models.

For the U-Net, the randomly cropped tiles (Figure 1a,c,d) were the candidate training datasets. The greater the variety of images used in the training data, the more robust the network and the better the detection of visible land boundaries. Data augmentation was applied to the provided images to increase the number of UAV images available for training the U-Net model. Of the data used for training, 30% was used for validation. Once the U-Net model was trained, we applied it to the test UAV images (Figure 1a).

The architecture was based on the original architecture of the U-Net, considering the number of layers (network depth) and the size of the convolutional filters. However, to avoid the resizing of the output image by the max-pooling operation, the padding was set to 'same'. In addition, a dropout rate of 0.8 was used as an optional function. The dropout rate aims to avoid overfitting the model, which means that the training and validation accuracy curves are less likely to diverge, then the model is more robust. To avoid under-fitting, the layer depth was set to 1024. The larger the layer size, the higher the probability that the curve for validation will be close to the training accuracy. We used sigmoid instead of softmax as the final activation layer to retrieve the predictions, which is good for binary classification. The main point is that when using sigmoid, the probabilities were independent and did not necessarily sum to one. This is because the sigmoid considers each raw output value separately. During training, the optimization algorithm stochastic gradient descent (SGD) was used as the optimizer, and the momentum was set to 0.9. The learning rate in the optimization defines the speed of learning, which makes the network training converge. We used an adjusted learning rate of 0.001. Table 3 shows the adjusted settings and parameters.

Table 3. Settings and adjusted parameters for our fine-tuned CNN based on the U-Net architecture.

	Settings	Parameters
Trainable layers	pooling layer	maxpooling 2D
	connected layer	layer depth = 1024 activation = ReLU
	dropout layer	dropout rate = 0.8
	logistic layer	activation layer = sigmoid
Learning optimizer	SGD optimizer	learning rate = 0.001 momentum = 0.9
Training	UAV images 256 × 256 × 3, data augmentation, validation split 0.3	number of epochs = 100 batch size = 32 steps per epoch = training samples/number of epochs

The model was trained with a batch size of 32 for 100 epochs. An early stop function was also used to monitor validation loss. The number of steps per epoch was calculated by dividing the total number of training images by the batch size. Deep learning by the U-Net model was performed in Google Collaboratory, which provided a GPU with 25 GB of RAM. A total of 4768 samples, i.e., augmented images, were used for training and 2044 samples for validation. Training the model for 100 epochs took 4 h. The best model was saved at epoch 92 by achieving an overall accuracy of 0.978 and a loss of 0.058. The training performance of the U-Net is visualized in Figure 5.

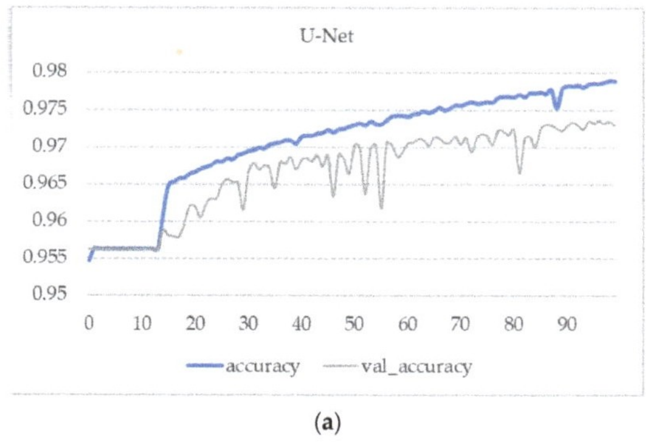

(a)

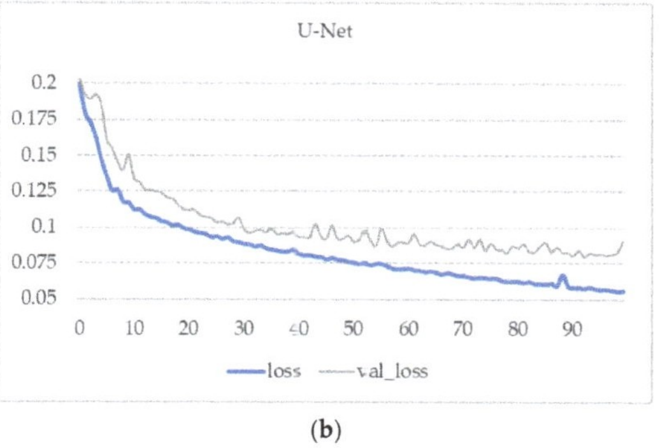

(b)

Figure 5. Model performance: (**a**) accuracy and (**b**) loss for our fine-tuned U-Net.

In this study, we also used ENVI deep learning to compare the results obtained with the U-Net model. In this study, ENVI deep learning is considered a 'black box'. The information we had is that ENVINet5 is based on U-Net architecture, and it uses the same layer size and the same number of convolution layers.

The ENVINet5 model was trained with a patch size equal to the total extent of the training UAV data. In addition, the training data shown in Figure 1a,c,d were also processed as UAV images for validation. The adapted training parameters of ENVINet5, namely patch size of 256 × 256, number of epochs 50 and class weights min. 1 and max. 2, data augmentation 'yes', resulted in a fine-tuned model for visible boundary detection. The values of the other parameters were automatically filled by ENVI deep learning as they are suggested to be left blank. The model with the best performance was saved at epoch 24, where the validation loss reached its lowest value. The overall accuracy of the model was 0.946 and with a loss of 0.234. The training performance of the CNN model ENVINet5 is shown in Figure 6.

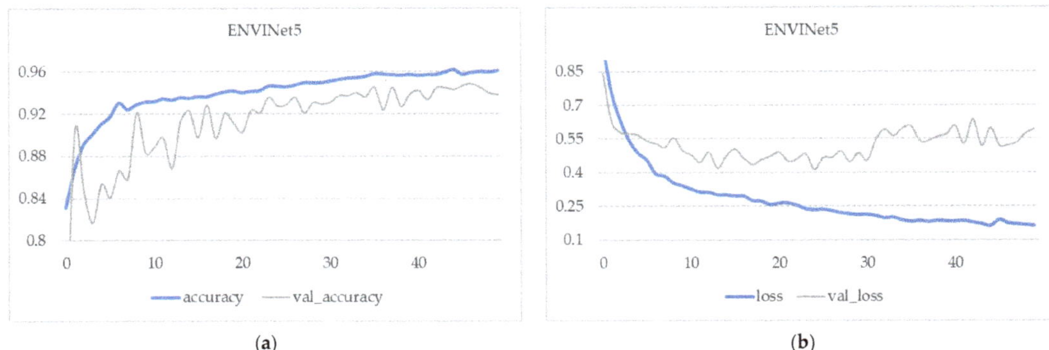

(**a**) (**b**)

Figure 6. Model performance: (**a**) accuracy and (**b**) loss for our fine-tuned ENVINet5.

All experiments with ENVI deep learning were performed on an Intel® Core ™ i7-4771 CPU 3.5 GHz machine with an NVIDIA GeForce GTX 650 GPU with 2 GB of RAM. The training time for 50 epochs was 6 h.

3.2. Detection of Visible Land Boundaries by U-Net

After training the CNN model, we evaluated its performance by applying it to the test area (Figure 1a). We applied the trained U-Net model to the test UAV tiles of size 256 × 256 to predict the visible land boundaries. Some results of the predicted boundary maps based on UAV tiles are shown in Figure 7.

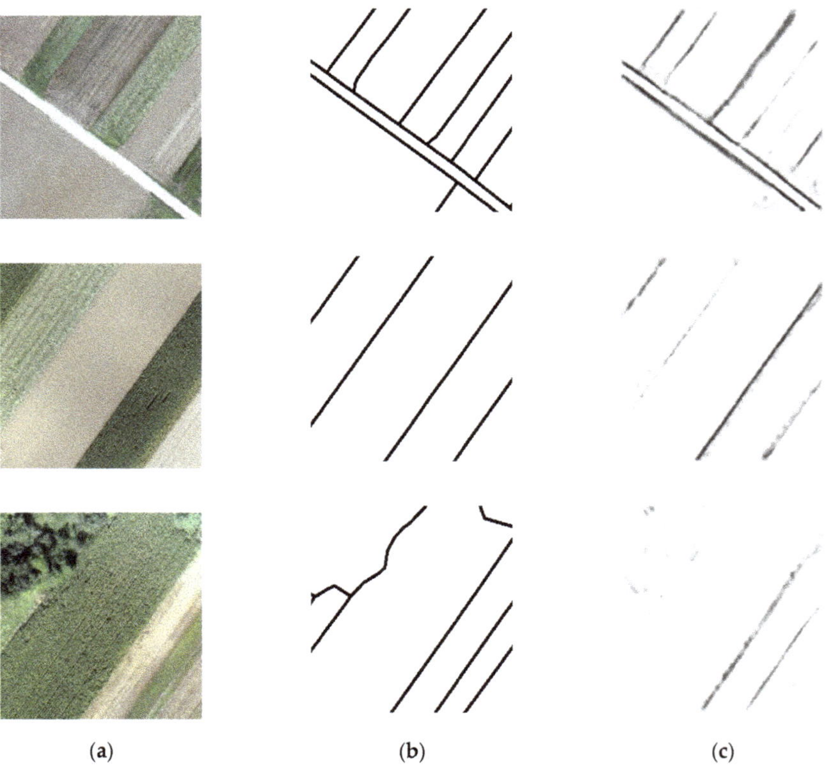

(**a**) (**b**) (**c**)

Figure 7. (**a**) Examples of UAV testing tiles; (**b**) label images; (**c**) predicted boundary maps with values 0–1.

The next step was to georeference the predicted visible land boundaries and merge them into a single land boundary map (Figure 8c). Considering that the predicted values were in the range of 0–1, in order to assess the accuracy and thus match the ground truth class values, it was necessary to reclassify the predicted values to 0 and 1, namely to 'boundary' and 'no boundary'. In this study, few boundary map reclassifications were performed, e.g., 'boundary' $\leq$ 0.9; 'boundary' $\leq$ 0.7; 'boundary' $\leq$ 0.5. The predicted boundary maps for the test area showed a good match with the labelling image (ground truth). The results of the georeferenced and merged predictions along with the reclassified boundary maps are shown in Figure 8c–f.

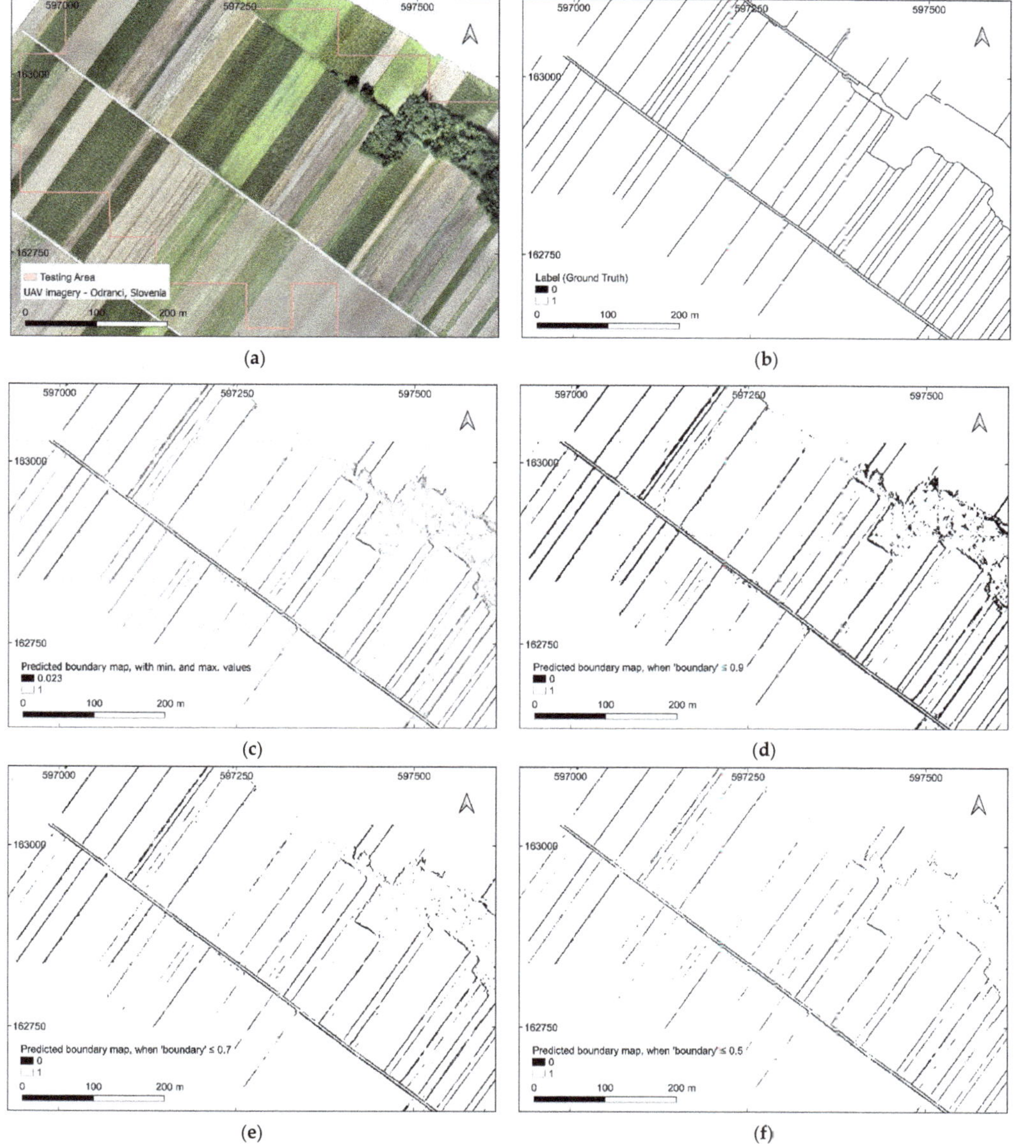

Figure 8. (**a**) Test UAV area; (**b**) label image; (**c**) predicted boundary map (0–1); reclassified boundary maps when (**d**) 'boundary'$\leq$ 0.9; (**e**) 'boundary'$\leq$ 0.7; (**f**) 'boundary'$\leq$ 0.5.

For a quantitative description of the predicted boundary maps, overall accuracy, F1 score, recall and precision are summarized in Table 4. Overall accuracy represents a general metric by counting true positives/negatives and false positives/negatives, i.e., it considers both 'boundary' and 'no boundary' classes. All predicted boundary maps resulted in an overall accuracy of over 94%. To get a better insight into the detection quality, F1 score, recall and precision were calculated for the class' boundary' or '0' as a positive class. The results showed that more relevant visible land boundaries were detected when the predicted boundary map was reclassified with the threshold 'boundary' $\leq$ 0.9, resulting in an F1 score of 0.51. More balanced scores were retrieved for the boundary map with 'boundary' $\leq$ 0.7, resulting in an F1 score of 0.52. Higher precision was obtained for the boundary map with the reclassification threshold 'boundary' $\leq$ 0.5, resulting in an F1 score of 0.46.

Table 4. Accuracy assessment of visible land boundary detection with U-Net.

Predicted Boundary Map	Overall Accuracy (%)	Recall	Precision	F1 *Score*
Boundary $\leq$ 0.9	94.5	0.654	0.413	0.506
Boundary $\leq$ 0.7	96.2	0.480	0.565	0.519
Boundary $\leq$ 0.5	96.5	0.348	0.675	0.459

3.3. Comparison with ENVI Deep Learning—ENVINet5

The predicted land boundary map for the test area (Figure 8a) retrieved using ENVINet5 model was already georeferenced, so no further post-processing step was required. In addition, the retrieved boundary map contained predicted values of 0 and 1, and no additional reclassification step was performed to compare the results to the ground truth map and to assess accuracy. The predicted boundary is visualized in Figure 9b.

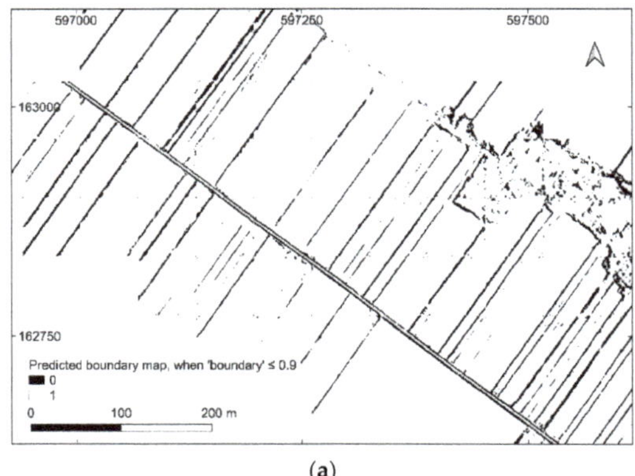

(a)

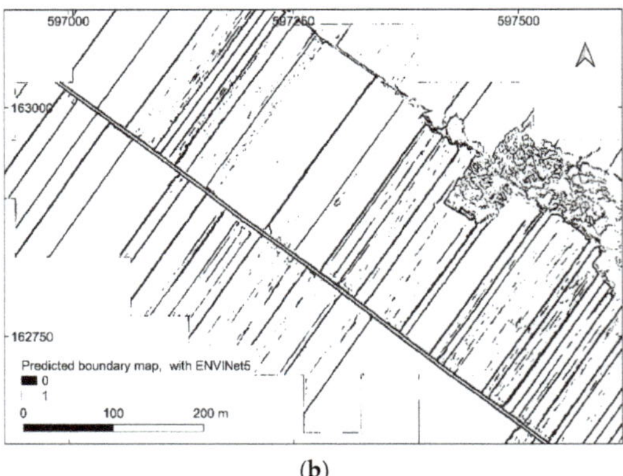

(b)

Figure 9. Comparison of predicted land boundary map: (**a**) predicted boundary map retrieved with U-Net, threshold 'boundary' $\leq$ 0.9; (**b**) predicted boundary map retrieved with ENVINet5.

Considering that all predictions retrieved with ENVINet5 were assigned the prediction value 0 for the class' boundary', we selected the boundary map for the comparison of results with U-Net, where all predictions $\leq$0.9, were reclassified as 0—'boundary'. With this, we wanted to compare predictions from U-Net that were as close as possible to the predictions of ENVINet5. The overall accuracy was 94.5% for U-Net and 96.2% for ENVINet5. However, in terms of detection quality for the 'boundary' class, ENVINet5 showed higher recall and lower *precision* than U-Net. In short, F1 score showed a slightly higher value for U-Net, i.e., 0.51 compared to ENVNet5, where the value was 0.49. The confusion matrices are shown in Table 5 and the quantitative results in Table 6.

Table 5. Confusion matrices based on the number of pixels.

		Ground Truth	
		Boundary	**No Boundary**
U-Net	Boundary	137,056	195,156
	No boundary	72,524	8,966,912
ENVINet5	Boundary	175,559	325,076
	No boundary	34,021	8,836,992

Table 6. Accuracy assessment and comparison with ENVINet5.

Predicted Boundary Map	Overall Accuracy (%)	Recall	Precision	*F*1 Score
U-Net	94.5	0.654	0.413	0.506
ENVINet5	96.2	0.838	0.351	0.494

4. Discussion

Deep learning is a relatively new research area and offers great potential for feature detection from remote sensing imagery [21,24]. The application of CNNs for detecting visible land boundaries is becoming increasingly important, especially for UAV-based cadastral mapping. In this work, we presented a deep learning application using Python with Keras to implement U-Net, and software-based ENVI deep learning for visible land boundary detection from UAV imagery. The research obtained encouraging and reasonable results that can help to automate the process of cadastral mapping.

4.1. CNN Architecture and Implementation

In both network models, the loss was constantly decreasing from the first epoch until the end. This indicated that the model was still learning on training samples. However, the training of the models was monitored with the validation loss to avoid overfitting. The training performance of the network models was shown in Figures 5 and 6. The validation loss for U-Net was decreasing until epoch 92 and for ENVINet5 until epoch 24. This was a good sign that the model did not lose the ability to generalize predictions for test datasets that were not seen by the model during training. The evaluation metric showed relatively high accuracies, 0.978 for U-Net and 0.946 for ENVINet5. The high accuracy of the network models, including the first epochs, is mainly due to the unbalanced pixels of the classes. The land boundaries occupy a minimal number of pixels compared to the background pixels.

In this study, we used a deep learning-based visible land boundary detector. Here, providing balanced pixels for 'boundary' and 'no boundary' is a bit challenging, especially for UAV imagery. UAV imagery usually has a small GSD (2–5 cm) and a limited coverage area beside the efficient and flexible data acquisition system [14]. Moreover, the number of background pixels in cadastral maps is always much higher than the number of pixels representing the course of the cadastral boundaries themselves (line-based). The imbalance of pixels per class is even more evident in randomly cropped tiles from UAV imagery. Resampling the original GSD to a larger GSD contributed somewhat to an increase in the field of view and balance between classes. However, the size of the GSD and the number of training tiles is limited by the coverage area. To increase the amount of training data, we applied data augmentation. Data augmentation has proven to be an efficient technique to supplement original UAV training data, especially when training the U-Net model from scratch. However, it remains a challenge to confirm what should be a sufficient variety of UAV training data to learn a robust network model for visible cadastral boundary detection.

The problem of unbalanced classes could be solved by rebalancing the class weights, using additional evaluation metrics besides overall accuracy, or performing deep learning with multiple classes for land cover (polygon-based). In addition, other remote sensing imagery can be used for the training data, e.g., aerial or satellite imagery; imageries can be cropped in a way to cover more balanced pixels for 'boundary' and 'no boundary' and may not be limited with the coverage area. This can be applied if the deep learning model is to be trained using only cadastral data that requires manual data preparation, such as the creation of image tiles and corresponding ground truths. Instead, the CNN model could be trained via transfer learning, similar to [17]. To avoid ambiguity, the detection quality for UAV test data in this study was evaluated using recall, precision and *F*1 score for the class 'boundary'. Thus, we had two indicators, overall accuracy, which includes both 'boundary' and 'no boundary' classes and one that is specific only to the 'boundary' class. Although both models performed well, there were significant differences in implementation and training, as one approach is customized, e.g., U-Net, and is offered as an API, while the other, e.g., ENVINet5, is software-based, where we have fewer parameters available but can still achieve good results.

Training a deep learning model requires more memory, a stronger GPU and efficient computation. Training of the U-Net model was performed in Google Collaboratory, which is open-source and can be considered as an alternative for the hardware costs to get more memory and a more powerful GPU. On the other hand, ENVI deep learning had some hardware and software requirements to perform the training of the network model. Google Collaboratory allowed faster training compared to our machine. For 100 epochs, the training time was 4 h with Google Collaboratory and three times the training time with ENVI deep learning since it was run on a local machine with less computational power. It should be emphasized that ENVI deep learning provided more stable training in terms of a training session interruption, which occasionally happened with Google Collaboratory.

4.2. Detection of Visible Land Boundaries

The network models, both U-Net and ENVINet5, generally performed well in detecting visible land boundaries, with some exceptions in the forest area. The results of the quality of visible land boundary detection are shown in Figures 8 and 9 and quantitatively in Tables 4 and 6. The results show that most visible land boundaries were correctly detected, which demonstrates the ability of the UAV imagery and network models to detect these types of land boundaries, especially in rural areas.

U-Net generated boundary maps with low recall and high precision when the threshold for 'boundary' was set $\leq$0.5. This resulted in a recall of 0.35 and a precision of 0.68. More balanced results and a higher F1 score were obtained when the threshold for 'boundary' was set $\leq$0.7, namely a recall of 0.48, precision of 0.57 and F1 score of 0.52. The boundary map with high recall and low precision was generated when the threshold was set almost to the maximum, namely 'boundary' $\leq$ 0.9. This boundary map was used for comparison with the new map obtained with ENVINet5, since nearly all predictions were reclassified to the 'boundary' class, which is in accordance with the output of ENVINet5.

The results show an overall accuracy of 94% and 96% for U-Net and ENVINet5, respectively. However, for the 'boundary' class, U-Net gave 0.51 F1 score and ENVINet5 0.49. This is mainly because U-Net provided more balanced scores, namely 0.65 in recall and 0.41 in precision. On the other hand, ENVINet5 provided higher recall (0.84) and lower precision (0.35), which means that the 'boundary' class is well detected, but the model also includes points of the background class in it.

U-Net provided boundary maps that were in the range of 0–1. This is due to the chosen sigmoid function as the activation function of the output layer, where the output values obtained are estimates of the probability that the input belongs to class 'boundary'. Then, we set a threshold to decide whether the input belongs to class 'boundary' or class 'no boundary'. The results maintain a balance; the lower the threshold, the lower the recall and the higher the precision. The significant point of the threshold is that the same can be used as a filtering method for boundary maps, depending on the need and purpose of the application. For example, a low threshold provided high precision, while a high threshold provided high recall. The recall is also referred to as completeness, while the precision is referred to as correctness [15]. Imbalanced classes are common in cadastral maps, and when it comes to specific use cases, more importance should be given to the metrics recall and precision, and how a balance between them can be achieved—which in our case was supported by filtering the predicted boundary maps (Figure 8e). Unlike U-Net, ENVINet5 provided all predictions with values 0 and 1, and no further thresholding or filtering could be applied.

In cadastral mapping, it is desirable that the relevant or candidate boundaries are correctly extracted since the correct determination of the location of the cadastral boundaries is the core of the cadastre itself (correctness). On the other hand, increasing the number of possible boundaries increases the cadastral coverage (completeness). Considering this, a model that provides a balance between recall and precision is preferable. In short, a model that provides a high F1 score.

The comparison of the results obtained with U-Net with other studies, in particular [15,17,25], which deal with the automation of cadastral mapping using different CNN architectures, is not possible at this stage. This is mainly because the training approach of the network models along with the input training data differs from study to study. Thus, a reliable and qualitative comparison is not possible.

4.3. Boundary Mapping Approach

This section refers to the visible land boundary detection workflows applied in custom-based U-Net and software-based ENVI deep learning. In general, boundary mapping approaches are quite different, starting from data preparation to the final predicted boundary map. However, these differences provide advantages and disadvantages for each boundary mapping approach used in this study.

In general, programming-based deep learning is open-source and offers a more flexible but complex approach compared to software-based deep learning. Software-based deep learning, e.g., ENVI deep learning, is simpler but at the same time more rigid. For example, U-Net can be trained in a machine and in online platforms such as Google Collaboratory, where the hyperparameters can be configured individually. In contrast, ENVI deep learning has no implementation choices, but it also requires no additional configuration. The latter can be considered a very important aspect as not all land administrators are experts in programming, and this can be an option for them to perform deep learning. The main challenge with CNNs is the preparation of a large amount of training data [26], especially when the goal is to train the network only cadastral data [17]. In order to increase the amount of training data for the U-Net, it was necessary to decompose the UAV orthoimages in tiles before data augmentation. Moreover, for each UAV tile, a corresponding label image (ground truth) was manually created using additional software for rasterisation. In contrast, training in ENVI deep learning was patch-based, and the entire extent or a larger UAV tile can be used as input for training. In addition, the labelling images were created quite quickly within the software—directly by uploading reference boundaries as ROIs. The boundary maps retrieved using U-Net were the same size as the input but were not georeferenced. Considering that georeferencing is the key element in cadastral mapping, it was necessary to georeference and merge predicted boundary maps from the test UAV tiles. In ENVI deep learning, the prediction boundary map was already georeferenced, and the predictions had values of 0 and 1. Therefore, further filtering of the predicted boundary maps was not possible. The advantages and disadvantages of the U-Net and ENVI deep learning mapping approaches used in this study are summarized in Table 7.

Table 7. Summarized advantages (pros) and disadvantages (cons) for boundary mapping approaches used in this study.

U-Net		ENVI Deep Learning	
pros	**cons**	**pros**	**cons**
• open-source	• programming	• no programming	• commercial
• impl. online or on machine	• additional georeferencing step	• georeferencing	• impl. on machine only
• hyper-parameter configuration	• label image manually	• label image by software	• hyper-parameter configuration
• prediction values in range; filtering of boundary maps			• fixed predictions

4.4. Application of Detected Visible Boundaries

Cadastral boundaries are often demarcated by objects visible in remote sensing imagery [2,8]. Automatic detection of cadastral boundaries based on remote sensing imagery, especially UAV imagery, has rarely been investigated. Automatic extraction of visible land boundaries, i.e., property boundaries, offers the potential to improve current approaches to cadastral mapping. The boundary mapping approaches investigated are based on deep learning and offer improvements in terms of time and cost.

Both boundary mapping approaches, i.e., U-Net and ENVI deep learning, can help to facilitate and accelerate cadastral mapping, especially in areas where large parts of the cadastral boundaries are continuous and visible. In terms of delineation effort per parcel, automatic delineation approaches (including post-alignments) require up to 40% less time in rural areas compared to manual delineation, based on [17]. However, in areas where cadastral boundaries are not visible in the image, manual delineation remains superior. Overall, it can be said that manual methods provide slower but more accurate delineations, while automatic methods are faster but less accurate (once the model is trained).

In countries with low cadastral coverage, deep learning-based mapping approaches can be used to produce cadastral maps. In countries with full cadastral coverage, the detected visible boundaries can be used to automate the process of revising the up-to-dateness of existing cadastral maps. In this way, areas requiring updating and improving cadastral boundary maps can be automatically identified. Notwithstanding the advances in cadastral mapping, the automation of cadastral boundary detection is still ongoing [15,17,18]. This is due to the nature of cadastral boundaries, which may have a simple geometry but are very complex to interpret. Consequently, automatically detected visible land boundaries should be considered as preliminary cadastral boundaries. Verification of automatically detected land boundaries should be aligned with the existing technical, legal and institutional framework of land administration. Moreover, not every cadastral boundary is demarcated with visible objects. In this study, boundary mapping approaches were tested in rural areas. It is argued that the number of visible cadastral boundaries is higher compared to urban areas [2].

Automating the detection of invisible cadastral boundaries remains a challenge in land administration, which has already been highlighted in [17]. Future work could investigate and analyze the applicability of deep learning for invisible cadastral boundaries that are marked prior to the UAV survey. It should be further investigated which type and size of land boundary markers are more appropriate for demarcating the invisible boundaries.

5. Conclusions

Deep learning is becoming increasingly important in cadastral applications as a state-of-the-art method for automatic boundary detection. The aim of this study was to investigate the potential of CNN architecture, namely U-Net, based on UAV imagery training samples—as a deep learning-based detector for visible land boundaries. The results and land boundary mapping approach using U-Net were compared with software-based ENVI deep learning. The overall accuracy for both CNN models was higher than 95%. This indicates that deep learning-based land boundary detection usually faces an unbalanced distribution of pixels per class, namely for 'boundary' and 'no boundary'.

Regarding the quality of recognition for the class 'boundary' in the case of U-Net, we obtained low recall and high precision when the threshold 'boundary' ≤ 0.5 was set. This resulted in a recall of 0.35 and a precision of 0.68. Prediction reclassification can be considered as a tool to filter the predicted boundary maps. For example, to compare the results with ENVINet5, the threshold had to be set almost to its maximum. Here, U-Net provided a recall of 0.65 and a precision of 0.41. For ENVI deep learning, we obtained a recall of 0.84 and a precision of 0.35. Based on the F1 score (U-Net 0.51 and ENVI deep learning 0.49), U-Net provided slightly better and more balanced results. The predicted land boundary maps obtained with U-Net were georeferenced and merged in an additional post-processing step. This was not an issue with ENVI deep learning—the output boundary maps were already georeferenced. Overall, U-Net is a programming-based solution and provides a more flexible boundary mapping approach in terms of hyperparameters and CNN model setting. On the other hand, it can be somewhat complex and demanding for the practice as not all land administrators are skilled in programming. In contrast, ENVI deep learning does not require any programming and deep learning is guided by the software process.

While programming-based deep learning is challenging due to the complexity of the processes and their control, commercial software-based deep learning brings some abstraction but at the same time has limitations in terms of influencing the processes flow. Both land boundary mapping approaches investigated in our study can be used to accelerate and facilitate cadastral mapping in rural areas. However, the automatically detected visible land boundaries should be considered as preliminary boundaries for cadastral map production and updating. The results should be further aligned with technical, legal and institutional framework of land administration.

Author Contributions: Conceptualization, B.F. and A.L.; methodology, B.F., M.R. and A.L.; software, B.F. and M.R.; validation, B.F., M.R. and A.L.; formal analysis, B.F.; investigation, B.F.; resources B.F., M.R. and A.L.; writing—original draft preparation, B.F.; writing—review and editing, B.F., M.R. and A.L.; visualization, B.F.; supervision, A.L. All authors have read and agreed to the published version of the manuscript.

Funding: This research was funded by Slovenian Research Agency, research core funding number P2-0406 Earth Observation and Geoinformatics. The APC was funded by P2-0406 and V2-1934 from the Slovenian Research Agency and the Surveying and Mapping Authority of the Republic of Slovenia.

Institutional Review Board Statement: Not applicable.

Informed Consent Statement: Not applicable.

Data Availability Statement: The data presented in this study are openly available in https://unilj-my.sharepoint.com/:f:/g/personal/bfetai_fgg_uni-lj_si/EhZieqQdkc5EkrdfjPwo5AYBQl9n_GOE-yM3Yux-wjmfwA?e=5mTYrH (accessed on 25 May 2021).

Acknowledgments: We thank the anonymous reviewers for their valuable comments and suggestions. We acknowledge Klemen Kozmus Trajkovski for capturing the UAV data and for the support during the fieldwork.

Conflicts of Interest: The authors declare no conflict of interest.

References

1. Enemark, S.; Bell, K.C.; Lemmen, C.; McLaren, R. *Fit-For-Purpose Land Administration*; International Federation of Surveyors (FIG): Copenhagen, Denmark, 2014; ISBN 978-87-92853-11-0.
2. Luo, X.; Bennett, R.; Koeva, M.; Lemmen, C.; Quadros, N. Quantifying the Overlap between Cadastral and Visual Boundaries: A Case Study from Vanuatu. *Urban Sci.* **2017**, *1*, 32. [CrossRef]
3. Zevenbergen, J. A systems approach to land registration and cadastre. *Nord. J. Surv. Real Estate Res.* **2004**, *1*, 11–24.

4. Simbizi, M.C.D.; Bennett, R.M.; Zevenbergen, J. Land tenure security: Revisiting and refining the concept for Sub-Saharan Africa's rural poor. *Land Use Policy* **2014**, *36*, 231–238. [CrossRef]
5. Williamson, I.P. *Land Administration for Sustainable Development*, 1st ed.; ESRI Press Academic: Redlands, CA, USA, 2010; ISBN 9781589480414.
6. Binns, B.O.; Dale, P.F. *Cadastral Surveys and Records of Rights in Land*; Food and Agriculture Organization of the United Nations: Rome, Italy, 1995; ISBN 9251036276.
7. Grant, D.; Enemark, S.; Zevenbergen, J.; Mitchell, D.; McCamley, G. The Cadastral triangular model. *Land Use Policy* **2020**, *97*, 104758. [CrossRef]
8. Crommelinck, S.; Bennett, R.; Gerke, M.; Nex, F.; Yang, M.; Vosselman, G. Review of Automatic Feature Extraction from High-Resolution Optical Sensor Data for UAV-Based Cadastral Mapping. *Remote Sens.* **2016**, *8*, 689. [CrossRef]
9. Zevenbergen, J.; Bennett, R. The visible boundary: More than just a line between coordinates. In Proceedings of the GeoTech Rwanda, Kigali, Rwanda, 18–20 November 2015; pp. 1–4.
10. Manyoky, M.; Theiler, P.; Steudler, D.; Eisenbeiss, H. Unmanned Aerial Vehicle in Cadastral Applications. *Int. Arch. Photogramm. Remote Sens. Spat. Inf. Sci.* **2011**, 57–62. [CrossRef]
11. Puniach, E.; Bieda, A.; Ćwiąkała, P.; Kwartnik-Pruc, A.; Parzych, P. Use of Unmanned Aerial Vehicles (UAVs) for Updating Farmland Cadastral Data in Areas Subject to Landslides. *ISPRS Int. J. Geo-Inf.* **2018**, *7*, 331. [CrossRef]
12. Koeva, M.; Muneza, M.; Gevaert, C.; Gerke, M.; Nex, F. Using UAVs for map creation and updating. A case study in Rwanda. *Surv. Rev.* **2018**, *50*, 312–325. [CrossRef]
13. Stöcker, C.; Nex, F.; Koeva, M.; Gerke, M. High-Quality UAV-Based Orthophotos for Cadastral Mapping: Guidance for Optimal Flight Configurations. *Remote Sens.* **2020**, *12*, 3625. [CrossRef]
14. Colomina, I.; Molina, P. Unmanned aerial systems for photogrammetry and remote sensing: A review. *ISPRS J. Photogramm. Remote Sens.* **2014**, *92*, 79–97. [CrossRef]
15. Xia, X.; Persello, C.; Koeva, M. Deep Fully Convolutional Networks for Cadastral Boundary Detection from UAV Images. *Remote Sens.* **2019**, *11*, 1725. [CrossRef]
16. Ramadhani, S.A.; Bennett, R.M.; Nex, F.C. Exploring UAV in Indonesian cadastral boundary data acquisition. *Earth Sci. Inf.* **2018**, *11*, 129–146. [CrossRef]
17. Crommelinck, S.; Koeva, M.; Yang, M.Y.; Vosselman, G. Application of Deep Learning for Delineation of Visible Cadastral Boundaries from Remote Sensing Imagery. *Remote Sens.* **2019**, *11*, 2505. [CrossRef]
18. Fetai, B.; Oštir, K.; Kosmatin Fras, M.; Lisec, A. Extraction of Visible Boundaries for Cadastral Mapping Based on UAV Imagery. *Remote Sens.* **2019**, *11*, 1510. [CrossRef]
19. Crommelinck, S.; Bennett, R.; Gerke, M.; Yang, M.; Vosselman, G. Contour Detection for UAV-Based Cadastral Mapping. *Remote Sens.* **2017**, *9*, 171. [CrossRef]
20. Arbeláez, P.; Maire, M.; Fowlkes, C.; Malik, J. Contour detection and hierarchical image segmentation. *IEEE Trans. Pattern Anal. Mach. Intell.* **2011**, *33*, 898–916. [CrossRef]
21. Ma, L.; Liu, Y.; Zhang, X.; Ye, Y.; Yin, G.; Johnson, B.A. Deep learning in remote sensing applications: A meta-analysis and review. *ISPRS J. Photogramm. Remote Sens.* **2019**, *152*, 166–177. [CrossRef]
22. Zhu, X.X.; Tuia, D.; Mou, L.; Xia, G.-S.; Zhang, L.; Xu, F.; Fraundorfer, F. Deep Learning in Remote Sensing: A Comprehensive Review and List of Resources. *IEEE Geosci. Remote Sens. Mag.* **2017**, *5*, 8–36. [CrossRef]
23. Persello, C.; Stein, A. Deep Fully Convolutional Networks for the Detection of Informal Settlements in VHR Images. *IEEE Geosci. Remote Sens. Lett.* **2017**, *14*, 2325–2329. [CrossRef]
24. Pan, Z.; Xu, J.; Guo, Y.; Hu, Y.; Wang, G. Deep Learning Segmentation and Classification for Urban Village Using a Worldview Satellite Image Based on U-Net. *Remote Sens.* **2020**, *12*, 1574. [CrossRef]
25. Park, S.; Song, A. Discrepancy Analysis for Detecting Candidate Parcels Requiring Update of Land Category in Cadastral Map Using Hyperspectral UAV Images: A Case Study in Jeonju, South Korea. *Remote Sens.* **2020**, *12*, 354. [CrossRef]
26. Ronneberger, O.; Fischer, P.; Brox, T. U-Net: Convolutional Networks for Biomedical Image Segmentation. 2015. Available online: http://arxiv.org/pdf/1505.04597v1 (accessed on 24 February 2021).
27. Diakogiannis, F.I.; Waldner, F.; Caccetta, P.; Wu, C. ResUNet-a: A deep learning framework for semantic segmentation of remotely sensed data. *ISPRS J. Photogramm. Remote Sens.* **2020**, *162*, 94–114. [CrossRef]
28. Flood, N.; Watson, F.; Collett, L. Using a U-net convolutional neural network to map woody vegetation extent from high resolution satellite imagery across Queensland, Australia. *Int. J. Appl. Earth Obs. Geoinf.* **2019**, *82*, 101897. [CrossRef]
29. Zhao, X.; Yuan, Y.; Song, M.; Ding, Y.; Lin, F.; Liang, D.; Zhang, D. Use of Unmanned Aerial Vehicle Imagery and Deep Learning UNet to Extract Rice Lodging. *Sensors* **2019**, *19*, 3859. [CrossRef]
30. Alshaikhli, T.; Liu, W.; Maruyama, Y. Automated Method of Road Extraction from Aerial Images Using a Deep Convolutional Neural Network. *Appl. Sci.* **2019**, *9*, 4825. [CrossRef]
31. Wierzbicki, D.; Matuk, O.; Bielecka, E. Polish Cadastre Modernization with Remotely Extracted Buildings from High-Resolution Aerial Orthoimagery and Airborne LiDAR. *Remote Sens.* **2021**, *13*, 611. [CrossRef]
32. Wani, M.A.; Bhat, F.A.; Afzal, S.; Khan, A.I. *Advances in Deep Learning*; Springer: Singapore, 2020; ISBN 978-981-13-6793-9.
33. GRASS Development Team. *GRASS GIS Bringing Advanced Geospatial Technologies to the World*; GRASS: Beaverton, OR, USA, 2020.
34. Google Colaboratory. Available online: https://colab.research.google.com (accessed on 29 April 2021).

35. Chollet, F.; et al. Keras. 2015. Available online: https://keras.io (accessed on 29 April 2021).
36. Martin, A.; Ashish, A.; Paul, B.; Eugene, B.; Zhifeng, C.; Craig, C.; Greg, S.C.; Andy, D.; Jeffrey, D.; Matthieu, D.; et al. TensorFlow: Large-Scale Machine Learning on Heterogeneous Systems. 2015. Available online: https://www.tensorflow.org/ (accessed on 29 April 2021).
37. Zhixuhao. Unet for Image Segmentation. Available online: https://github.com/zhixuhao/unet (accessed on 24 February 2021).
38. Gillies, S. Rasterio: Geospatial Raster I/O for Python Programmers. 2013. Available online: https://github.com/mapbox/rasterio (accessed on 30 April 2021).
39. GDAL/OGR Contributors. GDAL/OGR Geospatial Data Abstraction Software Library. 2021. Available online: https://gdal.org (accessed on 30 April 2021).
40. Harris, C.R.; Millman, K.J.; van der Walt, S.J.; Gommers, R.; Virtanen, P.; Cournapeau, D.; Wieser, E.; Taylor, J.; Berg, S.; Smith, N.J.; et al. Array programming with NumPy. *Nature* **2020**, *585*, 357–362. [CrossRef]
41. Exelis Visual Information Solutions. In *ENVI Deep Learning*; L3Harris Geospatial: Boulder, CO, USA, 1977.
42. Exelis Visual Information Solutions. ENVI Deep Learning—Training Background. Available online: https://www.l3harrisgeospatial.com/docs/BackgroundTrainDeepLearningModels.html (accessed on 9 March 2021).

Article

Application of a Hand-Held LiDAR Scanner for the Urban Cadastral Detail Survey in Digitized Cadastral Area of Taiwan Urban City

Shih-Hong Chio * and Kai-Wen Hou

Department of Land Economics, National Chengchi University, No. 64, Sec. 2, ZhiNan Rd., Wenshan District, Taipei 11605, Taiwan; 108257030@nccu.edu.tw

* Correspondence: chio0119@nccu.edu.tw or chio0119@gmail.tw; Tel.: +886-2-2939-3091 (ext. 51657)

Abstract: The cadastral detail data is used for overlap analysis with digitized graphic cadastral maps to solve the problem of inconsistencies between cadastral maps and the current land situation. This study investigated the feasibility of a handheld LiDAR scanner to collect 3D point clouds in an efficient way for a detail survey in urban environments with narrow and winding streets. Then, urban detail point clouds were collected by the handheld LiDAR scanner. After point cloud filtering and the ranging systematic error correction that was determined by a plane-based calibration method, the collected point clouds were transformed to the TWD97 cadastral coordinate system using control points. The land detail line data were artificially digitized and the results showed that about 97% error of the digitized detail positions was less than 15 cm compared to the check points surveyed by a total station. The results demonstrated the feasibility of using a handheld LiDAR scanner to perform an urban cadastral detail survey in digitized graphic areas. Therefore, the handheld LiDAR scanner could be used for the production of the detail lines for urban cadastral detail surveying for digitized cadastral areas in Taiwan.

Keywords: cadastral survey; detail survey; handheld LiDAR scanner; calibration

Citation: Chio, S.-H.; Hou, K.-W. Application of a Hand-Held LiDAR Scanner for the Urban Cadastral Detail Survey in Digitized Cadastral Area of Taiwan Urban City. *Remote Sens.* **2021**, *13*, 4981. https://doi.org/10.3390/rs13244981

Academic Editors: Mila Koeva, Rohan Bennett and Claudio Persello

Received: 14 October 2021
Accepted: 2 December 2021
Published: 8 December 2021

Publisher's Note: MDPI stays neutral with regard to jurisdictional claims in published maps and institutional affiliations.

1. Introduction

Depending on the surveying method and the way the cadastral maps are stored in Taiwan, the areas corresponding to the cadastral maps can be categorized into digital cadastral areas, graphic cadastral areas, and digitized graphic cadastral areas. Although graphic cadastral maps have been digitized as digitized graphic cadastral maps, they still preserve differential shrinkage, wrinkles, folds, and damage from the original maps, and give rise to problems such as the difference between maps and land details in the real world, map joins, and changes to the area subsequent to a map's registration [1]. In that case, the possible parcel points along the parcel boundaries must be surveyed for overlap analysis and map registration between the digitized cadastral maps and land detail data (possible parcel points) to eliminate inconsistencies between cadastral maps and the actual land details. The surveying of possible parcel points along the parcel boundaries or possible lines is called a cadastral detail survey. Therefore, an urban cadastral detail survey in the digitized cadastral area of Taiwan's urban city is to eliminate inconsistencies between cadastral maps and the actual land details in urban areas. However, most studies examining this issue focused on ground survey methods in a local area using ground instruments such as the GNSS (Global Navigation Satellite System) systems or total stations, and it is time-consuming. For an urban cadastral detail survey by GNSS, it is still necessary to consider the environmental sky view, multipath effects, and other errors, which are easily restricted by the urban construction environment. With this technological development, a terrestrial LiDAR scanner, a vehicle-based LiDAR system, and UAV aerial images were studied for an urban cadastral detail survey. For example, Chio and Chiang [2] investigated the feasibility of UAV aerial photogrammetry for a boundary verification survey of a digitized cadastral area in an urban city of Taiwan, the study demonstrated its feasibility in the accuracy of the urban cadastral detail data by UAV aerial photogrammetry.

The GeoSLAM Zeb-Horizon LiDAR scanner is a form of handheld mobile mapping system (HMMS) and the HMMS has been applied on many occasions due to its compact size, cost effectiveness, and high performance [3]. Because the HMMS uses a simultaneous localization and mapping (SLAM) algorithm [3–5] with IMU (inertial measurement unit) data for positioning without using GNSS data [6], it can avoid the environmental limits and be used in narrow and winding alleys, indoors and other areas where GNSS signals cannot be received. Compared to a total station and a terrestrial LiDAR scanner, it also shows high performance in collecting terrain data in general areas, such as in non-narrow alleys. Therefore, it has been employed in different fields, for example, cultural asset preservation in ancient cities [7], forest investigations [6,8–10], mine monitoring [3,11], disaster site reconstruction [12], tunnel surveying [13], topographic surveying [14], and the mapping of the building interior structures [7,15].

This study investigated the feasibility of GeoSLAM Zeb-Horizon LiDAR scanner for a detail survey in urban environments with narrow and winding streets. Such an application for a cadastral detail survey needs to consider the accuracy problem. To date, only two studies concerning the accuracy evaluation focused on forest investigations. Park and Um [9] employed GeoSLAM Zeb-Horizon to measure the diameter at breast height (DBH), and showed a deviation of less than 4 cm compared to the caliper measurement. Hunčaga et al. [10] used GeoSLAM Zeb-Horizon for diameter at DBH estimation by a circle-fitting method from point clouds, and a 1.62 cm root mean squared error (RMSE) was reached. The average RMSE of all cross sections was 1.26 cm. There is no relevant research on the application of HMMS to the cadastral survey. As the maximum scale of the cadastral map of the graphic digitized area in Taiwan's urban area is 1/500, and the maximum mapping accuracy is 0.3 mm, this corresponds to a 15 cm error in situ. Based on the accuracy demand, this study investigated the feasibility of a handheld LiDAR scanner, GeoSLAM Zeb-Horizon, which uses the SLAM algorithm for positioning without GNSS data, to collect 3D point cloud with as an efficient way to conduct a detail survey in urban environments with narrow and winding streets. First, a plane-based calibration method proposed in our previous study [16] was used to calibrate the handheld LiDAR scanner to improve the accuracy of the handheld LiDAR point cloud. The rough result was presented in [16] due to the paper page limit, therefore, the detailed derivation and result improvement were described in this study. Then, the urban detail point cloud in the test area was collected by the handheld LiDAR scanner. After point cloud filtering and ranging systematic error correction, the handheld LiDAR point cloud was transformed to the TWD97 cadastral coordinate system using ground control points. The land detail line data was artificially digitized from the handheld LiDAR point cloud. The accuracy was evaluated to investigate the feasibility of a handheld LiDAR scanner for the urban cadastral detail line survey in a digitized cadastral area of Taiwan's urban city. The following section will introduce the methodology used in this study.

2. Methodology

Figure 1 shows the study flowchart. This study was mainly divided into two parts: (1) the handheld LiDAR scanner calibration, as seen in the left part of Figure 1; and (2) the urban cadastral detail survey, as seen in the right part of Figure 1. A detailed description of a handheld LiDAR scanner calibration was explained in Section 2.1. Section 2.2 described how to apply a handheld LiDAR scanner on collecting the point data and correcting the system error to perform an urban cadastral detail survey with more details. Both results were verified by the check data.

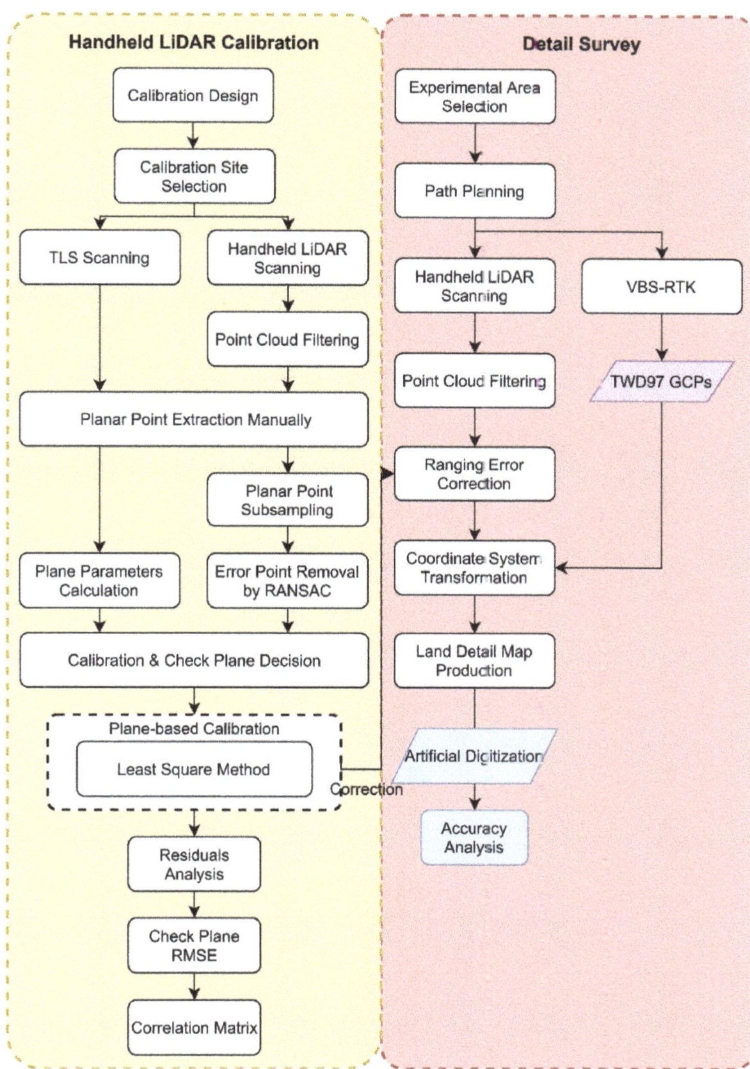

Figure 1. Study flowchart.

2.1. Handheld LiDAR Scanner Calibration

The calibration steps for the handheld LiDAR scanner were described as follows, and the description was more detailed than those presented by us in [16]. First, an indoor calibration field was selected. Then, the points of this calibration field were collected by a ground-based LiDAR scanner and the plane features in the calibration field were classified into the calibration planes and check planes. Subsequently, the points on the corresponding calibration planes were dynamically collected by a handheld LiDAR scanner, GeoSLAM Zeb-Horizon. The points corresponding to the calibration planes and check planes were selected manually and the blunder points were removed by a RANSC algorithm. The plane-based dynamic calibration method for GeoSLAM Zeb-Horizon proposed in our previous study [16] was used to calculate the ranging systematic errors, including the range scale factor (S) and the rangefinder offset (C), by the least-squares adjustment method.

The calibration results were investigated. Details for each step were described in the following subsections.

2.1.1. Selection of the Calibration Field

The calibration field had to be filled with plane features and be large enough to collect point clouds with various ranging measurements to extract planes as calibration reference data for calibration.

2.1.2. Acquisition and Extraction of Calibration Reference Data

Ground-Based LiDAR Point Cloud Acquisition

In this study, a Faro ground-based LiDAR scanner was used to capture the point cloud of the calibration site in one station instead of multiple stations to avoid the registration errors of point clouds and affecting the accuracy of calibration reference data. The specification of FARO Focus S350 ground-based LiDAR scanner was tabulated in Table 1. FARO Focus S350 is a phase-based 3D laser scanner and has better accuracy than a 3D laser scanner based on time of flight (TOF) scanning technique for the collection points in short distances under indoor conditions without interference.

Table 1. Technical information of FARO Focus S350. (https://echosurveying.com/3d-laser-scanner/faro-focus-s350-laser-scanner, accessed on 2 December 2021).

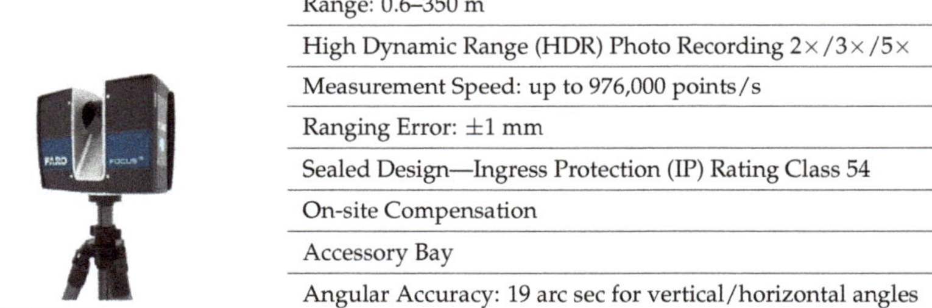

	Range: 0.6–350 m
	High Dynamic Range (HDR) Photo Recording 2×/3×/5×
	Measurement Speed: up to 976,000 points/s
	Ranging Error: ±1 mm
	Sealed Design—Ingress Protection (IP) Rating Class 54
	On-site Compensation
	Accessory Bay
	Angular Accuracy: 19 arc sec for vertical/horizontal angles

Extraction of Calibration Planes and Check Planes

This study used the plane features that were abundant in building interior environments as the calibration reference data, eliminating the need for a large number of construction procedures for the calibration targets. The plane equation of the plane feature k is shown in the following Equation (1).

$$a_k x + b_k y + c_k z + d_k = 0 \quad (1)$$

where (a_k, b_k, c_k) is a unit normal vector of plane feature k.

The plane parameters (a_k, b_k, c_k, d_k) were regarded as a priori parameters in the calibration process and served as the calibration reference data. As mentioned in Section 2.1.2, in order to improve the precision and efficiency of data acquisition, a FARO Focus S350 ground-based LiDAR scanner was operated to acquire point clouds of the calibration field, and the points located on the used plane features were manually extracted. The extracted plane points were used to determine the plane parameters by the least-squares method. Moreover, the extracted plane features were classified into the calibration planes and check planes. The calibration planes were used for the determination of the planar parameters for calibration, and the check planes were used for the verification of the calibration result.

2.1.3. Acquisition of Handheld LiDAR Pnt Cloud for Calibration Data

The GeoSLAM Zeb-Horizon handheld LiDAR scanner with the technical specification shown in Table 2 was used to collect the point cloud data of the calibration field in a mobile manner to obtain as many point clouds on various plane features as possible for calibration data. To compare with static calibration, dynamic calibration can obtain richer points with various ranging measurements, and there is no need to place the handheld LiDAR scanner in multiple stations to capture data separately, thus saving a lot of time [17]. According to the manual of GeoSLAM Zeb-Horizon, the scanned path must be closed to the starting point of the trajectory in order to employ the SLAM algorithm [3–5] together with IMU data to calculate the better scanning trajectory. The Velodyne multibeam LiDAR sensor was installed on the GeoSLAM Zeb-Horizon.

Table 2. Technical specification of GeoSLAM Zeb-Horizon. (https://microsolresources.com/wp-content/uploads/2019/06/GeoSLAM-Family-Brochure.pdf, accessed on 2 December 2021).

Technical specification	
Handheld I Backpack I UAV Ready	
Range	100 m
Protection Class	IP54
Scanner Weight	1.3 kg
Points per Second	300,000
Relative Accuracy	1–3 cm
Raw Data File Size	100–200 MB a minute
Processing	Point Processing
Battery Life	3.5 hrs

Filtering and Subsampling of Point Clouds

According to the studies of Glennie and Lichti [18] and Glennie [17], if the Velodyne multibeam LiDAR sensor has an excessively large incident angle to the object surface during scanning, it is likely to cause a significant increase in point cloud noise due to the decrease in reflection intensity. Although GeoSLAM Zeb-Horizon scans the data with VLP-16 in a rotating manner, it might also decrease the accuracy of the point cloud due to the large incident angle. In order to reduce the influence of errors caused by factors other than the ranging measurement of the handheld LiDAR scanner, the GeoSLAM Hub software was used to output the normal information of each point and the SLAM calculation quality (SLAM condition) to calculate the incident angle λ of each point and obtain the quality index of trajectory calculation for filtering point clouds. Points with a large incident and bad SLAM condition were filtered out.

After the point cloud was filtered, the point cloud located on the calibration planes and check planes were manually extracted for calibration data. In order to avoid a large difference in the point number on different planes and the excessive concentration in certain ranging measurements to affect the determination of the ranging system error parameters, including the range scale factor (S) and the rangefinder offset (C), the same number of point clouds was randomly subsampled on each plane so that the calibration point data could be evenly collected in various ranging measurements.

Blunder Point Filtering Using the RANSAC Algorithm

In order to avoid the error or noisy points being included in the determination of the range scale factor (S) and the rangefinder offset (C), the blunder points of the subsampled points were removed using the RANSAC algorithm [19]. RANSAC is an algorithm for estimating a specific mathematical model from a sample containing gross errors. A fixed amount of data is randomly sampled from the sample to calculate a mathematical model that matches the sampled data. The remaining data after sampling is substituted into this mathematical model and the residual is calculated. If the residual error is less than the threshold, the data are regarded as the inner group conforming to the mathematical model. If the residual error is greater than the threshold, the data is regarded as gross error or blunder. The above steps are repeated and the largest number of inner groups conforming to the mathematical model is regarded as the best model parameter to classify and locate gross error data.

2.1.4. Mathematical Model for Calibration

Based on our preliminary study [16], the ranging system error Δr described only by the range scale factor (S) and the rangefinder offset (C). They were regarded as the additional parameters (APs) of the adjustment system and solved in the least-squares adjustment. Due to paper page limits, the rough description was presented in [16], and the sequent subsections describe the plane-based dynamic calibration with more detail.

Scanning Center Determination of Each Point

GeoSLAM Zeb-Horizon cannot output the original ranging measurements. In order to obtain each ranging measurement of r_i, called pseudo-ranging measurement, the laser scanning center coordinates corresponding to each point should be obtained from the trajectory data. The trajectory data could be output by the GeoSLAM Hub software. The recording frequency in the trajectory data was 0.01 s. Therefore, the corresponding laser scanning center coordinates (x_{ic}, y_{ic}, z_{ic}) for point i were determined by the linear interpolation formula as follows:

$$x_{ic} = x_{c0} + (t_i - t_0)\frac{x_{c1} - x_{c0}}{t_1 - t_0};\ y_{ic} = y_{c0} + (t_i - t_0)\frac{y_{c1} - y_{c0}}{t_1 - t_0};\quad z_{ic} = z_{c0} + (t_i - t_0)\frac{z_{c1} - z_{c0}}{t_1 - t_0} \tag{2}$$

where

t_i is the scanning time of point i

$(x_{c0},\ y_{c0},\ z_{c0})$ is the trajectory coordinates that the scanning time of point i is less than t_i but is closest to t_i.

$(x_{c1},\ y_{c1},\ z_{c1})$ is the trajectory coordinates that the scanning time of point i is larger than t_i but is closest to t_i.

t_0 is the time of the calculated trajectory that is less than the scanning time t_i for point i and is closest to t_i.

t_1 is the time of the calculated trajectory that is larger than the scanning time t_i . for point i and is closest to t_i.

By using the laser center coordinates of point i, the space vector $(\Delta x_i, \Delta y_i, \Delta z_i)$ of point i was calculated by the following Equation (3).

$$\begin{bmatrix} \Delta x_i \\ \Delta y_i \\ \Delta z_i \end{bmatrix} = \begin{bmatrix} x_i \\ y_i \\ z_i \end{bmatrix} - \begin{bmatrix} x_{ic} \\ y_{ic} \\ z_{ic} \end{bmatrix} \tag{3}$$

The calculated pseudo-ranging measurement r_i, the horizontal angle α_i and vertical angle β_i should be calculated according to the following Equations (4)–(6) for subsequent derivation.

$$r_i = \sqrt{\Delta x_i^{\,2} + \Delta y_i^{\,2} + \Delta z_i^{\,2}} \tag{4}$$

$$\alpha_i = \tan^{-1}\frac{\Delta z_i}{\sqrt{\Delta x_i^{\,2} + \Delta y_i^{\,2}}} \tag{5}$$

$$\beta_i = \tan^{-1}\frac{\Delta y_i}{\Delta x_i} \tag{6}$$

Through the pseudo-ranging measurement r_i corrected by ranging error APs (S for the ranging scale factor and C for the rangefinder offset), the horizontal angle α_i, the vertical angle β_i, and the corresponding laser center coordinates (x_{ic}, y_{ic}, z_{ic}), the coordinates of point i could be reconstructed by Equation (7):

$$\begin{bmatrix} x_i \\ y_i \\ z_i \end{bmatrix} = \begin{bmatrix} (r_i \times S + C) * \cos\alpha_i * \sin\beta_i \\ (r_i \times S + C) * \cos\alpha_i * \cos\beta_i \\ (r_i \times S + C) * \sin\alpha_i \end{bmatrix} + \begin{bmatrix} x_{ic} \\ y_{ic} \\ z_{ic} \end{bmatrix} \quad (7)$$

Mathematical Model for Calibration

The data obtained by the ground-based LiDAR scanner and the handheld LiDAR scanner were respectively located in the ground-based LiDAR coordinate system and the handheld LiDAR coordinate system. Only when the LiDAR point cloud was converted to the ground-based LiDAR coordinate system, could the plane parameters obtained by the ground-based LiDAR scanner be used as the calibration reference data to solve the ranging APs. Therefore, the six rigid body conversion parameters (three translation parameters and three rotation parameters) were regarded as the unknowns and added to the adjustment equation for simultaneous determination. The following Equation (8) describes the conversion of the handheld LiDAR point (x_i, y_i, z_i) after the correction of the ranging system error to the ground-based LiDAR coordinate system through the rigid body six conversion parameters.

$$\begin{bmatrix} X_i \\ Y_i \\ Z_i \end{bmatrix} = R(\kappa)R(\varphi)R(\omega) * \begin{bmatrix} x_i \\ y_i \\ z_i \end{bmatrix} + \begin{bmatrix} X_t \\ Y_t \\ Z_t \end{bmatrix} = R * \begin{bmatrix} x_i \\ y_i \\ z_i \end{bmatrix} + \begin{bmatrix} X_t \\ Y_t \\ Z_t \end{bmatrix} \quad (8)$$

where R: rotational transformation matrix; (X_t, Y_t, Z_t): translation vector

Equation (8) is the main equation of the plane-based dynamic calibration method that was originally developed by us [16]. Equation (8) also means the point i after correction by range scale factor and rangefinder offset and conversion should be located on the corresponding calibration plane fitted by the point cloud from the ground-based LiDAR scanner. Due to the random error, the converted point could be not located on the calibration plane. Thus, to minimize the sum of distance squares from the converted points to their corresponding calibration plane, the mathematical model of adjustment was developed to determine the ranging APs. The mathematical model included the functional model and the stochastic model. If the observation equations were regarded as the identical weight. Equation (9) shows the functional model for the least-squares adjustment. The calibration plane parameters (a_k, b_k, c_k) used in this study were unit vectors, so the above-mentioned sum of distance squares from the converted points to their corresponding calibration plane could be regarded as the square sums of correction v_i (i.e., residuals), and the pseudo-observation equation was shown in Equation (9). Each handheld LiDAR point could establish a pseudo-observation equation, and simultaneously solve the ranging APs and the rigid body six conversion parameters according to the least-squares principle.

$$F_n = \begin{bmatrix} a_k & b_k & c_k \end{bmatrix} \begin{bmatrix} X_i \\ Y_i \\ Z_i \end{bmatrix} + d_k = 0 + v_i \quad (9)$$

where a_k b_k c_k d_k are plane parameters of plane k. (X_i, Y_i, Z_i) are the coordinates of point i on the ground-based LiDAR system after rigid body conversion.

The unknowns of this adjustment were two ranging APs (S and C) and six coordinate transformation parameters. Since Equation (9) is a non-linear equation, it should be linearized by Taylor expansion to establish the indirect observation adjustment matrix form, see Equation (10) required by the least-squares method. The pseudo-observation equations were regarded as equal weights, the corrections to the initial values of the unknowns were determined by Equation (11), and then the corrections were added to the initial value before iterations to reorganize the indirect observation adjustment matrix, also see Equation (10). During iterations, the termination condition was set as the ratio of the posterior variance change less than 0.000001; the ranging APs and the rigid body six conversion parameters could be solved until the posterior variance change ratio converges.

$$JX = K + V \tag{10}$$

$$J = \begin{bmatrix} \frac{\partial F_1}{\partial S} & \frac{\partial F_1}{\partial C} & \cdots & \frac{\partial F_1}{\partial Z_t} \\ \frac{\partial F_2}{\partial S} & \frac{\partial F_2}{\partial C} & \cdots & \frac{\partial F_2}{\partial Z_t} \\ \vdots & \vdots & \vdots & \vdots \\ \frac{\partial F_n}{\partial S} & \frac{\partial F_n}{\partial C} & \cdots & \frac{\partial F_n}{\partial Z_t} \end{bmatrix}, X = \begin{bmatrix} dS \\ dC \\ d\varphi \\ d\omega \\ d\kappa \\ dX_t \\ dY_t \\ dZ_t \end{bmatrix}, K = \begin{bmatrix} -F_{10} \\ -F_{20} \\ \vdots \\ -F_{n0} \end{bmatrix}$$

where: J is Jacobian matrix.

X is the correction vector to the initial value of the unknowns.

K is the difference vector between 0 and the value of substituting the initial value into pseudo observation equations.

V is the residual vector of the pseudo observation equations.

N is the number of pseudo observation equations.

$$X = \left(J^T J\right)^{-1}\left(J^T K\right) \tag{11}$$

2.1.5. Result Analysis

This subsection described the analysis of the calibration results.

Residuals Analysis

The influences on the residual distribution, the average value of the residuals, and the posterior unit weight standard deviation before and after the ranging system error correction were used to verify whether the residual error distribution has the trend of implicit system error or not; whether the average value of the residual error and the standard deviation have decreased or not in order to evaluate if the calibrated ranging system error parameters could correct the system error and improve the accuracy of the handheld LiDAR points.

Verification by the RMSE of Check Planes

The RMSE of the calibration data from corrected and uncorrected handheld point clouds to each corresponding check plane before and after the adjustment was calculated for evaluation of calibration results. The improvement ratio for each check plane was calculated by the following Equation (12) to verify the efficiency of the system error correction.

$$ratio = \frac{(RMSE_{with\ APs} - RMSE_{without\ APs})}{RMSE_{without\ APs}} * 100\% \tag{12}$$

where

$RMSE_{with\ APs}$ is the RMSE calculated by adding the ranging additional parameters (APs) to the adjustment

$RMSE_{without\ APs}$ is the RMSE calculated by not adding the ranging additional parameters (APs) to the adjustment.

Analysis of Correlation Matrix of the Unknowns

Through the correlation coefficient matrix, the quality and robustness of the calibration results [20] can be verified, and check if there is a high correlation between the parameters of the ranging system or the parameters of the ranging system and the conversion parameters.

High correlation means that the calibration method or calibration data is not enough to solve the calibration parameters well. The correlation coefficients after the adjustment were shown as discussed in this study.

Analysis of Ranging Systematic Error Parameters

In this study, two systematic errors were estimated, S for the range scale factor and C for the rangefinder offset. The influence of certain ranging measurements, for example, 20 m, 30 m, and 40 m, was investigated.

2.2. *Urban Cadastral Detail Survey*

2.2.1. Ground Control Survey

The ground control survey collected the 3D coordinates for converting the handheld LiDAR points to the cadastral coordinate system. In Taiwan, the TWD97 coordinate system was adopted for the cadastral coordinate system which is a horizontal coordinate system. However, the elevation for each control point was also surveyed for 3D coordinate conversion of the handheld LiDAR points. The TWVD2001 system was used for elevation systems in Taiwan.

The control points might be of the announced TWD97 coordinates and for the selected supplementary control points. Therefore, considering accuracy, cost, and ease of operation, VBS-RTK Technology was employed for the control survey to determine the 3D coordinates, which were in the GNSS coordinate system. The ground control points were surveyed twice for averaging. For each survey, the standard deviation in the horizontal coordinate component and the elevation component should not exceed 2 cm and 5 cm, respectively. At the same time, the differences in horizontal position and elevation between the two surveys should be less than 3 cm and 5 cm, respectively. However, the TWD97 coordinate system was used for horizontal coordinates, and the TWVD2001 system was used for the elevation system in Taiwan. Therefore, surveyed elevation was converted to the ortho-height system according to the geoid model announced by the Ministry of the Interior (Taiwan).

Subsequently, coordinate conversion had to be performed to estimate and eliminate the systematic errors between the TWD97 cadastral coordinate system and GNSS coordinate system, as well as to verify the correctness of coordinate conversion. The systematic error might be caused by the tension between the VBS-RTK coordinate system and the TWD97 cadastral coordinate system due to the control network tension, crustal changes, and projection deformation. Least-squares collocation is a combination of least-squares adjustment, estimation, and filtering. Compared with the traditional adjustment method that can only deal with random errors in observation [21], therefore, the least-squares collocation method was implemented in this study to estimate the systematic error between two kinds of coordinates, namely the TWD97 cadastral coordinate system and VBS-RTK coordinate system, where the observation was not made [22] to determine the cadastral coordinates for those control points without announced cadastral coordinates.

2.2.2. Path Planning for Data Collection

According to the manual of the GeoSLAM Zeb-Horizon, the scanned path must be closed to the starting point of the scanning in order to employ the SLAM algorithm [3–5] together with IMU data to calculate a better scanning trajectory. The start and end points of the scanned path should be the same point to create a closed route and reduce the deviation caused by error propagation in the trajectory [6,8].

If the scanned path passes through control points, the alignment device provided, as shown in Figure 2, could be used to align the control point with the hole in the front of the alignment device and stop for ten seconds. The GeoSLAM Zeb-Horizon scanner could record the coordinates of this control point in the handheld LiDAR coordinate system for subsequent coordinate conversion.

Figure 2. GeoSLAM control point alignment device.

Since the GeoSLAM Zeb-Horizon scanner uses the SLAM algorithm [3–5] together with IMU data to calculate a better scanning trajectory, the scanning environment will have a significant impact on the quality of its trajectory calculation. It is necessary to investigate the application environment and plan an appropriate path before scanning to understand the scan area and its situations, which might cause incorrect estimation of the SLAM algorithm, such as the lack of features, and good path planning could ensure the stability of the SLAM trajectory calculation. In this study, the test area was in the urban area. According to the user manual and past literature, the following guidelines organized the following items:

- Avoid scanning moving objects, the SLAM algorithm may recognize them as static features and cause calculation errors [23].
- The start and end points of the scanned path should be the same point to create a closed route and reduce the offset caused by error propagation in the trajectory [7,9].
- The moving speed should not be too fast which is not more than normal walking speed (1.1~1.5 m/s), and normal walking speed should be maintained to ensure a good point cloud density. When passing the building corner, because the scanning angle of view changes greatly, the speed and travel should be slowed down to obtain sufficient features to establish a trajectory [15,24].
- The scanning distance is recommended to be kept within 50 m to maintain good point cloud accuracy and point cloud density.
- The time of a single scanning task should be less than 30 min. When scanning a large field, the scanning task should be divided into several portions and the point cloud should be registered to reduce the probability of trajectory deviation [23].
- Avoid areas containing a lot of glass and windows. Glass and windows are prone to refraction of the laser beam and cause false point clouds [25].
- When scanning narrow passages, the scanned path should be in the middle of the passage so that the scanner can obtain the features of the walls on both sides. If it is too close to the wall, the scanning angle is too small and it will lack feature acquisition [23].

2.2.3. Point Cloud Processing

Filtering of Point Clouds

As mentioned in Section 2.1.3, after the point cloud data collection, the point cloud normal vector and the SLAM calculation quality were exported from GeoSLAM Hub software, and the point cloud was filtered to remove the point cloud with poor quality or poor observation conditions.

In addition to the above-mentioned filtering conditions, namely the angle of incidence and the SLAM calculation quality, the point cloud with a longer ranging measurement would be affected by the ranging measurement capability of the LiDAR scanner, and a point cloud with an abnormal distance usually has a larger error and is also more likely to be erroneous data in an outdoor environment. For mobile surveying and mapping of a LiDAR system, the ranging measurement data of a short distance could be retained to retain the point cloud with higher accuracy. According to the manual, the scanning distance is recommended to be kept within 50 m. In order to maintain a good point cloud accuracy, the scanning distance threshold was added to the filtering condition of the points to eliminate points of possible poor quality.

Ranging System Error Correction of Point Clouds

The corresponding laser center coordinates of each point by linear interpolation was determined, and the space vector $(\Delta x, \Delta y, \Delta z)$ of each point was calculated, then the pseudo-ranging measurement r, horizontal and vertical angle were calculated for each point according to Equations (4)–(6).

The ranging APs (S and C) of the ranging system obtained by the calibration method described in Section 2.1 were employed to correct the ranging system error, and the pseudo-ranging measurement r of each point was corrected and Equation (7) is used to reconstruct the point coordinates to complete the system error correction.

Coordinate Conversion

After the point cloud filtering and the ranging system errors were corrected, the control point coordinates in the handheld LiDAR coordinate system and their corresponding horizontal coordinates in the TWD97 coordinate system and the elevation in the TWVD2001 system were used to calculate the conversion parameters, and the corrected point cloud was converted from the handheld LiDAR coordinate system to the TWD97 cadastral coordinate system through the rigid body six conversion parameters.

2.2.4. Urban Cadastral Detail Line Data Production

The urban cadastral detail lines mean possible parcel points on parcel boundaries. The production of the urban cadastral detail line data is difficult to extract automatically because it contains some subjective factors to be identified and digitized with reference to actual building conditions, cadastral reconnaissance, existing cadastral maps, and land use zoning maps. Therefore, this study used PointCab software to digitize the detail line data manually. While digitizing manually, the following principles should be followed:

(1) For urban buildings along roads, most of them are bounded by the building line designated by the road centerline stake of the urban planning road or the boundary of the existing road boundary. Road boundaries are the detail lines for manual digitization.
(2) The detail lines of townhouses are mostly bounded by the center of the wall, but they still need to be judged by considering the difference in their structure or the decorative form of the exterior. The centerline of a wall on a building façade with the same building style except for the wall on the outer most boundary of a building.
(3) Where there are exposed steel bars on the walls of side houses or independent houses, the center of the wall shall be the boundary, otherwise, the outer edge of the wall shall be the boundary.
(4) The outer edge of a wall on the most outside boundary of a building with the same building style, and the outer edge of a wall of a single building except the wall attached with exposed steel reinforcing;
(5) The eaves of the building belong to the building itself.

Results Analysis

(1) Analysis of digitized detail lines.

The results analysis of the error of the urban cadastral detail survey using the handheld LiDAR points was divided into planar position error analysis (Figure 3a) and vertical distance error analysis (Figure 3b). If the points surveyed by a total station can be analyzed definitely by the junction points of the digitized lines, the planar position errors of those points could be calculated, as shown in Figure 3a. Otherwise, the vertical distances of the remaining points surveyed by a total station to the closest digitized detail line segment were calculated for analysis, as shown in Figure 3b.

(a) (b)

Figure 3. The demonstration of error calculation for an urban cadastral detail survey using the handheld LiDAR points where points mean the survey points measured by a total station and lines were digitized manually from the handheld LiDAR points. (**a**) Planar position error analysis. (**b**) Vertical distance error analysis.

(2) Analysis on the effect of ranging system error correction.

The effect analysis of error correction of the ranging system was represented by the difference, as expressed in Equation (13), of digitized detail lines using the handheld LiDAR point clouds after and before ranging system error correction, compared with the detail points surveyed by a total station.

$$Difference = DE_{before\ correction} - DE_{after\ correction} \tag{13}$$

where,

$DE_{before\ correction}$ is the digitized error using uncorrected point cloud for digitization compared with the detail points surveyed by a total station.

$DE_{after\ correction}$ is the digitized error using corrected point cloud for digitization compared with the detail points surveyed by a total station.

3. Results and Discussion

3.1. Handheld LiDAR Scanner Calibration

3.1.1. Selection of the Calibration Field

The size of the calibration site was about 35 m by 27 m by 3 m, located in the underground parking lot of Research and Innovation-Incubation Center at National Chengchi University in Taiwan. The calibration site provided a large variety of planar features with different ranges for calibration, as shown in Figure 4.

Figure 4. Calibration site.

3.1.2. Acquisition and Extraction of Calibration Reference Data

As mentioned in Section 2.1.2, a Faro ground-based LiDAR scanner was used to capture the point cloud of the calibration site in one station instead of multiple stations to avoid the point cloud registration errors. The collected point cloud data are shown in Figure 5.

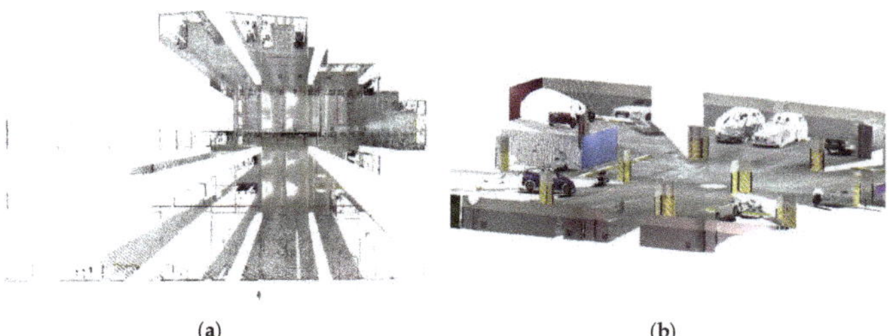

(a) (b)

Figure 5. Top and side views of the collected points by a FARO Focus S350 in a single station. (**a**) Top view. (**b**) Side view.

Seventeen plane locations, labeled A to Q, were manually extracted for calibration reference data, and the location of each plane is shown in Figure 6. The size of each plane was about 0.7 m by 1.2 m and its planar parameters were determined by the least-squares method. The parameters and fitting RMSEs are shown in Table 3. All plane fitting RMSEs were not greater than 0.001 m, indicating that the point cloud data of the FARO Focus S350 was of a certain accuracy and reliability as the calibration planes and the check planes.

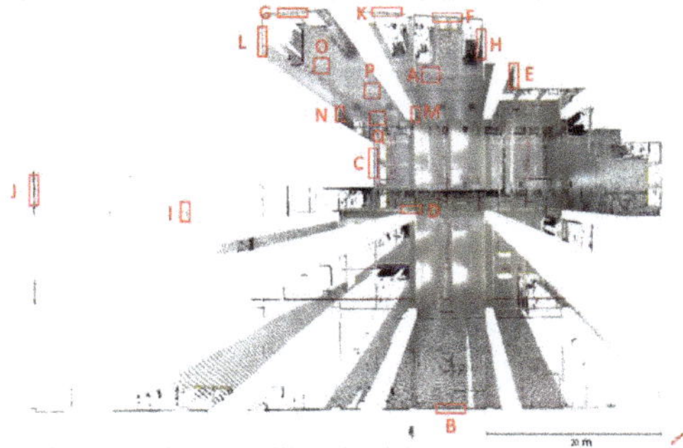

Figure 6. Locations of the selected 17 planes (labeled A to Q) for calibration reference data.

Table 3. Fitting plane information.

Plane	a	b	c	d	Fitting RMSE (m)	DIP(°)
A	0.012	0.007	0.999	−44.416	0.0009	0
B	0.982	0.190	0.002	−23.319	0.0009	89
C	−0.196	0.981	−0.001	9.670	0.0008	89
D	0.982	0.190	0.001	−1.516	0.0006	89
E	−0.183	0.983	−0.003	−5.552	0.0008	89
F	0.993	0.121	0.002	19.513	0.0008	89
G	0.964	0.265	0.002	21.626	0.0006	89
H	−0.193	0.981	−0.002	−2.357	0.0007	89
I	−0.194	0.981	0.001	30.209	0.0006	89
J	−0.191	0.982	0.001	46.958	0.0010	89
K	0.994	0.112	0.003	19.384	0.0009	89
L	−0.189	0.982	−0.005	22.097	0.0006	89
M	−0.191	0.982	−0.001	5.064	0.0008	89
N	−0.187	0.982	−0.004	13.458	0.0005	89
O	−0.004	0.005	0.999	−44.549	0.0009	0
P	−0.003	−0.002	0.999	−44.667	0.0006	0
Q	0.002	0.003	0.999	−44.545	0.0010	0

In order to solve simultaneously the ranging APs together with the six-parameter rigid body conversion parameters. three planes with orthogonal normal were included in the plane selection [20]. The horizontal and vertical planes were identified from the DIP in Table 3. Table 3 shows the planes including 4 horizontal (DIP 0°) and 13 vertical (DIP 89°) planes. The triple pair (a, b, c) indicated the unit normal vector in Table 3.

3.2. Acquisition of Handheld LiDAR Point Cloud for Calibration Data

A plane-based dynamic calibration developed in our previous study [11] and mentioned in Section 2.1.4 was employed to investigate the calibration results. The dynamic or kinematic calibration means to capture the point cloud data for calibration in a mobile way. The calibration data was collected on 26 January 2021 at 9:30 a.m., the scanning time took about 87 s, the number of point clouds was about 160,000,000 points. Figure 7 shows the collected calibration data and the scanning trajectory, the colors of points in Figure 7a were determined based on the SLAM quality (SLAM condition), the best quality was blue (R = 0; G = 0; B = 255), the closer to red, the worse was the quality. In an indoor environment with rich features, the SLAM quality was noticeably stable, and the majority of the point clouds were dark blue indicating no significant problem of SLAM solution or failure of the SLAM solution. The color of the point clouds in Figure 7b was given according to the scanning time of the scanning trajectory, and the color of the point clouds gradually changed from red to blue according to the scanning time, where red was the scanning time at the beginning, and blue was the scanning time at the end. The scanned path was planned to walk around the center of the parking lot and close to the starting point at a normal walking speed to ensure that the complete calibration field was scanned. It could be found that the starting point represented by red and the scanning end point represented by blue were closed, which met the scanning requirements.

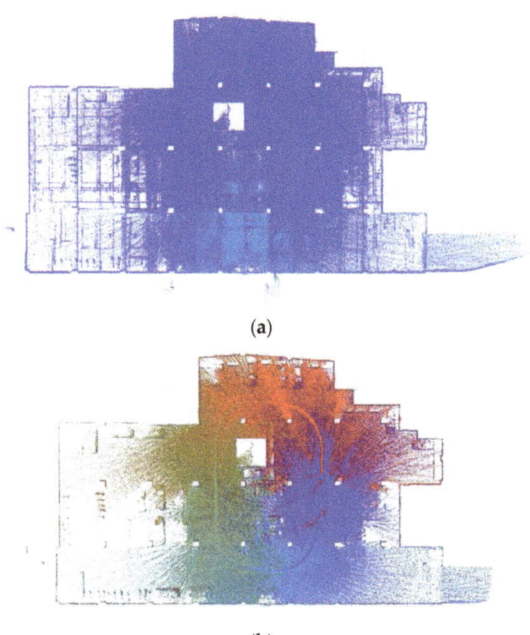

Figure 7. Point cloud and scanning trajectory of calibration data. (**a**) Colored by SLAM condition. (**b**) Colored by scanning time.

3.2.1. Filtering and Subsampling of Point Clouds

In this study, the filtering conditions of the handheld LiDAR point cloud were performed by the SLAM quality and the incident angle of the point. In order to remove the influence caused by the poor SLAM quality, the threshold value was set to R = 0, G = 0, and B = 255, namely the blue points colored by the best SLAM solution were used for the calibration data. Based on the study of Glennie and Lichti [18] and Glennie [17], the thresholds of the incidence angle for the effectiveness of filtering with the incident angle were 60° and 70°, respectively. However, the point cloud filtered at 60°, the plane features in the horizontal direction were relatively insufficient. Considering that a certain number of planes in the horizontal direction must be extracted, this study used 70° as the filter threshold for the incident angle.

In order to evaluate the effectiveness in filtering point clouds with poor quality by the thresholds, a point cloud on one plane was taken for analysis to discuss whether the RMSE of the plane fitting after filtering was reduced or not. The analysis results are shown in Table 4. After filtering using the SLAM quality or incident angle separately, the plane fitting RMSE was reduced from 0.0110 m to 0.0108 m. If both the SLAM quality and incident angle were used for filtering, the plane fitting RMSE was further reduced to 0.0106 m. The results showed that the plane fitting RMSE of the point cloud by the three filtering conditions were all better than the original point cloud used for plane fitting. It could say that the used conditions and thresholds could retain the point cloud with better observation conditions, therefore this study used these combined conditions for filtering. Figure 8 shows the point cloud data after filtering. After the point cloud filtering, the number of point clouds was 1,401,803, and the filtering ratio was about 91%. The remaining good quality point cloud data was employed as calibration data for calibration adjustment.

Table 4. Analysis results of filtering conditions using a planar point cloud

Filter Condition	Plane Fitting RMSE (m)	Number of Points	Filtering Points
None	0.0110	21,650	——
SLAM quality	0.0108	13,525	8125
Incidence angle	0.0108	19,620	2024
SLAM quality and incidence angle	0.0106	12,969	8678

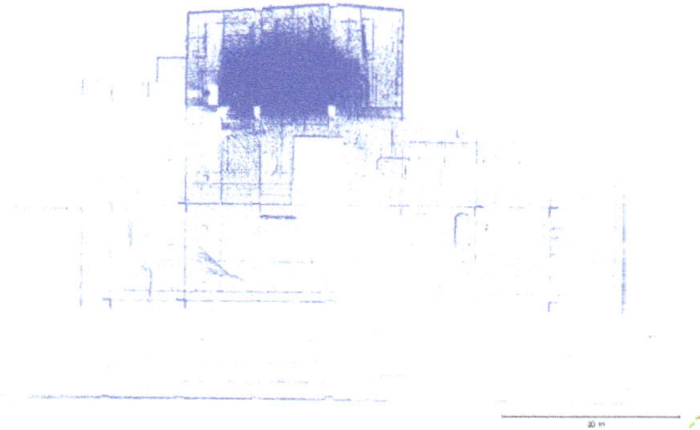

Figure 8. Handheld LiDAR point cloud after point cloud filtering.

The handheld LiDAR point cloud corresponding to each plane, see Figure 6, was selected manually, and the point number contained on each plane and the ranging measurement between each point relative to their corresponding laser center was calculated. The coordinates of each laser center could be seen in Equation (2). In order to increase the calculation efficiency and allow the calibration calculation to include uniform and various ranging measurements, the points on each plane were randomly subsampled to select the same number of point clouds as the calibration data, except for planes B, E, I, and J, which were of a lower number of points, the least number of points on the other planes was plane D, and the point number was 613. Therefore, this study randomly subsampled 600 points for the remaining planes, and evaluated whether there was a significant difference in the calculated pseudo-ranging measurements or not before and after subsampling and blunder removing by the RANSAC algorithm.

The statistical results of calculated pseudo-ranging measurements for each plane of calibration data before and after subsampling and blunder removing by the RANSAC algorithm are shown in Table 5. The statistics for pseudo-ranging measurements were divided into minimum, maximum, median, and average to evaluate the distance measurements provided by the points in each plane for calibration adjustment calculation. As seen in Table 5, planes B, I, and J were of relatively fewer points due to the longer scanning distances. Plane E was blocked by a wall, the point number was also relatively insufficient. The box diagram of the calculated pseudo-ranging measurements before and after subsampling and blunder removing by the RANSAC algorithm is shown in Figure 9. According to Table 5 and Figure 9, there was no significant difference between the minimum, maximum, median, and average values of the calculated pseudo-ranging measurements in each plane before and after subsampling and blunder removing by the RANSAC algorithm. The difference of statics of the calculated pseudo-ranging measurements in each plane before and after subsampling and blunder removing by the RANSAC algorithm was less than 1 m, indicating that the same rich calculated pseudo-ranging measurements could be retained after subsampling and blunder removing by the RANSAC algorithm could be used for calibration. Therefore, 600 points could be used as the number of subsampling points, taking into account the richness of the calibration data and improving the calculation efficiency.

Table 5. The statistics of pseudo-ranging measurements in each plane for calibration data before and after subsampling and blunder removing by the RANSAC algorithm for calibration data.

Plane	No. of Point before/after	Calculated Pseudo-Ranging Measurement (m)			
		Minimum before/after	Maximum before/after	Median before/after	Average before/after
A	3024/586	2.012/2.043	4.166/4.043	2.714/2.715	2.745/2.738
B	**233/211**	**34.100/34.100**	**37.353/37.230**	**35.474/34.982**	**35.444/35.313**
C	660/595	6.953/6.953	9.137/9.137	8.137/8.146	8.138/8.138
D	613/567	12.471/12.471	15.977/15.919	13.775/13.768	13.962/13.923
E	**322/297**	**9.226/9.226**	**10.454/10.328**	**9.495/9.474**	**9.737/9.691**
F	862/593	8.713/8.735	13.018/12.964	9.125/9.125	9.347/9.337
G	847/579	8.846/8.846	18.200/18.200	10.954/10.696	11.536/11.449
H	1420/581	7.832/7.860	11.380/11.380	9.227/9.161	9.354/9.317
I	**80/58**	**22.515/22.515**	**28.518/28.518**	**24.317/24.350**	**24.985/25.192**
J	**91/62**	**39.675/39.757**	**45.099/45.099**	**41.515/41.605**	**41.833/41.836**
K	1595/592	7.916/7.919	12.618/12.618	8.396/8.357	9.029/8.949
L	864/569	9.722/9.722	20.655/20.655	15.538/14.993	15.484/15.391
M	1110/599	2.310/2.310	2.909/2.906	2.542/2.536	2.546/2.542
N	1417/562	3.333/3.555	10.170/10.170	5.961/5.987	6.335/6.391
O	818/600	2.695/2.695	12.473/12.473	3.485/3.493	3.722/3.706
P	2119/599	1.456/1.466	7.881/7.295	2.082/2.088	2.342/2.348
Q	736/593	3.186/3.199	7.054/7.054	3.698/3.694	3.873/3.866

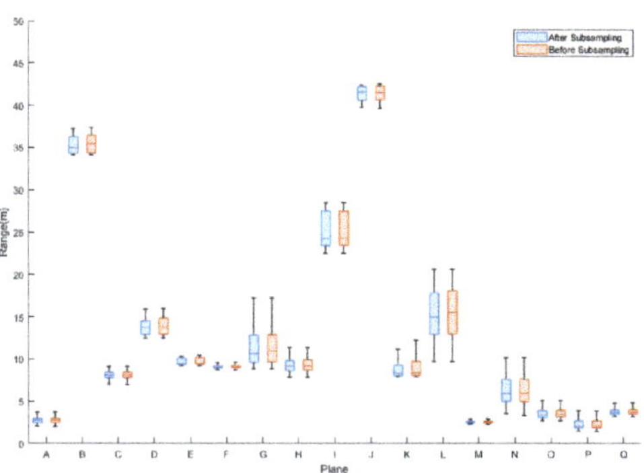

Figure 9. Box diagram of the calculated pseudo-ranging measurements of each plane before and after subsampling and blunder removing by the RANSAC algorithm.

3.2.2. Blunder Point Filtering Using the RANSAC Algorithm

The RANSAC algorithm was used to remove the gross errors of the subsampled points in each plane for the calibration data. According to the point cloud precision announced by the GeoSLAM Zeb-Horizon manufacturer, the allowable error threshold is set to 0.03 m. The results are shown in Table 6, where the red points were evaluated as a gross error by the RANSAC algorithm.

Table 6. RANSAC result for calibration data.

Plane	RANSAC Results	Number of Outliers	Outliers %
A		14	2.33%
B		22	9.44%
C		5	0.83%
D		33	5.5%
E		25	7.76%
F		7	1.17%
G		21	3.5%
H		19	3.17%
I		22	27.5%
J		19	20.88%
K		8	1.33%
L		31	5.17%
M		1	0.17%
N		38	6.33%
O		0	0%
P		1	0.17%
Q		7	1.17%

3.2.3. Results Analysis

A plane-based dynamic calibration proposed in our previous study [16] was performed. The location distribution of the calibration planes in this study is shown in Figure 10. The selection of the calibration planes should enclose the entire calibration field as much as possible and be evenly distributed. In order to solve the rigid body conversion parameters at the same time during the calibration adjustment calculation, in addition to the vertical planes, the calibration planes must also include the horizontal planes (such as planes A, O, Q) [20]. In particular, although plane E seems to be closer to the periphery of the calibration field than plane H, plane H was of a richer scanning distance than plane E from the maximum and minimum values in Table 6, plane H instead of plane E was selected as a calibration plane. The following subsections investigate the calibration results.

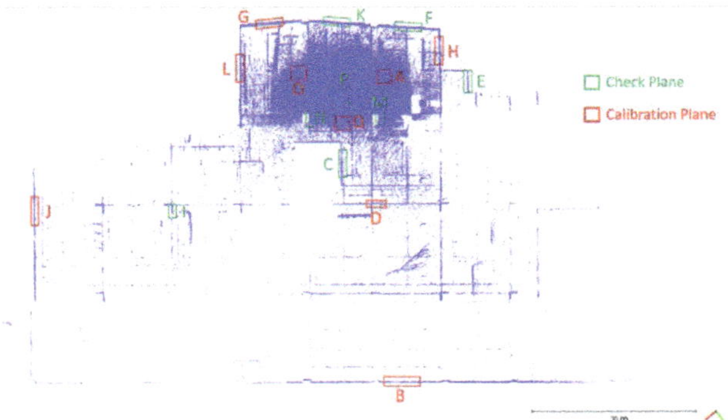

Figure 10. The locations of calibration planes (labeled A, B, D, G, H, J, L, O, Q) and check planes (labeled C, E, F, I, K, M, N, P).

Residuals Analysis

Figure 11 shows the residual distribution plots and Figure 12 shows the residual scatter plots after adjustment with or without determination of the ranging system error, respectively, where Figures 11a and 12a are the residual distribution plots and the residual scatter plots of the adjustment results by incorporating the ranging system error into the adjustment system as an additional parameter (referred to with APs) for determining together with the rigid body six conversion parameters, Figures 11b and 12b are the residual distribution plots and the residual scatter plots of the adjustment results without adding the ranging system error as the additional parameter (referred to without APs)) and only the rigid body six conversion parameters retained as the unknowns for determination.

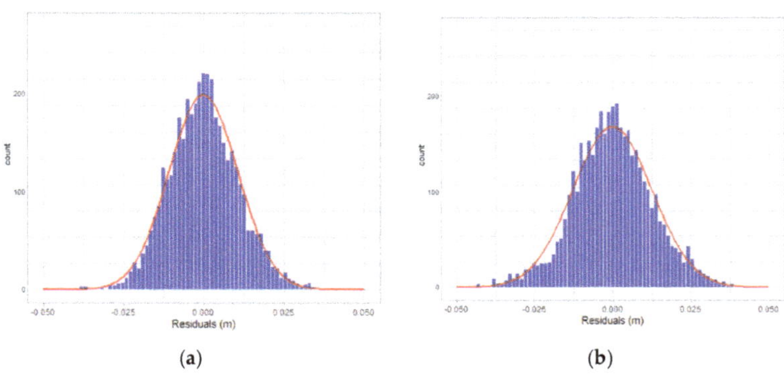

Figure 11. The residual distribution plots of calibration data. (**a**) With APs. (**b**) Without APs.

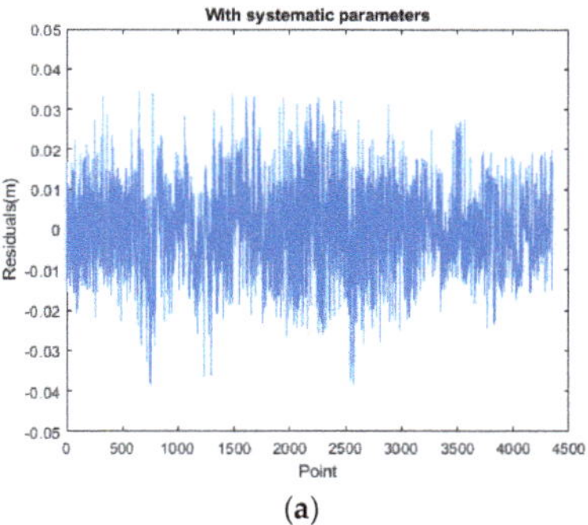

(a)

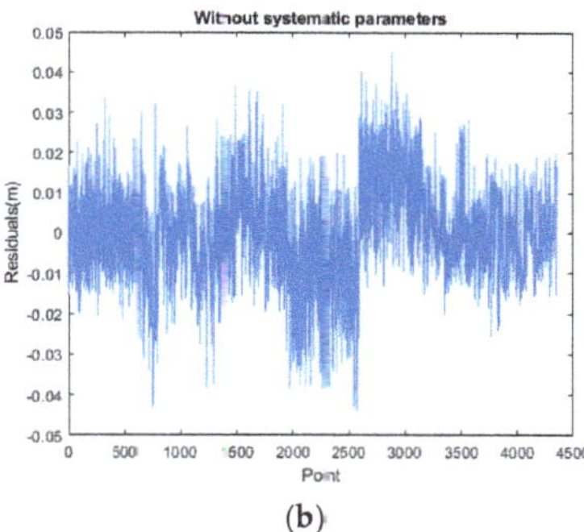

(b)

Figure 12. The residual scatter diagram of calibration data. (**a**) With APs. (**b**) Without APs.

Comparing Figure 11a,b, the residual value distribution was much more concentrated at 0 and was more in line with the normal distribution curve. Regarding further analysis by the residual scatter diagram, as shown in Figure 12, the residuals in Figure 12a were stably and evenly dispersed within ±0.03 m, which conformed to the 3 cm precision of point cloud announced by the manufacturer. The residuals in Figure 12b were affected by more significant systematic errors, which caused the residual dispersion to present an unstable undulation. After adding the ranging APs, the average residual was closer to 0 (from −0.000045195 m to −0.000014845 m), and the posterior unit weight standard deviation became smaller (from ±0.01278 m to ±0.01077 m), both of which were improved compared to those without adding the ranging APs into the adjustment. There might be no significant statistical difference in the figures. However, it could be clearly understood from the change of the relevant figures that adding the ranging APs in this study could eliminate most of the ranging system errors.

Verification by the RMSE of Check Planes

By evaluating the calibration results, the RMSE of each check plane was calculated for calibration data using least-squares adjustment with and without ranging APs. Table 7 shows the RMSE results of each check plane for three datasets. Among them, the RMSE of all check planes were all improved after the correction of the ranging system error, also see the RMSE bar chart of each check plane in Figure 13. Among them, the RMSE of the check planes E, F, K, M, and N were significantly improved. Up to 72.12% in plane F, an increase of about 2.4 cm and an improvement of 1.6 cm in plane E were reached. The overall average improvement was 32.61%, which demonstrated that the proposed calibration approach could effectively improve the overall point cloud accuracy.

Table 7. RMSE of each check plane.

Check Plane	$RMSE_{withAPs}$ (m)	$RMSE_{withoutAPs}$ (m)	Difference (m)	Improvement (%)
C	0.0121	0.0125	0.0004	2.98%
E	**0.0129**	**0.0290**	**0.0161**	**55.35%**
F	**0.0092**	**0.0329**	**0.0237**	**72.12%**
I	0.0236	0.0287	0.0051	17.75%
K	0.0083	0.0164	0.0081	49.25%
M	0.0057	0.0105	0.0048	45.27%
N	0.0102	0.0118	0.0016	13.82%
P	0.0066	0.0069	0.0003	4.32%
Mean	0.0111	0.0186	0.0075	32.61%

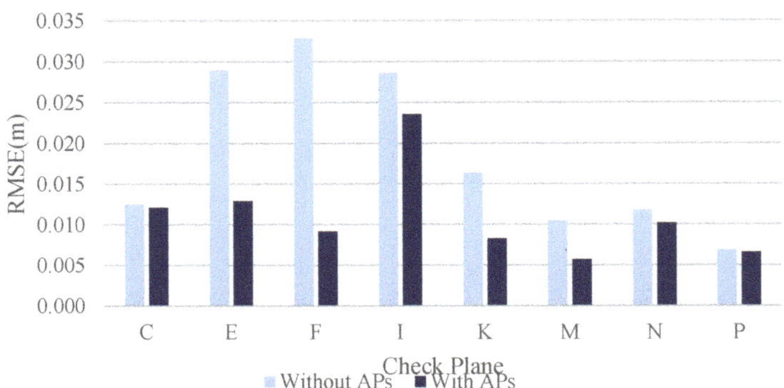

Figure 13. The bar chart of RMSE for check planes.

Analysis of Correlation Matrix of the Unknowns

Table 8 shows part of the matrix of correlation coefficients of the unknowns of the least-squares calibration solution. The correlation coefficients between the ranging APs and the coordinate conversion parameters were maintained at a low correlation, and the absolute values of the correlation coefficients were mostly not larger than 0.7. This result is consistent with Glennie and Lichti [18], that the calibration reference and calibration data collected in different coordinate systems did not significantly affect the calculation of the system error parameters; however, there was a high negative correlation between the ranging APs (S and C), −0.82. The lower negative correlation between the ranging APs made the solution results of the ranging APs more reliable. The result was more reliable than the test in our previous study [16], the correlation between the ranging APs (S and C) was −0.985. However, in the calibration data of this test, only points in the planes B and J were with long pseudo-ranging measurements for calibration, as shown in Figure 10, and the calculated pseudo-ranging measurements from the points on these two planes ranged from about 35.3 m to 41.8 m, as shown in Table 5. In the future, if a larger calibration site or suitable plan for scanning could be found to collect the handheld LiDAR points to obtain more calculated pseudo-ranging measurements for calibration, the negative correlation between the ranging APs could be reduced and the solutions of S and C could be more reliable.

Table 8. The matrix of correlation coefficients of the unknowns for calibration data.

	S	C	X_t	Y_t	Z_t	φ	ω	κ
S	1	−0.82	−0.68	0.65	−0.04	−0.06	0.12	−0.61
C	−0.82	1	0.30	−0.28	0.08	0.06	−0.14	0.23

Analysis of Ranging Systematic Error Parameters

The estimated ranging systematic parameters S and C were 0.9996 and −0.0088, the corresponding standard deviations were ±0.0000397 m and ±0.00055 m. Table 9 indicates the correction for different distances. If the ranging measurement was 10 m, the correction was 1 cm; the ranging measurement was 30 m, the correction is 2 cm; the ranging measurement was 40 m, the correction is 2 cm. Even it was 2 m, the correction would be 1 cm. When using a handheld LiDAR scanner for precise surveying, for example, cadastral surveying, this ranging system error should be corrected to obtain a more accurate result.

Table 9. The different distance values after correction (unit: m).

Distance	**1**	**2**	**5**	**10**	**20**	**30**	**40**
Distance after correction	0.99	1.99	4.99	9.99	19.98	29.98	39.98

3.3. Urab Cadastral Detail Survey

The test area of an urban cadastral detail survey was located near National Cheng-Chi University in Taiwan, as shown in the red box in Figure 14. The aerial photographs illustrate that the area was densely built and filled with narrow lanes that were difficult to use GNSS for a detail survey and was time-consuming to use total stations for a detail survey.

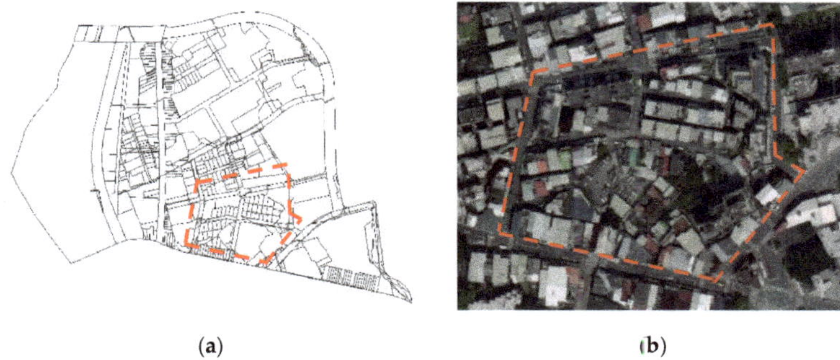

Figure 14. The test area of an urban cadastral detail survey. (**a**) Cadastral map. (**b**) Aerial image.

3.3.1. Ground Control Survey

The locations of four cadastral control points (red points) used in this study and two supplementary control points (blue points) are shown in Figure 15. Four cadastral control points were of the known announced control points in the TWD97 cadastral coordinate system, including point nos. 100005, NA0591, NA0657, and GA0477; and two supplementary control points, including point no. 1 and no. 2 were not. The cadastral coordinates of these two points were obtained by the process of least-squares collocation adjustment, as described in Section 2.2.3.

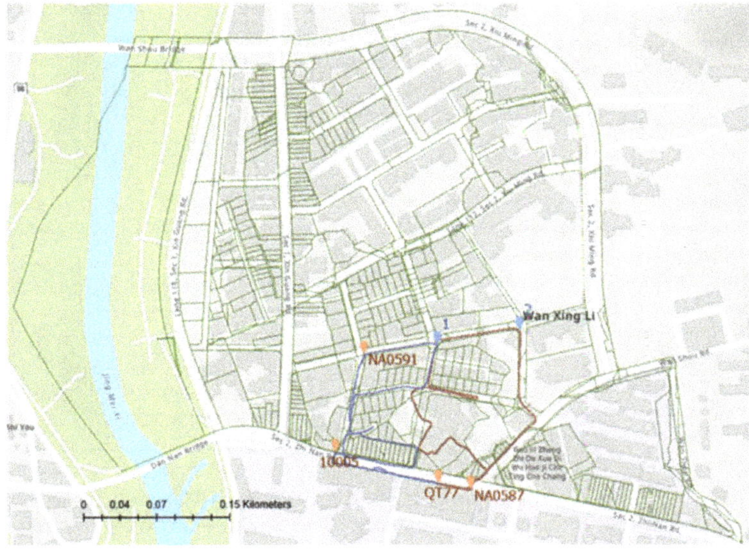

Figure 15. Two independent scanning operation areas and scanned paths.

Thee cadastral control point nos. QT77 and NA0587 were under the building eave, the ortho-height could not be surveyed by VBS-RTK after conversion. The ortho-height of these two points was surveyed by levelling from the known height point 10005, and their

N, E coordinates were adopted by the announced TWD97 cadastral coordinates. After the other six points were surveyed by VBS-RTK, the TWD97 cadastral control coordinates of point nos. 1 and 2 were obtained using least-squares collocation, as shown in Table 10. The ortho-height was converted by the geoid undulation. The final cadastral (N, E) coordinates and ortho-height of all control points are organized in Table 10.

Table 10. The cadastral TWD97 coordinates and ortho-height of used control points.

Point No.	N Coordinate (m)	E Coordinate (m)	Ortho-Height (m)
1	2,764,678.099	308,072.529	17.996
2	2,764,693.030	308,157.005	18.589
NA0591	2,764,668.076	307,997.099	18.491
100005	2,764,569.887	307,967.553	17.213
NA0657	2,764,853.892	308,221.181	19.417
GA0477	2,764,471.261	308,372.474	18.782
QT77	2,764,538.592	308,073.022	18.017
NA0587	2,764,532.553	308,106.479	17.991

3.3.2. Path Planning for Data Collection

After evaluating the area and scanning time of the test area, it was divided into two independent scanned paths. The area of these two scanned paths and the distribution of control points are shown in Figure 15. The scanned path 1 in red started from the control point no. QT77, passed through control point no. NA0587, 1, 2, and NA0587, then it closed at the starting point; the scanned path 2 in blue started from the control point no. QT77, passed through control point no. 100005, 1, NA0591, and closed at the starting point.

3.3.3. Point Cloud Filtering, System Error Correction, and Coordinate Conversion

In order to retain more point clouds for an urban cadastral detail survey to digitize the detail line data, the SLAM quality threshold value was changed to $R \leq 50$ and $G \leq 50$ and $B \geq 250$, and the point cloud data that was closer to the blue (good SLAM solution) was retained. The incident angle threshold was consistent with the calibration operation, set to $70°$. In addition, according to the manual, the scanning distance is recommended to be kept within 50 m. The calibration results in the previous section indicate that the point cloud larger than the calibration calculation range would tend to increase the error. Therefore, this study set the scanning distance threshold to 50 m. The conditions and thresholds of point cloud filtering for a land detail survey are summarized in Table 11.

Table 11. Conditions and thresholds of point cloud filtering for a cadastral detail survey.

Filtering Condition	Thresholds
SLAM quality	$R \leq 50$ & $G \leq 50$ & $B \geq 250$
Incident angle	$\leq 70°$
Scanning distance	≤ 50 m

By using the calculated ranging system error parameters, the pseudo-ranging measurement between the coordinates of each point and the corresponding laser center coordinates were corrected, see Equation (14), and the point coordinates after the correction of the ranging system error were recalculated based on Equation (7).

$$r_{icorrect} = r_i * 0.9996 - 0.0088 \tag{14}$$

The recalculated point cloud data of scanned paths 1 and 2 after filtering, merge, and conversion are shown in Figure 16. The control points used in scanned path 1 included four control point nos. QT77, NA0587, 1, and 2. The RMSE of the coordinate conversion was 0.0354 m, and the number of point clouds after point cloud filtering was 68118612 points, and the filtering ratio was about 40%. The control points used in the scanned path 2 included four control point nos. QT77, 100005, NA0591, and 1, the RMSE of coordinate conversion was 0.0309 m, the number of point clouds after point cloud filtering was 93961336 points, and the filtering ratio was about 31%. The points after filtering were used for an urban cadastral detail survey. The result will be demonstrated in the next subsection.

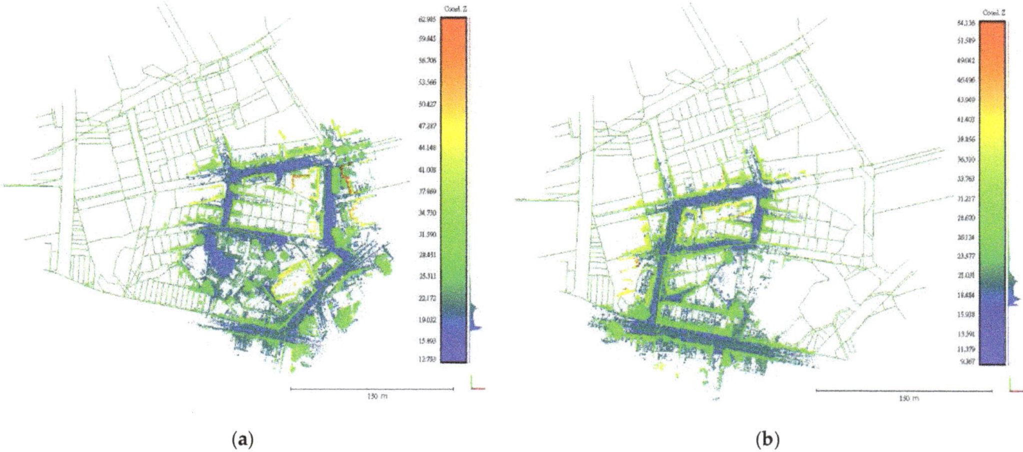

(**a**) (**b**)

Figure 16. Point cloud after filtering. (**a**) Point cloud after filtering for scanned path 1. (**b**) Point cloud after filtering for scanned path 2.

3.3.4. Urban Cadastral Detail Data Production

Detail Line Data by Manual Digitization

According to the principles described in Section 2.2.4 to digitize the cadastral detail lines, the types of possible cadastral detail lines in the test area of this study could be roughly divided into the boundary between townhouses and the existing road boundaries. The two types of possible cadastral detail lines are shown in Figures 17 and 18 after manual digitization.

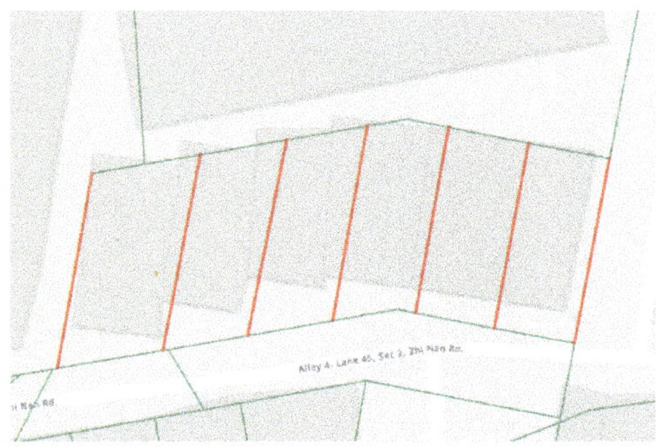

Figure 17. Possible cadastral detail line on the boundary between townhouses.

Figure 18. Possible cadastral detail line on an existing road boundary.

- Possible cadastral detail line on the boundary between townhouses

The possible cadastral detail line of the townhouses was mostly located at the center of the wall or the center of the stairwells. Taking the selected possible cadastral detail line in Figure 19a as an example, Figure 19b shows the overlapping results of the point cloud data and the cadastral map. As seen, the cadastral line was located at the center line of the stairwell of the townhouses and was roughly parallel to the plane on both sides of the stairwell. The point cloud data and the image of the townhouse are shown in Figure 20.

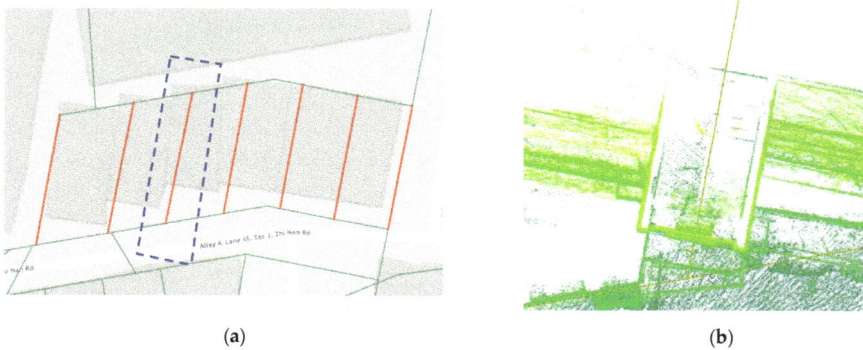

(**a**) (**b**)

Figure 19. An example of the possible cadastral detail line between townhouses. (**a**) The digitized possible cadastral detail lines. (**b**) Overlap of point cloud and the cadastral map.

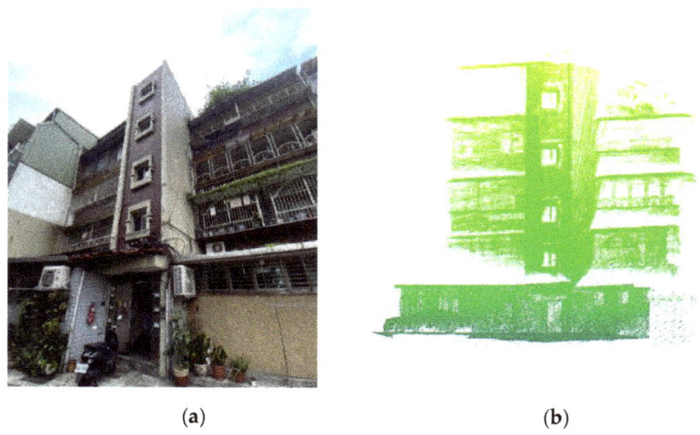

(**a**) (**b**)

Figure 20. The image of a stairwell and its corresponding point cloud. (**a**) Image of a stairwell. (**b**) Point cloud of a stairwell.

The digitized process of the possible cadastral detail line between townhouses is shown in Figure 21. The blue structure in the figure is the stairwell front façade. First, line AB was digitized according to the width of the stairwell, and a line CD from one of the two wall sides of the stairwell was also digitized. The line CD was translated to the midpoint of line AB to become the line C′D′ parallel to the line CD, the line C′D′ is the position of the possible cadastral detail line.

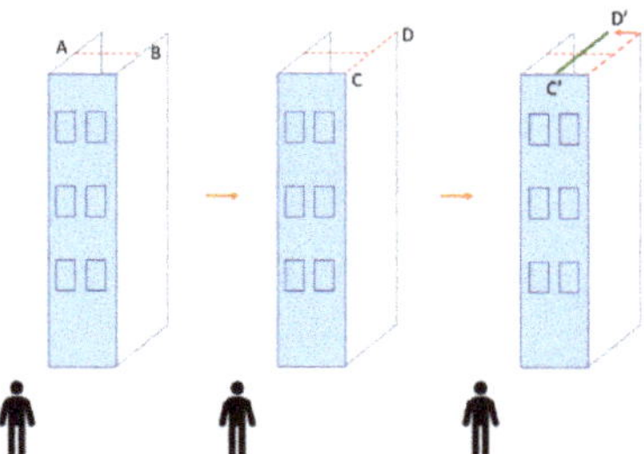

Figure 21. The digitized process of the possible cadastral detail line between townhouse.

If there were a gap between different townhouses, as shown in Figure 22, the midpoint of the gap space was digitized as the possible cadastral detail line. The digitization process was similar to the above, the line digitized at one side of the gap was translated to the midpoint of the gap width, and the translated line was regarded as a possible cadastral detail line.

Figure 22. The image of two different townhouses with a gap.

- Possible cadastral line on an existing road boundary

The possible cadastral detail line on an existing road boundary is as shown as the red line in Figure 23a. Figure 23b shows the overlapping results of the point cloud data and the cadastral map. The possible cadastral detail line was located at the junction of the outermost wall of the building area and the road, namely the boundary of an existing road boundary. The point cloud data and image of the wall are shown in Figure 24.

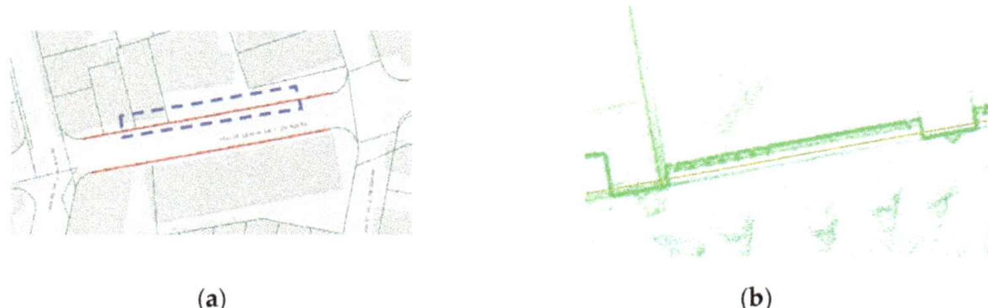

(**a**) (**b**)

Figure 23. Possible cadastral detail line on an existing road boundary. (**a**) The digitized possible cadastral lines. (**b**) Overlap of point cloud and the cadastral map.

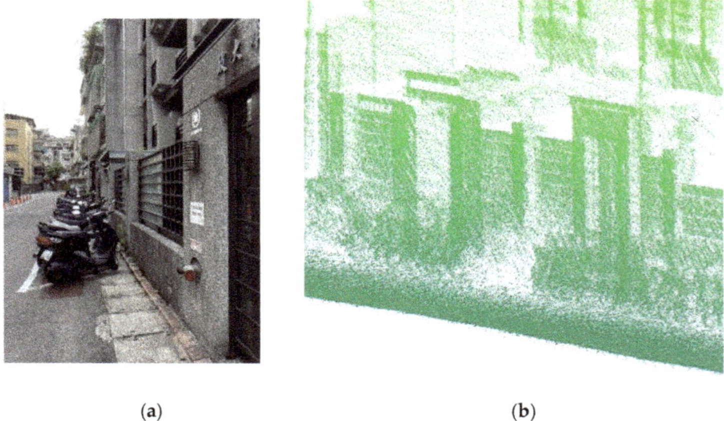

(**a**) (**b**)

Figure 24. The image of a road boundary and the corresponding point cloud. (**a**) Image of a road boundary. (**b**) Point cloud of a road boundary.

The digitization of a possible cadastral detail line on an existing road boundary is illustrated in Figure 25 where blue lines indicated the outermost boundary of the building top view. The outer boundary of the wall located on the road boundary was digitized, and then those line segments were connected as the possible cadastral detail line, as the line EF shown in Figure 25.

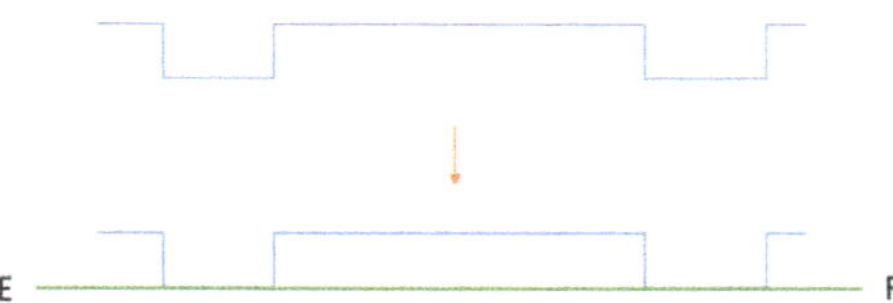

Figure 25. The digitized process of the possible cadastral line close to the road boundary.

According to the obtained hand-held LiDAR point cloud, the possible cadastral boundaries between the townhouses and on the existing roads were digitized manually and joined in AutoCAD software to produce the results of the urban cadastral detail lines of the test area. The final results are shown in Figure 26.

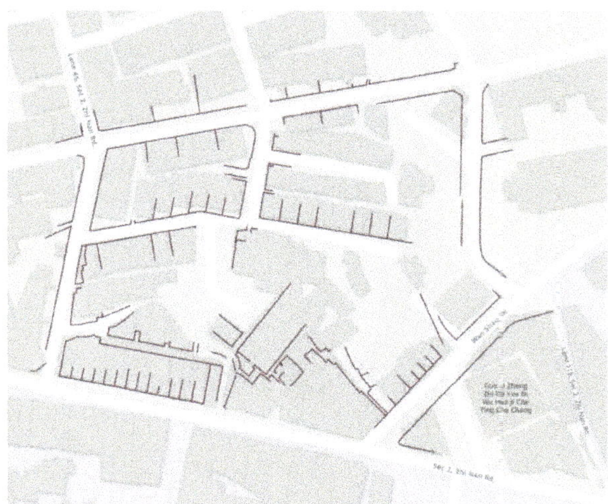

Figure 26. The result of digitized detail lines from the corrected handheld LiDAR point cloud.

3.3.5. Results Analysis

(1) Analysis of detail line data

The digitized data of the corrected hand-held LiDAR point cloud and the results survey by a total station are shown in Figure 27. The orange points were the detail points surveyed by a total station. By visual inspection, there was no significant difference between the digitized detail lines and detail points.

Figure 27. The digitized results of the corrected handheld LiDAR points and detail points surveyed by a total station.

This test area contained 10 detail points surveyed by a total station which could be used for planar position analyses (see Figure 3a) and 32 detail points surveyed by a total station which were used for vertical distance analyses (see Figure 3b). The calculated error value is shown in Figure 28. The maximum plane position error of the current point is about 8.3 cm, most of which is between 2 to 4 cm; the maximum error of the vertical distance between the digitized detail lines and the detail points was about 16.5 cm, and most of them were below 5 cm. In Taiwan, the maximum map scale in graphic digitized areas is 1/500, the accuracy is 15 cm on-site. Therefore, 5 cm was used for the first level for the analysis, then 5 cm to 10 cm was the second level, 10 cm to 15 cm was the third level, the other level was greater than 15 cm. Figure 29 shows the error statistics of the integration of the two error analyses based on the four levels.

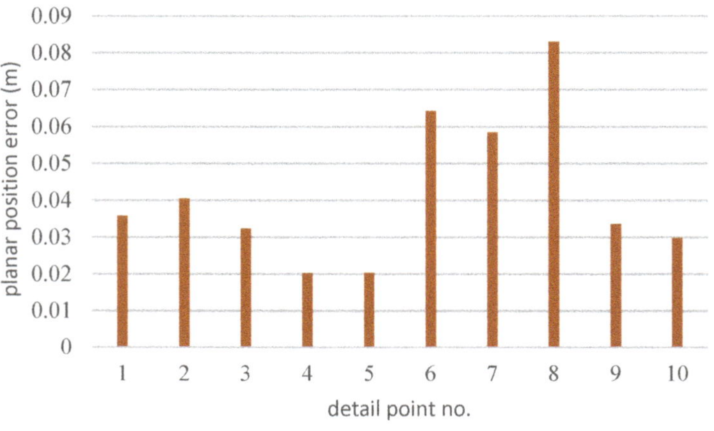

(a)

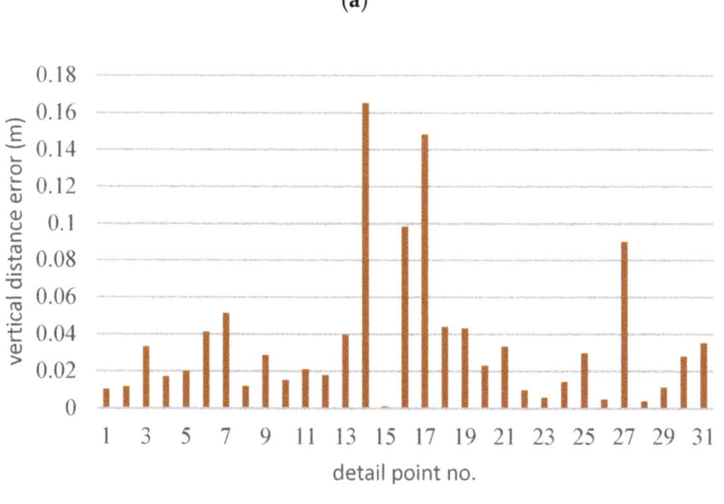

(b)

Figure 28. The error analysis of digitized detail line data. (**a**) Planar position analysis. (**b**) Vertical distance error analysis.

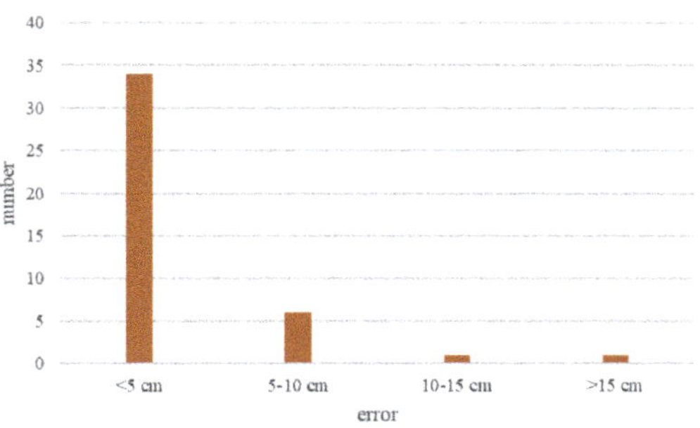

Figure 29. The statistics of two error analyses of the digitized line data.

The error of the digitized detail data using the point cloud after the error correction of the ranging system verified by the detail points surveyed by a total station, the average error was 3.44 cm. About 97.62% of the digitized detail line data was less than 15 cm. There was still one digitized detail error greater than 15 cm, but the selection of the detail point location obtained by a total station implied the subjective judgment of the surveyor.

In areas with more complex building types, the selection of the detail point location by the surveyor might be slightly different from the digitized location. The digitized detail line data was the centerline of the stairwell in the townhouses. The surveyor could only visually observe the approximate centerline position when surveying by a total station instrument. Compared with the corner points of the townhouses, the detail point surveyed by a total station was more susceptible to the influence of subjective judgment of the surveyor. Therefore, these test results verified that digitized detail line data using the corrected hand-held LiDAR point cloud was sufficient for the detail line data production in the cadastral graphic digitized area. With the advantages of its fast scanning speed and mobile mapping, it could scan a larger amount of detail point cloud on the ground in a short time, and the required detail line data could be digitized for subsequent cadastral overlap analysis.

(2) Analysis of the effect of ranging system error correction.

Based on the description in Section 2.2.4, the error difference of digitized detail lines using the handheld LiDAR point clouds after and before ranging system error correction is shown in Figure 30. Among 42 detail points, the error of 25 detail points, compared with the digitized line data using uncorrected point clouds, was greater than the digitized line data using the corrected point cloud. The positive values shown in Figure 30 indicate that the accuracy was improved. The error was reduced by a maximum of 2.7 cm. The error of 17 detail points compared with the digitized line data using uncorrected point clouds was less than the digitized line data using corrected point cloud. The negative values shown in Figure 30 indicate that the accuracy was not improved.

Regardless of the error difference within 5 mm, about 73% error of the digitized detail line data was reduced from the detail line data digitized using the uncorrected collected handheld LiDAR points. Although 27% error of the digitized detail line data was still higher than the digitized detail line data using the uncorrected collected handheld LiDAR points, only less than 5% of the digitized detail data was of an error difference greater than 2 cm. This suggests that the correction of the ranging system error could improve the accuracy of most of the digitized detail data. An error difference greater than 2 cm was not only affected by the quality of the trajectory solution of the handheld LiDAR but also affected by the detail points from the subjective judgment of the surveyor.

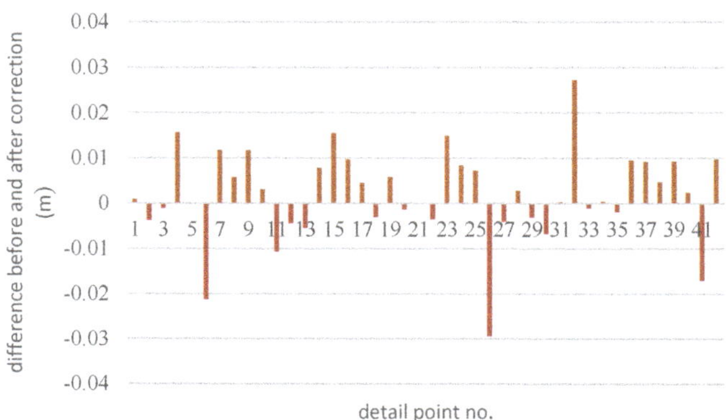

Figure 30. Error difference of digitized detail line data using the corrected and uncorrected handheld LiDAR point clouds.

4. Conclusions

In this study, the feasibility of using a hand-held LiDAR scanner for the urban cadastral detail survey was studied. Before performing the urban cadastral detail survey by the handheld LiDAR scanner, named the GeoSLAM Zeb-Horizon scanner, the scanner calibration was conducted by using the ground LiDAR scanner to collect the planar point cloud in the selected indoor calibration field for calibration planes to calculate the planar parameters for calibration. The ranging system error parameters, including the range scale factor (S) and the rangefinder offset (C), of the VLP-16 multi-beam sensor carried by the GeoSLAM Zeb-Horizon handheld LiDAR scanner were determined by the plane-based calibration method proposed in our previous study [16].

After calibration, the distribution of residuals was more concentrated at 0 and the residual distribution was more in line with the normal distribution curve. The average residual was much closer to 0, and the posterior unit weight standard deviation became smaller, both of which were improved compared to those without adding the ranging system error parameter into the adjustment. Therefore, the plane-based dynamic calibration method proposed in our previous study [16] used in this study could eliminate most of the ranging system errors of the GeoSLAM Zeb-Horizon handheld LiDAR scanner.

From the analysis of the RMSE results of the check planes, the RMSE of all the check planes was improved after the correction of the ranging system error for calibration data. Up to 72.12% in one plane, an increase of about 2.4 cm was reached. The overall average improvement was 32.61%. From the improvement of the RMSE of the check planes, it demonstrated again that the proposed calibration approach could effectively improve the overall point cloud accuracy of the GeoSLAM Zeb-Horizon handheld scanner.

From the investigation of the correlation between the additional ranging parameters S and C, the negative correlation between the ranging additional parameters S and C was −0.82. The lower negative correlation between the ranging additional parameters makes the solution results of the ranging additional parameters S and C more reliable. Meanwhile, the calibration data with about 40 ~45 m longer pseudo-calculated ranging measurements for calibration, the calibration data used in our previous study [16] with about only 20 m calculated pseudo-ranging measurements for calibration. Therefore, the negative correction in [16] was extremely high to −0.985. However, −0.82 was also high, so if a larger calibration site or a suitable scanning could be planned to collect the handheld LiDAR points with longer calculated pseudo-ranging measurements for calibration in the future, the negative correlation would expect to be reduced.

For the analysis of ranging systematic error parameters S and C, this study concluded that scanning by a handheld LiDAR scanner, even if it was 2 m, the correction would be 1 cm for the ranging measurement. When using a handheld LiDAR scanner for precise surveying, for example, cadastral surveying, the errors of this ranging system should be corrected to obtain more accurate results.

For the urban cadastral detail survey test, two independent scannings were performed in the test area. Each scanning task took about 15 min and each scanned path was all closed paths to ensure the accuracy of the trajectory calculations. Those scanning points were corrected by the calibration ranging system error and they were used to manually digitize the urban detail line data. According to the test results, the use of a handheld LiDAR scanner could collect the 3D detail point clouds more globally and is easy to use in narrow lanes where it is not easy for GNSS to receive the satellite signals and a total station or a ground LiDAR scanner is difficult to set up.

Using the detail points surveyed by a total station to verify the detail line data digitized from the corrected handheld LiDAR point cloud, 97% error of the digitized detail data was less than 15 cm. It demonstrated the digitized detail data was sufficient for the urban cadastral detail survey in the cadastral graphic digitization area in Taiwan.

Compared with the digitized detail data from uncorrected handheld LiDAR points, regardless of the 5 mm difference between the digitized detail line data before and after the system error correction, about 73% error of the digitized detail line data were reduced against the detail line data digitized using the uncorrected collected handheld LiDAR points. Although 27% error of the digitized detail line data was still higher than the digitized detail line data using the uncorrected collected handheld LiDAR points, only less than 5% of the digitized detail data was of a difference greater than 2 cm, indicating that the correction of the ranging system error could improve the accuracy of most of the digitized detail data.

The results demonstrated the feasibility of using a handheld LiDAR scanner to perform an urban cadastral detail survey in digitized graphic areas. Therefore, the handheld LiDAR scanner could be used for the production of the detail lines for an urban cadastral detail survey for digitized cadastral areas in Taiwan. In the future, it is possible that it could also be used as a handheld scanner to create a 3D cadaster for land management.

Author Contributions: S.-H.C. conceived, designed the experiments; S.-H.C. and K.-W.H. K.-W.H. simultaneously performed the test, analyzed the data, and wrote the paper. All authors have read and agreed to the published version of the manuscript.

Funding: This research received no external funding.

Acknowledgments: This study was financially supported under the project number of the Ministry of Science and Technology, Taiwan: MOST 109-2221-E-004-001. A special thanks to the Ministry of Science and Technology, Taiwan.

Conflicts of Interest: The founding sponsors had no role in the design of the study, in the collection, analyses, or interpretation of data, in the writing of the manuscript, or in the decision to publish the results.

References

1. Wu, T.P.; Kao, S.P.; Ning, F.S. Study on Computer Registering for Land Revision of Digitalized Graphic Maps. *J. Cadastr. Surv.* **2003**, *22*, 1–23. (In Chinese)
2. Chio, S.-H.; Chiang, C.-C. Feasibility Study Using UAV Aerial Photogrammetry for a Boundary Verification Survey of a Digitalized Cadastral Area in an Urban City of Taiwan. *Remote Sens.* **2020**, *12*, 1682. [CrossRef]
3. Park, J.; Lee, K.-W. Analysis of the Status of Mine and Methods of Mine Geospatial Information Construction Technology for Systematic Mine Management. *J. Korea Acad.-Ind. Coop. Soc.* **2018**, *19*, 355–361.
4. Bosse, M.; Zlot, R. Continuous 3D scan-matching with a spinning 2D laser. In Proceedings of the 2009 IEEE International Conference on Robotics and Automation, Kobe, Japan, 12–17 May 2009; pp. 4312–4319. [CrossRef]
5. Bosse, M.; Zlot, R.; Flick, P. Zebedee: Design of a spring-mounted 3-D range sensor with application to mobile mapping. *IEEE Trans. Robot.* **2012**, *28*, 1104–1119. [CrossRef]
6. Chen, S.; Liu, H.; Feng, Z.; Shen, C.; Chen, P. Applicability of personal laser scanning in forestry inventory. *PLoS ONE* **2019**, *14*, e0211392. [CrossRef] [PubMed]
7. Mudicka, S.; Matolak, M.; Kapica, R.P. Application of handheld scanner in documentation of historical buildings. *Sofia Surv. Geol. Min. Ecol. Manag. (SGEM)* **2019**, *10*, 39–46.

8. Bauwens, S.; Bartholomeus, H.; Calders, K.; Lejeune, P. Forest Inventory with Terrestrial LiDAR: A Comparison of Static and Hand-Held Mobile Laser Scanning. *Forests* **2016**, *7*, 127. [CrossRef]
9. Park, J.K.; Um, D.Y. Application of Handheld Scanner to Investigate Diameter at Breast Height and Tree Height. *Int. J. Mob. Device Eng.* **2019**, *3*, 1. [CrossRef]
10. Hunčaga, M.; Chudá, J.; Tomaštík, J.; Slámová, M.; Koreň, M.; Chudý, F. The Comparison of Stem Curve Accuracy Determined from Point Clouds Acquired by Different Terrestrial Remote Sensing Methods. *Remote Sens.* **2020**, *12*, 2739. [CrossRef]
11. Dewez, T.; Plat, E.; Degas, M.; Richard, T.; Pannet, P.; Thuon, Y.; Meire, B.; Watelet, J.-M.; Cauvin, L.; Lucas, J. Handheld Mobile Laser Scanners Zeb-1 and Zeb-Revo to map an underground quarry and its above-ground surroundings. In Proceedings of the 2nd Virtual Geosciences Conference: VGC, Bergen, Norway, 21–23 September 2016.
12. Chiabrando, F.; Sammartano, G.; Spanò, A. A Comparison among Different Optimization Levels in 3D Multi-Sensor Models: A Test Case in Emergency Context: 2016. *Int. Arch. Photogramm. Remote Sens. Spat. Inf. Sci.* **2017**, *XLII-2/W3*, 155–162. [CrossRef]
13. Makkonen, T.; Heikkilä, R.; Kaaranka, A.; Naatsaari, M. The Applicability of the Rapid Handheld Laser Scanner to Underground Tunnel Surveying. In Proceedings of the 32nd International Symposium on Automation and Robotics in Construction, Oulu, Finland, 15–18 June 2015.
14. James, M.R.; Quinton, J.N. Ultra-rapid topographic surveying for complex environments: The hand-held mobile laser scanner (HMLS). *Earth Surf. Process. Landf.* **2014**, *39*, 138–142. [CrossRef]
15. Sršan, A. Mogućnosti Primjene GeoSLAM Tehnologije za Izmjeru Zatvorenih Prostora. Undergraduate Thesis, Polytechnic of Međimurje in Čakovec, Čakovec, Croatia, 2019. Available online: https://urn.nsk.hr/urn:nbn:hr:110:324934 (accessed on 2 December 2021).
16. Hou, K.-W.; Chio, S.-H. Plane-based range calibration method for geoslam zeb-horizon handheld lidar instrument. In Proceedings of the International Symposium on Remote Sensing (ISRS) (ISRS 2021), Virtual Conference, 26–28 May 2021.
17. Glennie, C.; Lichti, D.D. Static Calibration and Analysis of the Velodyne HDL-64E S2 for High Accuracy Mobile Scanning. *Remote Sens.* **2010**, *2*, 1610–1624. [CrossRef]
18. Glennie, C. Calibration and Kinematic Analysis of the Velodyne HDL-64E S2 Lidar Sensor. *Photogramm. Eng. Remote Sens.* **2012**, *78*, 339–347. [CrossRef]
19. Fischler, M.A.; Bolles, R.C. Random sample consensus: A paradigm for model fitting with application to image analysis and automated cartography. *Commun. ACM* **1981**, *24*, 381–395. [CrossRef]
20. Chan, T.; Lichti, D.D.; Roesler, G.; Cosandier, D.; Durgham, K. Range scale-factor calibration of the velodyne vlp-16 lidar system for position tracking applications. In Proceedings of the 11th International Conference on Mobile Mapping, Shenzhen, China, 6–8 May 2019; Volume 70–77, pp. 350–355.
21. Moritz, H. *Advanced Least-Squares Methods*; Reports of the Department of Geodetic Science, Report No. 175; Ohio State University Research Foundation: Columbus, OH, USA, 1972.
22. Ruffhead, A. An introduction to least-squares collocation. *Surv. Rev.* **1987**, *29*, 85–94. [CrossRef]
23. Jana, Š. Testování Přístroje GeoSLAM ZEB-REVO RT. Master's Thesis, Czech Technical University in Prague, Prague, Czech Republic, 2020.
24. Maboudi, M.; Bánhidi, D.; Gerke, M. Evaluation of indoor mobile mapping systems. In Proceedings of the GFaI Workshop 3D North East, Berlin, Germany, 7–8 December 2017.
25. Russhakim, N.; Mohd Ariff, M.F.; Darwin, N.; Majid, Z.M.; Idris, K.; Abbas, M.; Zainuddin, K.; Yusoff, A. The suitability of terrestrial laser scanning for strata building. *ISPRS—Int. Arch. Photogramm. Remote Sens. Spat. Inf. Sci.* **2018**, *42*, 67–76. [CrossRef]

Article

Using LiDAR System as a Data Source for Agricultural Land Boundaries

Natalia Borowiec * and Urszula Marmol

Department of Photogrammetry Remote Sensing of Environment and Spatial Engineering, Faculty of Mining Surveying and Environmental Engineering, AGH University of Science and Technology, Al. Mickiewicza 30, 30-059 Krakow, Poland; entice@agh.edu.pl

* Correspondence: nboro@agh.edu.pl

Abstract: In this study, LiDAR sensor data were used to identify agricultural land boundaries. This is a remote sensing method using a pulsating laser directed toward the ground. This study focuses on accurately determining the edges of parcels using only the point cloud, which is an original approach because the point cloud is a scattered set, which may complicate finding those points that define the course of a straight line defining the parcel boundary. The innovation of the approach is the fact that no data from other sources are supported. At the same time, a unique contribution of the research is the attempt to automate the complex process of detecting the edges of parcels. The first step was to classify the data, using intensity, and define land use boundaries. Two approaches were decided, for two test fields. The first test field was a rectangular shaped parcel of land. In this approach, pixels describing each edge of the plot separately were automatically grouped into four parts. The edge description was determined using principal component analysis. The second test area was the inner subdivision plot. Here, the Hough Transform was used to emerge the edges. Obtained boundaries, both for the first and the second test area, were compared with the boundaries from the Polish land registry database. Performed analyses show that proposed algorithms can define the correct course of land use boundaries. Analyses were conducted for the purpose of control in the system of direct payments for agriculture (Integrated Administration Control System—IACS). The aim of the control is to establish the borders and areas of croplands and to verify the declared group of crops on a given cadastral parcel. The proposed algorithm—based solely on free LiDAR data—allowed the detection of inconsistencies in farmers' declarations. These mainly concerned areas of field roads that were misclassified by farmers as subsidized land, when in fact they should be excluded from subsidies. This is visible in both test areas with areas belonging to field roads with an average width of 1.26 and 3.01 m for test area no. 1 and 1.31, 1.15, 1.88, and 2.36 m for test area no. 2 were wrongly classified as subsidized by farmers.

Keywords: LiDAR system; segmentation; edge detection; agricultural land boundary

Citation: Borowiec, N.; Marmol, U. Using LiDAR System as a Data Source for Agricultural Land Boundaries. *Remote Sens.* **2022**, *14*, 1048. https://doi.org/10.3390/rs14041048

Academic Editor: Pinliang Dong

Received: 27 October 2021
Accepted: 17 February 2022
Published: 21 February 2022

1. Introduction

Technological progress and development methods of processing spatial data have popularized the use and increased the availability of various products presenting information about the area. Laser scanning became a dynamically developing technology at the turn of the 21st century and is finding an increasing number of applications in various fields of science. Data from an airborne laser scanning system (ALS) significantly facilitate and accelerate the collection of information about the topography and terrain, which leads to both cooperation and competition with the technology and products of classical photogrammetry and geodesy. Currently, the construction of digital terrain models (DTMs) and digital surface models (DSMs) using LiDAR data is common, but the identification of agricultural land boundaries based only on point clouds is a complex issue.

Light detection and ranging (LiDAR) is an active remote sensing system that first generates a laser pulse and then records the energy reflected from a given surface. Knowledge of the time of signal generation and the moment of its reception, as well as the properties of the generated light wave can be used to determine the distance to the object. Airborne laser scanning, which is performed using a flying plane or helicopter, works according to this principle. The system uses two main components: a laser scanner, which collects information about the distance between the scanner and a point on the ground surface, and a combination of Global Positioning System (GPS) and the inertial navigation system (INS), whose task is to measure the position and orientation of the system. As a result, data are acquired in the form of a point cloud [1].

The point cloud is not the final product. It is a set of data (points), defined by spatial coordinates, with stored information about intensity, RGB color, and echo. This representation reveals a wealth of information, and when processed into numerical models, it enables subsequent applications.

Airborne laser scanning and its products in the form of point clouds and digital terrain models and digital surface models are increasingly used. Precise terrain models contain a large amount of detailed information. Compared to photogrammetry, they enable the study of terrain overshadowed by vegetation. Therefore, this study focuses on the possibility of using LiDAR sensor to identify places of land use changes—agricultural boundaries.

Identification of the course of agricultural boundaries is important to control in the direct agricultural subsidies system (Integrated Administration Control System—IACS) [2]. The Land Parcel Identification System (LPIS), which is a part of IACS, is a system supporting direct subsidies to farmers, which depend on the area of crops. Farms with a minimum agricultural area of at least 1.0 ha, consisting of agricultural parcels of at least 0.1 ha, qualify for direct payments. The procedure is based on the farmer filling in a declaration, which involves specifying the area of crops intended for payments. The purpose of the control is to determine whether the submitted declaration is correct, i.e., whether the land declared by the farmer is indeed eligible for subsidies. Discrepancies are evidence of irregularities in the declarations, which should be corrected by the farmers. So far control of applications within the framework of direct payments for land is carried out by two methods, i.e., field inspection, most often carried out with the use of GPS technology, and the so-called "photo" method, based mainly on high-resolution satellite images or aerial images.

This study attempts to answer the question of whether ALS data can be used as a basis for inspecting agricultural land boundaries. LiDAR data are a powerful source of spatial information that includes not only coordinates but also intensity, return numbers, and point cloud classification data. Due to the increasing density of acquired data, ALS is increasingly used in new fields.

The purpose and novelty of this study was the attempt to automate the detection of agricultural land edges by using only LiDAR data in the analysis. The innovation of the method is the use of only airborne laser scanning data to indicate the course of agricultural land boundaries. Determination of the agricultural land boundary is important in the process of checking and updating the reference databases of the Land Parcel Identification System. The proposed algorithm—based only on free LiDAR data—was able to detect inconsistencies in farmers' declarations.

2. Literature Review

The data coming from the LiDAR sensor are characterized by high measurement accuracy. The resulting digital terrain models and digital surface models depict the surrounding reality in detail. These features determine the multidirectional use of airborne laser scanning in various fields of science, such as engineering solutions—calculation 3D displacements of bridges [3], 3D object detection along the road [4,5], building extraction [6,7], land cover change detection, and forest succession monitoring [8,9] for heterogeneous land use urban mapping [10], coastal monitoring [11,12], or archeological research [13,14].

The subject of the LPIS is widely discussed in many publications. Among other things, researchers compare the LPIS to the national cadastre. Reference [15] evaluated the extent to which the reference data from the cadastral register are modified in the

LPIS in Poland. Reference [16], using the LPIS of the Republic of Ireland, demonstrated significant differences in cropland/grassland reporting between an inter-annual based reporting schema and a land use history approach. Research has also been conducted on a data model for the collaboration between land administration systems and LPIS [17]. Study [18] focuses on a conceptual model of a large Turkish rural SDI design that combines the sensor usage and attribute datasets for all types of rural lands. In India, a government program used high-resolution aerial and satellite orthophotomaps, Global Positioning System, and electronic total stations (ETSs) to create and update land cadastres in a short time [19]. Reference [20] also analyzes the quality characteristics of orthoimages for visual identification of agricultural fields. Reference [21] presents a field boundary detection technique based on deep learning and a variety of image features which was combined with the graph-based growing contours (GGCs) method to extract agricultural fields in a study area in Northern Germany.

Airborne laser scanning is increasingly used in research to identify the type of land cover [22]. Reference [23] evaluated the use of high-resolution LiDAR for classification of native and tame grasslands and compared these classifications to the best available landcover mapping product that is currently available for this area. Reference [24] evaluated the effectiveness of integrating LiDAR data with high spatial resolution near-infrared digital imagery for object-based classification of land cover types and dominant tree species, using decision tree analysis. In [25] the authors proposed a process for objective and automated identification of agricultural parcel features based on processing and combining Sentinel-2 data (to sense different types of irrigation patterns) and LiDAR data (to detect landscape elements). Another example of combination cadastral data and remote sensing is in article [26] where high-resolution multi-spectral WorldView-2 satellite images were used with the object-oriented approach to image classification and image classification algorithm creation. The main objective is to compare the results obtained with the traditional methods of cadastral land evaluation and the results obtained by the methods of remote sensing. Analyses were performed for an area of Butmir Municipality in Sarajevo. Remote sensing data such as Sentinel-1 radar images were also used for mapping the different crops in the Camargue region in Southern France. In this study, deep machine learning was used to perform land classification [27]. Reference [28] proposed a geographic object-based image analysis approach to enable semiautomatic land classification and mapping using LiDAR elevation and intensity data. In [29], the authors used a hybrid capsule network for land cover classification using multispectral light detection and ranging data. Reference [30] focused on the extraction of uncultivable trails, ditches, and cultivated field parcels within farmland on the basis of a LIDAR high-resolution gridded DEM. Reference [31] discussed the impact that the quality of the digital elevation model has on the final result of landslide susceptibility modeling. The landslide map was developed on the basis of the analysis of archival geological maps and the light detection and ranging digital elevation model.

3. Specification of Test Data

The test area is located in Lubelskie Voivodeship in the village of Zimno and includes only the agricultural lands on which this study focuses. The test area was characterized by varied terrain. The laser data used in the study were taken from the ISOK project [32]. ISOK is a Polish information security system created with the aim of improving the protection of the economy, environment, and society against extraordinary threats, in the first place against floods. The basic specifications of the ISOK project are: 12 points per square meter, average distance between the points is 0.3 m, overlap between scans $\geq$20%, scan cross angle $\leq\pm25^{\circ}$, min scan lane width $\geq$100 m, and laser beam diameter $\leq$0.5 m. The obtained data are in the PL-1992 situation system and PL-KRON86-NH elevation system, with additional information, i.e., intensity and min 4 echo. In the study area, in accordance with the Regulation of the Minister of Regional Development and Construction on land and building cadastres [33], the following agricultural land was distinguished: arable crop land, marked with the symbol R; permanent grassland, marked by the symbol Ł; and permanent pastures, marked with the symbol Ps. In the selected area, the agricultural land should correspond to the cadastral parcels. This would make it possible to check whether there is any anomalous information in the area, indicating a discrepancy between the declarations in the payment system. Two study samples were selected for analysis in order to determine the actual land use status. A visualization of the area is shown in Figure 1.

Figure 1. The figure shows a point cloud acquired from the airborne laser scanning system. The cloud is displayed in plan view. (**a**) displays natural RGB colors, (**b**) the displayed point cloud uses the intensity, (**c**) the color corresponds to the height of the point, that is the Z coordinates.

4. Scheme of the Proposed Algorithm

The proposed algorithm consists of two stages. In the first stage, a point cloud is interpolated into a regular grid with a resolution of 0.3 m. Nearest neighbor interpolation was used. Two algorithms are used with the obtained rasters: edge detection and segmentation, on the basis of which the approximate location of agricultural land boundaries is estimated. In the next stage, we returned to the raw point cloud for pre-selected areas from the first part. Two methods were used for the original data: PCA and Hough Transform, which allowed for precise determination of agricultural land boundaries. The scheme of the algorithm is shown in Figure 2.

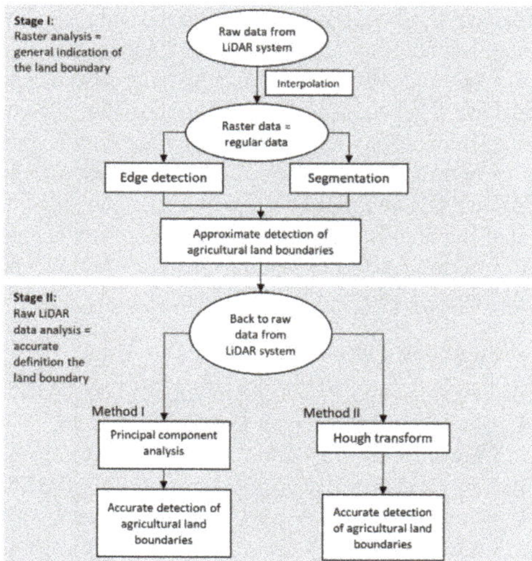

Figure 2. The scheme of the proposed algorithm.

5. Detection of Agricultural Land Boundaries from Raster Data

As a first step, an initial detection of agricultural land was undertaken using a point cloud stored as a raster. For this purpose, rasters with a resolution of 0.3 m were generated—images of intensity, classification, height differences, and RGB values.

An edge detection and segmentation process were performed in the MATLAB environment.

5.1. Edge Detection

Edge detection was carried out using the combined operators of Prewitt and Canny [34], based on intensity and classification rasters. It was decided to combine the two methods in order to obtain a more reliable agricultural land boundary. The Prewitt operator, based on gradients, more accurately determined the edges, but the information about them was point-wise. The Canny operator, on the other hand, introduced linear information but detected undesirable ploughing traces in the fields. Ploughing traces were recorded as lines, which made it difficult to eliminate them in subsequent steps. In the next step, simple morphological operators (erosion and dilation) were applied to remove noise in the image. The performed work made it possible to eliminate many errors and better illustrate land use boundaries in the analyzed area. However, errors are still visible. The result of this stage is presented in Figure 3.

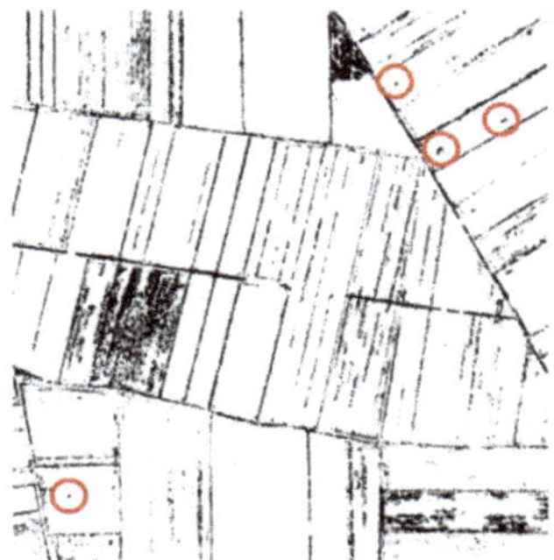

Figure 3. Edge detection of agricultural lands along with the visible errors—noise. The selected noise is marked with red circles.

5.2. Segmentation

Segmentation was used as a further supporting step. The aim was to roughly determine the areas of individual agricultural lands. Segmentation was carried out on the intensity image.

It was decided to use the multi-resolution algorithm (MRS). This is one of the most widely used segmentation models. It is based on the minimalization of the average heterogeneity in a single object extracted from an image [35]. This algorithm was chosen because a literature review suggested that this segmentation method would be more accurate [36,37].

The first key parameter of the algorithm is scale, i.e., the size of objects (segments) in the analyzed area. In this study, the value of 200 is used. In addition to the scale parameter, two additional parameters are required: shape and compactness [38]. The best results were obtained for values of 0.9 and 0.1, respectively. The segmentation result is shown in Figure 4.

The performed work allowed for a preliminary identification of the areas of agricultural fields, as well as their boundaries. The effect of the work is illustrated in Figure 5.

As can be seen from Figure 5, noise—incorrect edges in object detection—is still present. The integration of edges and segments can be used to identify some of them. Incorrect edges, related to ploughing traces in the fields, were located by correct segments in this area—this mainly concerns three parcels in the central-western part of the study and one parcel in the south-eastern part. Furthermore, in a few parts the segmentation did not work properly. This was due to the similarity of intensities in the neighboring parcels (see Figure 1b). The different edges in these areas indicate the locations of possible errors (mainly the central and central-southern part of the study). The stage described in this subsection is considered as a prelude to further work aimed at more accurate identification of agricultural land boundaries.

Figure 4. Segmentation of agricultural lands in the study area.

Figure 5. Overlaying of the segmentation and edge detection steps.

6. Accurate Detection of Agricultural Land Boundaries

After identification of initial regions of agricultural land boundaries, there was a return to the original point cloud, but this was limited to the indicated sub-areas defining the course of agricultural land boundaries. It was decided to return to distributed data because there are several airborne laser scanning points per image pixel, which will significantly increase the accuracy of the edge location (Figure 6). Therefore, the original data increase the amount of information that can define the agricultural land with higher accuracy.

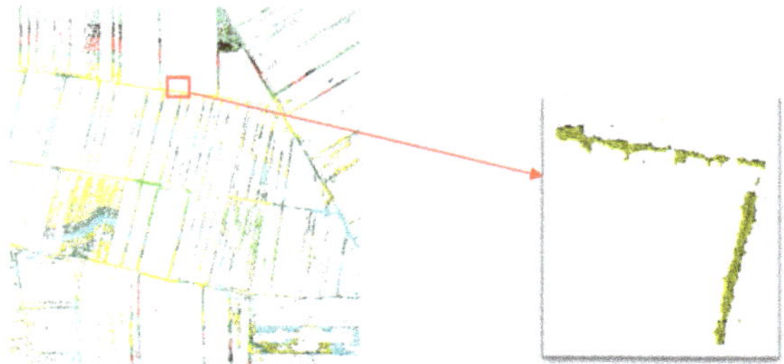

Figure 6. Scattered points superimposed on a raster representing parcel boundaries.

In this study, precise detection of agricultural boundaries was presented for two test areas (segments). In the first area, the segment boundaries coincided entirely with a single area of agricultural land. In contrast, the second selected segment included additional agricultural boundaries within its area. Therefore, two approaches for precise identification of agricultural boundaries with the use of scattered data (irregular point cloud) were applied in this study. In the first one, a solution based on principal component analysis was used to determine land parcel boundaries. On the other hand, the second approach used Hough Transform.

6.1. Boundary Detection Using Principal Component Analysis—Test Area No. 1

The first test field is an area of agricultural use whose extent overlaps with the rough edges delineated in stage 1 (Figure 7).

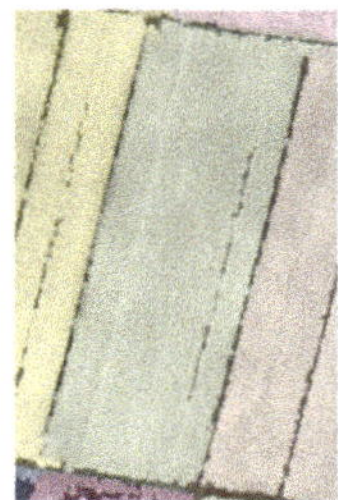

Figure 7. Test area no. 1—segment, coinciding with approximate edges.

Analyses began by indicating the test area (segment) from the segmentation image (Figure 8a), then a product was performed with the raster depicting the edges of the agricultural (Figure 8b), resulting in an image showing only the edges of the segment (Figure 8c).

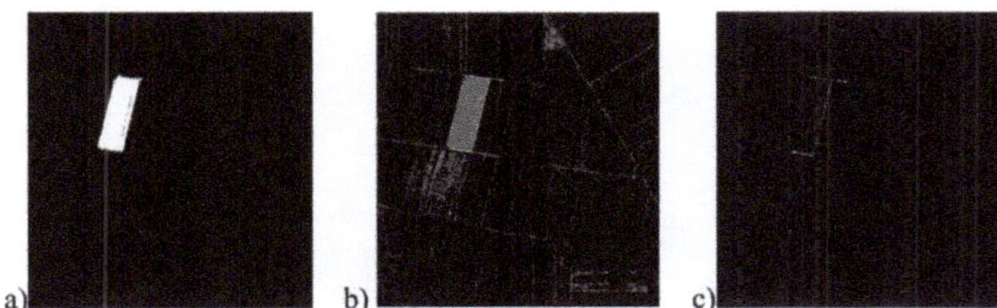

Figure 8. The sequence steps (**a**–**c**) leading to a raster edge representation of one segment.

The next stage leading to a correct determination of the edges was grouping the set of points into four parts containing points describing each parcel edge separately. For this purpose, a solution based on principal component analysis (PCA) was used. This solution allows us to determine the scatter model of the dataset in the form of a probability ellipse. Each pixel in the image is defined by two variables, the X variable and the Y variable (column number, row number). The ellipse is plotted based on the assumption that the given two variables follow a two-dimensional normal distribution (Gaussian distribution). The orientation of ellipse depends on the sign of correlation coefficient between variables, the size of the ellipse is determined by the intervals and its center is defined by the averages of the variables X and Y. The term intervals refers to the root of the eigenvalues multiplied by the user-selected value. Eigenvalues can be interpreted as proportions of the variance explained by correlations between relevant variables. Therefore, in solving this task, the values of variance and covariance were calculated for the variables X and Y to build the designated perimeter of the object.

The value of variance, which determines the diversity of the community, is equal to the sum of the arithmetic mean of squares of deviations of individual feature values from the arithmetic mean of the community. The unconstrained variance estimator for X and Y coordinates was calculated from the following formulas [39]:

$$s_x{}^2 = \frac{\sum_{i=1}^{n} (x - \overline{x_i})^2}{n-1} s_y{}^2 = \frac{\sum_{i=1}^{n} (y - \overline{y_i})^2}{n-1}, \quad (1)$$

S_x^2/S_y^2—the variance for X/Y calculated from the sample—unbiased variance estimator;

cov(X,Y)—covariance of a variables X, Y set;

x/y—sample mean value for X/Y;

$\overline{x}_i, \overline{y}_i/$—value of the X/Y variable for the i-th point

n—sample size.

The covariance value, which determines the linear relationship between the random variables X and Y, can be calculated from the following formula:

$$\text{cov}(X, Y) = \frac{1}{n} \sum_{i=1}^{n} (x - \overline{x_i}) \cdot (y - \overline{y_i}), \quad (2)$$

As a result of the calculations performed, the variance–covariance matrix can be constructed $\begin{bmatrix} S_x^2 & \text{cov}(X, Y) \\ \text{cov}(X, Y) & S_y^2 \end{bmatrix}$ for X, Y coordinates. From the obtained matrix, the eigenvalues of the matrix can be calculated.

The obtained variance–covariance matrix and eigenvalues were used to calculate the ellipse parameters, i.e., orientation and size of the ellipse according to Hausbrandt's formulas [40].

$$\Omega = \frac{\arctan\left(\frac{2 \cdot \text{cov}(X,Y)}{s_x^2 - s_y^2}\right)}{2}, \quad (3)$$

Ω—the omega angle between the horizontal line and the direction of the eigenvector with the larger eigenvalue;

S_x^2/S_y^2—the variance for X/Y calculated from the sample—unbiased variance estimator;

cov(X,Y)—covariance of a variables X, Y set.

The lengths of the minor (a) and major (b) half-axes of the confidence area ellipse were calculated from the formulas below:

$$\text{e}_1 = 2 \cdot \sqrt{\text{a}_1}, \ \text{e}_2 = 2 \cdot \sqrt{\text{a}_2}, \quad (4)$$

e_1—the length value for the minor half-axis of the ellipse;

e_2—the length value for the major half-axis of the ellipse;

a_1—minor eigenvalue of the test area object;

a_2—major eigenvalue of the test area object.

In the above formula, the square root of the eigenvalues was multiplied by 2, which means that 95.5% of the feature values lie at a distance ≤ 2 from the expected value. The center of gravity was calculated as the arithmetic mean of the X, Y coordinates, which determined the midpoint of the ellipse. The obtained parameters can be used to determine the span of the probability ellipse and to determine the main directions along which the points defining the boundaries of the parcel are arranged (Figure 9).

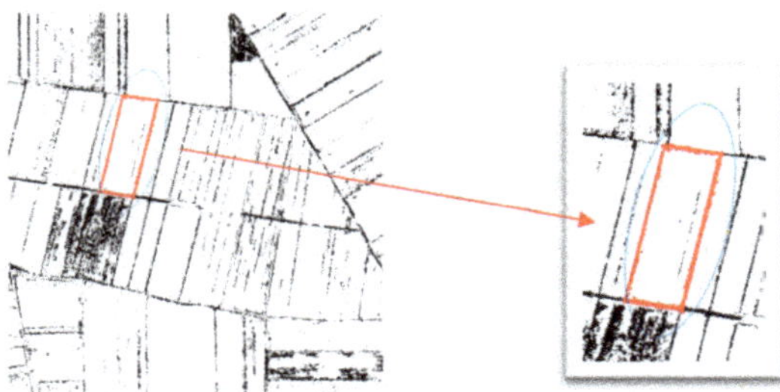

Figure 9. Example of an error ellipse calculated for one of the parcels of land.

The parameters of the ellipse also made it possible to divide the points into four sets of data that define four corresponding boundaries. In each of the four sets of points, a simple least squares approximation was performed (Figure 10). The approximation is an iterative process that, in successive iterations, discards outlier points from the straight line until the agricultural edges are accurately determined. Outlier points are those points that lie further than the average distance of the points from the straight line obtained in successive iterations. The iterative process ends when the sum of the squares of the outliers reaches a minimum.

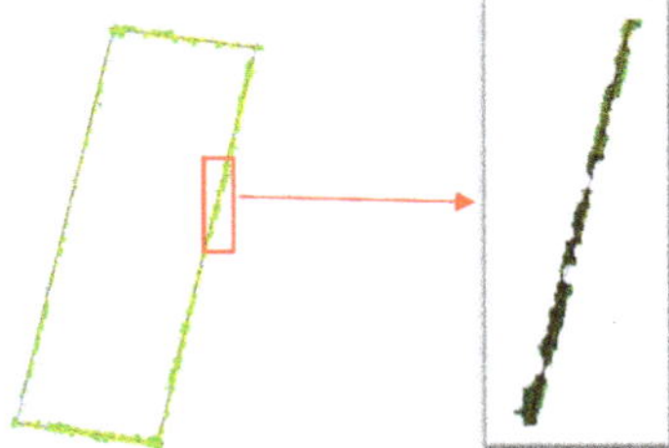

Figure 10. Approximation of straight lines based on scattered points that define utility boundaries.

The intersections of the detected lines determined the vertices of the sought parcel.

6.2. Boundary Detection Using Hough Transform—Test Area No. 2

The second test field was an erroneously delineated segment. The rough edges produced in the first stage indicate that there is probably more agricultural land here (Figure 11).

Figure 11. Test area no. 2—segment with more approximate edges.

Similar to test area no. 1, a segment was first selected on which a raster representing the edges of the utilities was overlaid, resulting in a binary image containing only the edges of the segment (Figure 12a–c).

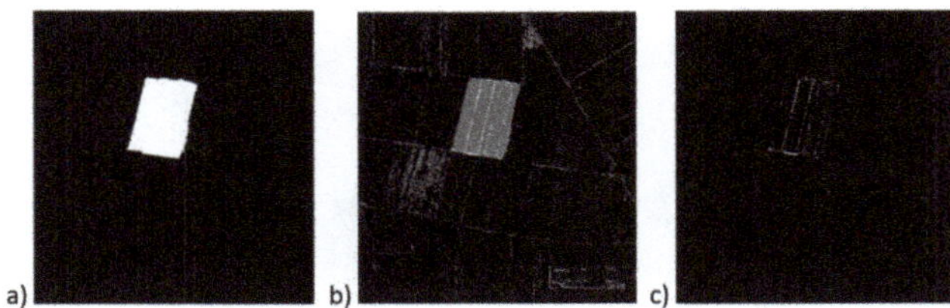

Figure 12. The next steps (**a–c**) detected the outer and inner edges of the selected segment.

The selected segment contains not only edges describing its perimeter but also several edges inside the area. The study showed that the principal components method cannot automatically write to separate sets of points representing each edge (those inside and outside). Therefore, Hough Transform was used to select and identify all the edges of the analyzed area. Hough Transform enables fast detection of straights in a binary image [41]. When detecting collinear pixels in an image in this method, it is possible to indicate the number of straights that need to be detected. In this study, nine lines were detected (Figure 13).

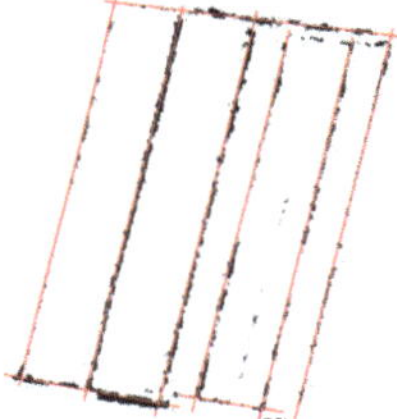

Figure 13. A binary raster representing the edges with the detected lines by Hough Transform.

Based on the detected lines, it was possible to group the scattered points into appropriate sets, defining the course of the edge. The condition determining points belonging to a given edge was the distance of the point from the straight line. Thus, nine sets of points were defined, and in each set a straight line was approximated by the least squares method (Figure 14).

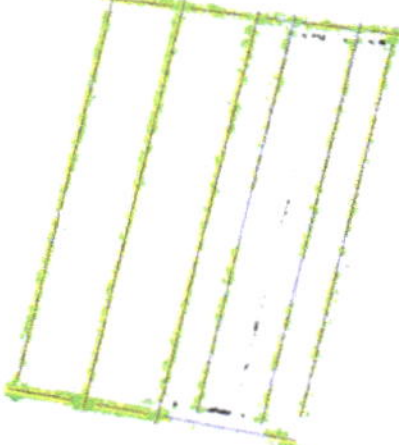

Figure 14. Approximation of straight lines defining the course of agricultural edges.

In the next step, the vertices of the areas where the lines intersect were determined. Next, the vector cadastral data were overlaid on the raster representing the edges of the agricultural land (Figure 15). According to IACS, there can be several types of agricultural land in each parcel of land. However, after visual verification it was found that in the third segment there is one agricultural area of land in the whole cadastral parcel. Therefore, the inclusion of two straights in further analyses was discontinued. The straights are indicated by arrows in Figure 15.

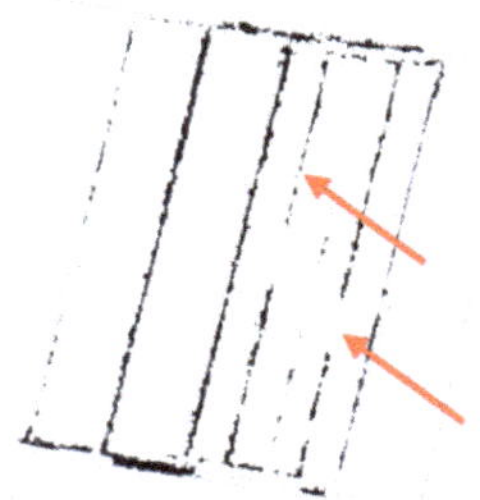

Figure 15. Land use boundaries and overlaid vectors from land records. Arrows indicate boundaries that were not considered in further analyses.

7. Analysis and Discussion of the Obtained Results

7.1. Test Area No. 1

Test area no. 1 was the arable crop land (R) corresponding to cadastral parcel no. 700. The analysis of the obtained results consisted of comparing the borders of the agricultural land obtained by means of the PCA (Section 6.1) with the borders of the agricultural land from the cadastral records. For this purpose, additional points were inserted on both lines at corresponding intervals of 0.5 m. Next, deviations were determined as distances between corresponding points.

Western boundary (W): Conformity of agricultural land boundaries with the data recorded in the land cadastre was observed. Obtained deviations are within the range from 0.23 to 0.35 m. The average value is 0.29 m.

Southern boundary (S): Much larger differences are obtained. Deviations range from 0.37 to 2.16 m, with an average value of 1.26 m. Anomalous information is observed, showing a discrepancy between the declaration and the actual state of use (Figure 16a). There is a road running in the area, and its boundary is not clear, resulting in changes in the actual state of use for these parts of the parcel.

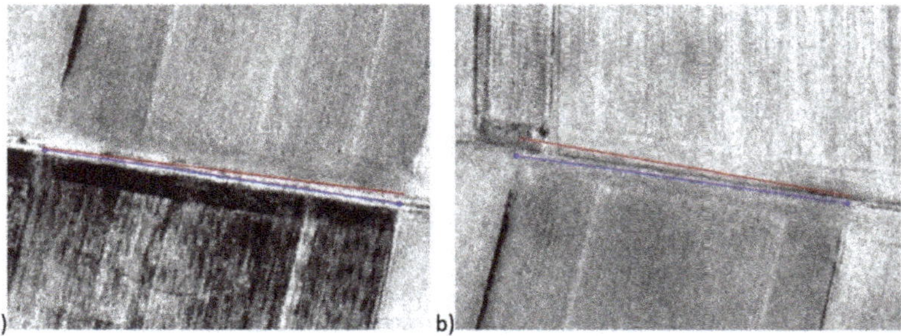

Figure 16. Comparison of the southern boundary of the land parcel (S) (**a**) (red line—boundary from the land cadastre, blue—from the LiDAR) and the northern boundary (N) (**b**) (red line—boundary from the land cadastre, blue—from the LiDAR data).

Eastern boundary (E): Conformity of agricultural land boundaries with the data recorded in the land cadastre was observed. Deviations range from 0.13 to 0.39 m. The average value is 0.26 m.

Northern boundary (N): The largest differences were obtained for the northern boundary of the agricultural land (Figure 16b). LiDAR determined a completely different course of the agricultural land boundary than indicated by the land cadastre. Deviations range from 1.81 to 4.21 m. The average value of deviations is as high as 3.01 m. Analyzing the data against the intensity map, it can be observed that both the LiDAR data and the cadastre data show discrepancies in relation to the actual land use.

The analysis is presented in two tables. Table 1 gives the summarized results for the analyzed agricultural land.

Table 1. Basic statistics for agricultural land R (parcel no. 700).

Agricultural Land	Parcel	Boundary	Minimum Deviation (m)	Maximum Deviation (m)	Average Deviation (m)	Remarks
arable crop land—R	700	boundary W	0.23	0.35	0.29	conformity of the declaration with the actual state
		boundary S	0.37	2.16	1.26	non-conformity of the declaration with the actual state—field road
		boundary E	0.13	0.39	0.26	conformity of the declaration with the actual state
		boundary N	1.81	4.21	3.01	non-conformity of the declaration with the actual state—field road

It is noted that the largest amount of anomalous information occurs for areas of agricultural land boundaries with field roads. Table 2 is a comparison carried out for two variants: The first considers the situation of the agricultural land–agricultural land boundary and the second the agricultural land–field road.

Table 2. Anomalous information for variant 1, land–land boundary, and variant 2, land–field road boundary.

Boundary	Minimum Deviation (m)	Maximum Deviation (m)	Average Deviation (m)
land–land boundary	0.13	0.39	0.27
land–field road boundary	0.37	4.21	2.01

7.2. Test Area No. 2

After the analysis of the cadastral data, it was revealed that the selected segment consists of three agricultural lands—R, Ł, corresponding to cadastral parcels with the numbers: 702 (R), 705 (R), and 707 (Ł). The algorithm based on Hough Transform correctly extracted these three areas.

As a first step, a detailed analysis of the agricultural land R for parcel 702 was carried out. Western boundary (W): Conformity of agricultural land boundaries with the data recorded in the land cadastre was observed. Deviations range from 0.23 to 0.40 m. The average value is 0.31 m.

Southern boundary (S): Much larger differences were obtained. The deviations range from 1 to 1.61 m, with an average deviation of 1.31 m. In this area we are dealing with a field road, the course of which is unclear. A discrepancy between the declaration and the actual state of use can be seen in the form of a systematic shift between the LiDAR data and the agricultural land from the land cadastre (Figure 17a).

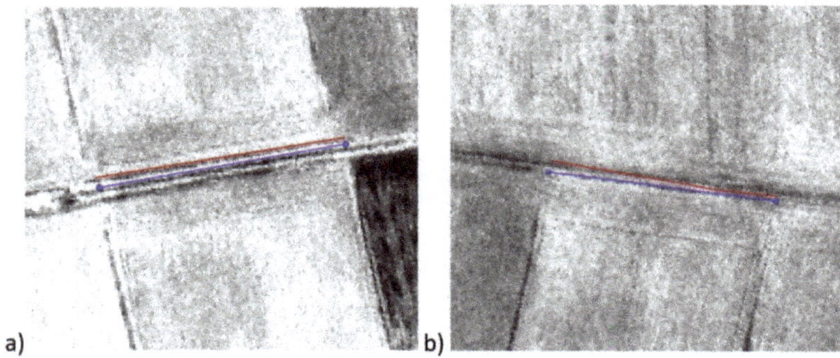

Figure 17. Comparison of the southern boundary of the land parcel (S) (**a**) (red line—boundary from the land cadastre, blue—from the LiDAR) and the northern boundary (N) (**b**) (red line—boundary from the land cadastre, blue—from the LiDAR).

Eastern boundary (S): For the eastern boundary, deviations of 0.40–0.70 m were obtained, with a mean value of 0.55 m.

Northern boundary (N): Much larger differences were obtained (Figure 17b). The deviations range from 0.66 to 1.96 m, and the mean value of the deviations is 1.15 m. In this case a field road is also located in the area.

Due to the repeated spatial situation, the results for parcels 705 and 707 are grouped together in Table 3. For both parcels similar types of differences to parcel 702 appeared, as discussed above. The results for plot 702 are also included in the table.

Table 3. Basic statistics for parcels 702, 705, and 707.

Agricultural Land	Parcel	Boundary	Minimum Deviation (m)	Maximum Deviation (m)	Average Deviation (m)	Remarks
arable crop land—R	702	boundary W	0.23	0.40	0.31	conformity of the declaration with the actual state
		boundary S	1.0	1.61	1.31	non-conformity of the declaration with the actual state—field road
		boundary E	0.40	0.70	0.55	conformity of the declaration with the actual state
		boundary N	0.66	1.96	1.15	non-conformity of the declaration with the actual state—field road
arable crop land—R	705	boundary W	0.40	0.70	0.55	conformity of the declaration with the actual state
		boundary S	1.72	2.33	1.88	non-conformity of the declaration with the actual state—field road
		boundary E	0.55	1.22	0.88	conformity of the declaration with the reality state
		boundary N	0.72	0.77	0.74	conformity of the declaration with the actual state—field road

Table 3. *Cont.*

Agricultural Land	Parcel	Boundary	Minimum Deviation (m)	Maximum Deviation (m)	Average Deviation (m)	Remarks
grassland—Ł	707	boundary W	0.55	1.22	0.88	conformity of the declaration with the reality state
		boundary S	1.86	2.87	2.36	non-conformity of the declaration with the actual state—field road
		boundary E	0.00	0.42	0.16	conformity of the declaration with the reality state
		boundary N	0.31	0.64	0.47	conformity of the declaration with the actual state—field road

A similar rule was observed as in the previous analyses, i.e., a much higher agreement of boundaries for the land–land boundary variant than for the land–field road boundary variant. This is presented in Table 4.

Table 4. Anomalous information for variant 1, land–land boundary, and variant 2, land–field road boundary.

Boundary	Minimum Deviation (m)	Maximum Deviation (m)	Average Deviation (m)
land–land boundary	0.00	1.22	0.47
land–field road boundary	0.31	2.87	1.31

8. Conclusions

In summary, the proposed algorithms make it possible to carry out controls in the system of direct payments to agriculture. The use of laser data makes it possible to determine specific agricultural land and to determine the size of anomalous information. Correctness has been noted in relation to the conformity of the actual statue with the declaration in the case of boundaries of adjacent agricultural land (variant 1: land—land). However, in areas where there are borders of agricultural land with field roads, there are visible discrepancies (variant 2: land—field road). In these areas, anomalous information connected with the difference between the actual use of a given area and the legal status recorded in the declaration should be noted.

The obtained results were considered satisfactory. LiDAR proved to be very useful technology in the process of detecting agricultural boundaries. Most of the boundaries were readable in the laser data.

The advantage of using a point cloud over traditional aerial images is that an additional elevation information can be used. In the case of aerial images, analyses are carried out on 2D data stored as a raster. In the developed algorithm, the second step returns to the raw point cloud. The knowledge of an additional Z coordinate may allow for more precise edge detection in areas where the 2D information is ambiguous.

It was noted that the introduction of higher resolution data would certainly contribute to an increase in accuracy. The use of laser data from a UAV flight would allow a more precise determination of boundaries in doubtful cases, not fully legible in the case of the data used in this study (average distance between points 0.3 m).

The two-stage approach to analysis also proved to be a valuable solution. Edge detection and segmentation algorithms used in the first stage allowed us to roughly estimate the area and boundaries of individual agricultural lands. In the second stage, we returned to the original data for the locations presenting the contours of the agriculture boundaries. In the developed method of detecting straights, two approaches were used. Two approaches were chosen because the first area is a rectangular parcel of land and the second test area is a rectangle with an internal land boundary, and precise determination of the land boundary is possible if only one boundary is displayed in the raster image. Therefore, in each approach, the aim was to divide the point cloud into datasets representing only one land use boundary. With such datasets it is possible to approximate straights with higher accuracy than by using raster data. Detected land use boundaries are described by the equation of the straight line, determined in an iterative process, wherein subsequent iterations' outliers are rejected. Thus, the obtained straight line reliably reflects the course of land use boundaries detected based on ALS data.

The created algorithm allowed the detection of inconsistencies in farmers' declarations. These were related to areas of field roads that were incorrectly declared by farmers as donated land, when in fact they should be excluded from subsidies. It was detected that both test areas, test field 1, areas belonging to field roads with an average width of 1.26 and 3.01 m, and test field 2, areas belonging to field roads with an average width of 1.31, 1.15, 1.88, and 2.36 m, were incorrectly classified by farmers as donated land.

In this study the authors focus on identification of land boundaries between land uses covered with low vegetation. In the next research, the authors intend to analyze the boundaries of parcels that are farmed and covered with different species of plants. Such diversity can help at the stage of segmentation because each species can have a different intensity value. Different intensity value will contribute to easier initial identification of parcel edges. At the stage of precise identification of plot boundaries, on the other hand, it may be necessary to select only points reflected from the ground. Then it will be necessary to perform a filtering of the lidar data to be used for further analysis.

In conclusion, the method used in this study based on LiDAR data is useful for automatic verification and monitoring of anomalous information showing inconsistency of the declaration with the actual state of land use.

Author Contributions: Conceptualization, N.B. and U.M.; methodology, N.B. and U.M.; software, N.B. and U.M.; validation N.B. and U.M.; formal analysis N.B. and U.M.; writing—original draft preparation, N.B. and U.M.; writing—review and editing, N.B. and U.M.; visualization, N.B. and U.M.; supervision, N.B. and U.M.; project administration, N.B. and U.M. All authors have read and agreed to the published version of the manuscript.

Funding: Research project supported by program „Excellence initiative—research university" for the AGH University of Science and Technology no. 501.696.7996. The project title: Integration of remote sensing data for control in the system of direct agricultural subsidies (IACS).

Institutional Review Board Statement: Not applicable.

Informed Consent Statement: Not applicable.

Data Availability Statement: The data presented in this study (.las files) are available in ISOK project—https://isok.gov.pl/index.html, accessed on 31 July 2021.

Acknowledgments: The authors thank the editor and anonymous reviewers for their helpful comments and valuable suggestions.

Conflicts of Interest: The authors declare no conflict of interest.

References

1. Vosselman, G.; Mass, H.-G. *Airborne and Terrestrial Laser Scanning*; Whittles Publishing: Dunbeath, UK, 2010.
2. Prawa i Obowiązki Rolników w Procesie Kontroli na Miejscu | Agencja Restrukturyzacji i Modernizacji Rolnictwa. Available online: https://www.arimr.gov.pl/kontrole-beneficjentow/prawa-i-obowiazki-rolnikow-w-procesie-kontroli-na-miejscu.html (accessed on 22 August 2021).
3. Cha, G.; Sim, S.-H.; Park, S.; Oh, T. LiDAR-Based Bridge Displacement Estimation Using 3D Spatial Optimization. *Sensors* **2020**, *20*, 7117. [CrossRef] [PubMed]

4. Chen, Z.; Lin, Q.; Sun, J.; Feng, Y.; Liu, S.; Liu, Q.; Ji, Y.; Xu, H. Cascaded Cross-Modality Fusion Network for 3D Object Detection. *Sensors* **2020**, *20*, 7243. [CrossRef] [PubMed]
5. Imad, M.; Doukhi, O.; Lee, D.-J. Transfer Learning Based Semantic Segmentation for 3D Object Detection from Point Cloud. *Sensors* **2021**, *21*, 3964. [CrossRef] [PubMed]
6. Dey, E.K.; Tarsha Kurdi, F.; Awrangjeb, M.; Stantic, B. Effective Selection of Variable Point Neighbourhood for Feature Point Extraction from Aerial Building Point Cloud Data. *Remote Sens.* **2021**, *13*, 1520. [CrossRef]
7. Huang, J.; Zhang, X.; Xin, Q.; Sun, Y.; Zhang, P. Automatic building extraction from high-resolution aerial images and LiDAR data using gated residual refinement network. *ISPRS J. Photogramm. Remote Sens.* **2019**, *151*, 91–105. [CrossRef]
8. Szostak, M. Automated Land Cover Change Detection and Forest Succession Monitoring Using LiDAR Point Clouds and GIS Analyses. *Geosciences* **2020**, *10*, 321. [CrossRef]
9. Gu, C.; Zhai, C.; Wang, X.; Wang, S. Cmpc: An innovative lidar-based method to estimate tree canopy meshing-profile volumes for orchard target-oriented spray. *Sensors* **2021**, *21*, 4252. [CrossRef]
10. Megahed, Y.; Shaker, A.; Yan, W.Y. Fusion of Airborne LiDAR Point Clouds and Aerial Images for Heterogeneous Land-Use Urban Mapping. *Remote Sens.* **2021**, *13*, 814. [CrossRef]
11. Terefenko, P.; Zelaya Wziątek, D.; Dalyot, S.; Boski, T.; Pinheiro Lima-Filho, F. A High-Precision LiDAR-Based Method for Surveying and Classifying Coastal Notches. *ISPRS Int. J. Geo-Inf.* **2018**, *7*, 295. [CrossRef]
12. Swirad, Z.M.; Young, A.P. Automating coastal cliff erosion measurements from large-area LiDAR datasets in California, USA. *Geomorphology* **2021**, *389*, 107799. [CrossRef]
13. Vilbig, J.M.; Sagan, V.; Bodine, C. Archaeological surveying with airborne LiDAR and UAV photogrammetry: A comparative analysis at Cahokia Mounds. *J. Archaeol. Sci. Rep.* **2020**, *33*, 102509. [CrossRef]
14. Balsi, M.; Esposito, S.; Fallavollita, P.; Melis, M.G.; Milanese, M. Preliminary Archeological Site Survey by UAV-Borne Lidar: A Case Study. *Remote Sens.* **2021**, *13*, 332. [CrossRef]
15. Kocur-Bera, K. Understanding information about agricultural land. An evaluation of the extent of data modification in the Land Parcel Identification System for the needs of area-based payments—A case study. *Land Use Policy* **2020**, *94*, 104527. [CrossRef]
16. Zimmermann, J.; González, A.; Jones, M.B.; O'Brien, P.; Stout, J.C.; Green, S. Assessing land-use history for reporting on cropland dynamics-A comparison between the Land-Parcel Identification System and traditional inter-annual approaches. *Land Use Policy* **2016**, *52*, 30–40. [CrossRef]
17. Inan, H.I.; Sagris, V.; Devos, W.; Milenov, P.; van Oosterom, P.; Zevenbergen, J. Data model for the collaboration between land administration systems and agricultural land parcel identification systems. *J. Environ. Manag.* **2010**, *91*, 2440–2454. [CrossRef]
18. Iban, M.C.; Aksu, O. A model for big spatial rural data infrastructure in Turkey: Sensor-driven and integrative approach. *Land Use Policy* **2020**, *91*, 104376. [CrossRef]
19. Parida, P.K.; Sanabada, M.K.; Tripathi, S. The Digital Cadastral Map/Layer Generation and Conclusive Titling of Land Parcels Using Hybrid Technology (Aerial/High-Resolution Image (HRSI) and DGPS and ETS Survey) Adopted by Govt. of Odisha Under Digital India Land Record Modernization Programme (DIL). In *Lecture Notes in Civil Engineering*; Springer: Singapore, 2020; Volume 33, pp. 439–459.
20. Tarko, A.; de Bruin, S.; Fasbender, D.; Devos, W.; Bregt, A.K. Users' assessment of orthoimage photometric quality for visual interpretation of agricultural fields. *Remote Sens.* **2015**, *7*, 4919–4936. [CrossRef]
21. Wagner, M.P.; Oppelt, N. Deep Learning and Adaptive Graph-Based Growing Contours for Agricultural Field Extraction. *Remote Sens.* **2020**, *12*, 1990. [CrossRef]
22. Kamal, M.; Phinn, S.; Johansen, K. Object-based approach for multi-scale mangrove composition mapping using multi-resolution image datasets. *Remote Sens.* **2015**, *7*, 4753–4783. [CrossRef]
23. Fisher, R.J.; Sawa, B.; Prieto, B. A novel technique using LiDAR to identify native-dominated and tame-dominated grasslands in Canada. *Remote Sens. Environ.* **2018**, *218*, 201–206. [CrossRef]
24. Sasaki, T.; Imanishi, J.; Ioki, K.; Morimoto, Y.; Kitada, K. Object-based classification of land cover and tree species by integrating airborne LiDAR and high spatial resolution imagery data. *Landsc. Ecol. Eng.* **2012**, *8*, 157–171. [CrossRef]
25. Estrada, J.; Sánchez, H.; Hernanz, L.; Checa, M.; Roman, D. Enabling the Use of Sentinel-2 and LiDAR Data for Common Agriculture Policy Funds Assignment. *ISPRS Int. J. Geo-Inf.* **2017**, *6*, 255. [CrossRef]
26. Mulahusić, A.; Topoljak, J.; Tuno, N.; Ajvazović, K. The Possibilities of the Cadastral Land Use Assessment by the Methods of Remote Sensing. In *Lecture Notes in Networks and Systems*; Springer Nature: Cham, Switzerland, 2019; Volume 42, pp. 452–458.
27. Ho Tong Minh, D.; Ndikumana, E.; Baghdadi, N.; Courault, D.; Hossard, L. Applying deep learning for agricultural classification using multitemporal SAR Sentinel-1 for Camargue, France. In *Proceedings of the Image and Signal Processing for Remote Sensing XXIV*; Bruzzone, L., Bovolo, F., Benediktsson, J.A., Eds.; SPIE: Berlin, Germany, 2018; Volume 10789, p. 39.
28. Zhou, X.; Li, W. A Geographic Object-Based Approach for Land Classification Using LiDAR Elevation and Intensity. *IEEE Geosci. Remote Sens. Lett.* **2017**, *14*, 669–673. [CrossRef]
29. Yu, Y.; Guan, H.; Li, D.; Gu, T.; Wang, L.; Ma, L.; Li, J. A Hybrid Capsule Network for Land Cover Classification Using Multispectral LiDAR Data. *IEEE Geosci. Remote Sens. Lett.* **2020**, *17*, 1263–1267. [CrossRef]
30. Qian, T.; Shen, D.; Xi, C.; Chen, J.; Wang, J. Extracting Farmland Features from Lidar-Derived DEM for Improving Flood Plain Delineation. *Water* **2018**, *10*, 252. [CrossRef]

31. Kamiński, M. The Impact of Quality of Digital Elevation Models on the Result of Landslide Susceptibility Modeling Using the Method of Weights of Evidence. *Geosciences* **2020**, *10*, 488. [CrossRef]
32. ISOK. Available online: https://isok.gov.pl/index.html (accessed on 31 July 2021).
33. Regulation of the Minister for Regional Development and Construction of 29 March 2001 on the land and building cadastre. Journal of Laws, No. 38 item 454 as Amended. 2001.
34. Gonzalez, R.C.; Woods, R.E. *Digital Image Processing*, 3rd ed.; Pearson: Upper Saddle River, NJ, USA, 2008; ISBN 978-0-13-168728-8.
35. Baatz, M.; Schape, A. Multiresolution segmentation—An optimization approach for high quality multi-scale image segmentation. In Proceedings of the AGIT Symposium, Salzburg, Austria, 5–7 July 2000.
36. Kavzoglu, T.; Tonbul, H. A comparative study of segmentation quality for multi-resolution segmentation and watershed transform. In Proceedings of the 2017 8th International Conference on Recent Advances in Space Technologies RAST 2017, Istanbul, Turkey, 19–22 June 2017; pp. 113–117. [CrossRef]
37. Kavzoglu, T.; Tonbul, H. An experimental comparison of multi-resolution segmentation, SLIC and K-means clustering for object-based classification of VHR imagery. *Int. J. Remote Sens.* **2018**, *39*, 6020–6036. [CrossRef]
38. Benz, U.C.; Hofmann, P.; Willhauck, G.; Lingenfelder, I.; Heynen, M. Multi-resolution, object-oriented fuzzy analysis of remote sensing data for GIS-ready information. *ISPRS J. Photogramm. Remote Sens.* **2004**, *58*, 239–258. [CrossRef]
39. Sobczyk, M. *Statystyka*; Wydawnictwo Naukowe PWN: Warsaw, Poland, 2007.
40. Hausbrandt, S. *Rachunek Wyrównawczy i Obliczenia Geodezyjne*; Państwowe Przedsiębiorstwo Wydawnictw Kartograficznych: Warsaw, Poland, 1970.
41. Marmol, U.; Borowiec, N. Detection of Line Objects by Means of Gabor Wavelets and Hough Transform. *Arch. Civ. Eng.* **2020**, *66*, 339–363. [CrossRef]

Article

Polish Cadastre Modernization with Remotely Extracted Buildings from High-Resolution Aerial Orthoimagery and Airborne LiDAR

Damian Wierzbicki *, Olga Matuk and Elzbieta Bielecka

Institute of Geospatial Engineering and Geodesy, Faculty of Civil Engineering and Geodesy, Military University of Technology, 00-908 Warsaw, Poland; olga.matuk@wat.edu.pl (O.M.); elzbieta.bielecka@wat.edu.pl (E.B.)
* Correspondence: damian.wierzbicki@wat.edu.pl

Abstract: Automatic building extraction from remote sensing data is a hot but challenging research topic for cadastre verification, modernization and updating. Deep learning algorithms are perceived as more promising in overcoming the difficulties of extracting semantic features from complex scenes and large differences in buildings' appearance. This paper explores the modified fully convolutional network U-Shape Network (U-Net) for high resolution aerial orthoimagery segmentation and dense LiDAR data to extract building outlines automatically. The three-step end-to-end computational procedure allows for automated building extraction with an 89.5% overall accuracy and an 80.7% completeness, which made it very promising for cadastre modernization in Poland. The applied algorithms work well both in densely and poorly built-up areas, typical for peripheral areas of cities, where uncontrolled development had recently been observed. Discussing the possibilities and limitations, the authors also provide some important information that could help local authorities decide on the use of remote sensing data in land administration.

Keywords: image segmentation; deep-learning; building outlines; cadastre modernization; FCN; high-resolution aerial orthoimages; LiDAR data

Citation: Wierzbicki, D.; Matuk, O.; Bielecka, E. Polish Cadastre Modernization with Remotely Extracted Buildings from High-Resolution Aerial Orthoimagery and Airborne LiDAR. *Remote Sens.* **2021**, *13*, 611. https://doi.org/10.3390/rs13040611

Academic Editors: Mila Koeva, Rohan Bennett and Claudio Persello

Received: 24 January 2021
Accepted: 6 February 2021
Published: 8 February 2021

1. Introduction

Modern cadastral systems, being the part of land administration systems, constitute an indisputable tool for sustainable land management [1]. Over the past several decades, the cadastre, both as a concept and system, has significantly developed and changed its role from a simple land register to a technologically advanced multidimensional and multi-functional system, supporting effective and sustainable land management [2]. Moreover, the continuous evolution of land administration systems, as well as the cadastre as part of them, results from the increasing human pressure on the environment [3,4]. Particularly, changes are driven by and heavily dependent on processes such as economic and political reform, urbanization, agricultural intensification and deforestation, and on the other hand, concern for nature protection, human well-being and sustainable development [1–4].

For three decades, care for the environment has been one of the key global trends in land use and management. This has led to the emergence of many global initiatives related to environmental protection and care for the present and future human well-being in friendly natural and socio-economic environments. The most important of them are Agenda 21 [5] and Agenda 2030 [6], which aim to define care for the environment and sustainable development to a large extent and at an earlier stage of planning. Monitoring towards the 2030 Agenda Sustainable Development Goals (SDGs) involves the availability of high-quality, timely and disaggregated data, which are of great importance for evidence-based decision making and for ensuring accountability for the implementation of the 2030 Agenda. Many concerns about the state of the environment reflected in the SDGs are related to unprecedented urban growth. Urbanization, as a complex and continuous process worldwide, has been going on for hundreds of years, although it has significantly accelerated in the last few decades. As stated by UN Secretary General [7] "from 2000 to 2015, in all regions of the world, the expansion of urban land outpaced the growth of urban populations". This results in uncontrolled urban growth and a decrease in city density. Remote sensing data are undoubtedly one of the most important data sources for monitoring urban sprawl and updating data in cadastral systems, as they provide not only information on the geographic location but also some characteristics of buildings and associated artificial infrastructure [8,9]. Since the 1990s, the implementation of remote sensing for land management and cadastre updating has evolved significantly due to technological advances, particularly, high-resolution and multispectral images, advances in aerial imaging technologies, image processing algorithms as well as internet and mobile technologies [9–11].

In Poland, the register of buildings along with the land and mortgage registers is a coherent part of the cadastre, which is ultimately to be kept in accordance with the assumptions of the INSPIRE (INfrastructure for SPatial InfoRmation in Europe) Directive [12] and Land Administration Domain Model [13]. These registers, being public registers, form part of the Polish land administration system and provide support to local authorities in their decision-making [12]. A significant incentive to the commencement of the Polish cadastre modernization, resulting in its transition from analogue to electronic form, was initiated in 2004 with Poland's accession to the European Union [12]. Since that year, intensified works of the Head Office of Geodesy and Cartography, aimed at modernizing the existing cadastral system, have been observed [13–15]. The amendment to the Geodetic and Cartographic Law of 2010 and the Ministry Regulation on the land and building register (denoted as EGiB) [16] introduce comprehensive administrative procedures for updating, verifying and modernizing the cadastre to ensure data accuracy and reliability. The necessity to adapt the Polish cadastre to European requirements (including the INSPIRE directive and the Land and Parcel Identification System (LPIS)) was the driving force for launching the two nationwide projects aimed at converting analogue parcel boundaries and building outlines to digital form, namely the Integrated System of Real Estate Information [14] and the Polish LPIS [17].

In a few years, the modernized cadastre covered almost the entire country's territory. As documented by the Head Office of Geodesy and Cartography at the end of 2017, building outlines were still recorded in an analogue (paper) form in 3% of urban and 14% rural areas, and respectively 1% and 6% in the raster (scanned analogue map) supplemented by buildings' centroids [18]. Moreover, § 63.1. of the latest update of the land and buildings register regulation [19] allowed for the registration of building outlines in the form of a point representing the building center. The regulation also requires cadastral authorities to periodically verify the land and building register (§ 44), as well as to eliminate existing inconsistencies by updating or modernizing cadastral data. It should be noted, however, that the modernization of the cadastre has been defined by law [19] as a set of technical, organizational and administrative activities undertaken to adapt the existing cadastral data to the requirements of a modern, up-to-date and fit-for-purpose IT system. Hence, the motivation of this study is a thorough analysis of the possibilities and limitations of verification and modernization of the Polish cadastre by remotely extracted buildings from high-resolution aerial orthoimages and LiDAR data. We proposed a three-stage end-to-end methodology of building rooftop outlines extraction. An inseparable part of the method constituted the geo-processing of vectors of buildings' rooftop surfaces, derived from deep learning orthoimage segmentation and LiDAR data processing. Therefore, the research makes both a scientific and a practical contribution. The main goal of this study is to automate the extraction of building rooftop outlines based on the deep learning algorithm enriched with the LiDAR segmentation and geoprocessing to map the rooftop outline of buildings. The deep learning algorithm is based on a modified U-Shape Network (U-Net) so it provides precise buildings segmentation with a relatively small number of training images. In the proposed U-Net architecture, the size of the feature map was empirically set-up as 416 × 416 × 1. The analysis is supported by publicly available data (orthoimagery and LiDAR densely classified data (DSM)) from the official geoportal of the Head Office of Geodesy and Cartography, the National Mapping Agency in Poland.

The remainder of the paper is structured as follows. Section 2 gives an overview of building extraction algorithms based on high-resolution images and LiDAR data. Study area, materials and methods descriptions are presented in Section 3. Section 4 provides a concise description of the experimental results, while a discussion with the results, and hitherto achievements in the field are given in Section 5. The paper ends with the concluding remarks, Section 6.

2. Deep Learning-Based Building Extraction-Related Works

Automatic building extraction from remotely sensed data has been a major research topic for decades due to the importance of building data in many areas of economy and science, inter alia land administration, topographic mapping, urban planning and sustainable development, natural hazards risk management and mitigation or humanitarian aid [20–24]. Moreover, building extraction algorithms enable cost- and time-effective approaches to 3D data acquisition, maintenance and analysis [25]. At the beginning of the 21st century, Baltsavias [26] observed some tendencies in the development of image analysis methods for building extraction. These included the increasing use of holistic and rule-based approaches to the problem, such as semantic and Bayesian networks or artificial neural networks (ANN) and fuzzy logic. Furthermore, increased use of a priori knowledge (e.g., from vector data) and multi-image and multi-sensor 3D methods have become standard in both image processing and object modelling. However, in those days, as stated by Baltsavias [26], "reliability and completeness of automated results together with their automatic evaluation remain the major problems". The issue of reliability and accuracy of building models has improved significantly in recent years through multi-sensor fusion-based building detection methods [23,27–30] and the possibility of applying deep learning techniques. Ma et al. [31] provided a meta-analysis of deep learning methods in remote sensing applications and highlighted the pioneering achievement of Zhuo et al. [32] on increasing the OpenStreetMap (OSM) buildings' location accuracy derived from deep learning-based semantic segmentation of oblique Unmanned Aerial Vehicle (UAV) images.

Convolutional neural networks (CNNs), being a typical deep learning method, are widely used in building extraction, especially for object detection and image semantic segmentation [24,33,34]. Wang et al., [21] found that the drawback of building extraction algorithms, due to the lack of global contextual information and careless up-sampling methods, may be overcome by a U-shaped network and an adjusted non-local block called the Asymmetric Pyramid Nonlocal Block (APNB). The test provided by authors [21] showed that the accuracy of the established Efficient Non-local Residual U-shape Network (ENRU-Net) gives a remarkable improvement against commonly used semantic segmentation models (e.g., U-Net, FCN-8s, SegNet, or Deeplab v3). Shao et al., [34] have introduced a new, two-module deep learning network named BRRNet for complete and accurate building extraction from high-resolution images. The prediction module, based on encoder–decoder structure, is aimed at building extraction, while the residual refinement module improved the accuracy of building extraction. The experiment of Shao et al. [34] in Massachusetts showed superiority over other state-of-the-art methods (e.g., USPP, EU-Net and MC-FCN) [35] in terms of building integrity (wholeness) and building footprint accuracy. USPP introduced by Liu Y. et al. [35] denotes a U-shaped encoder-decoder structure with spatial pyramid pooling. It contains four encoder blocks, one spatial pyramid pooling module and four decoder blocks. In the encoder phase, the VGG-11 architecture is used as the backbone. EU-Net is an effective fully convolutional network (FCN)-based neural network consisting of three parts: encoder, dense spatial pyramid pooling (DSPP) bloc and decoder. The network, developed by Kang et al. [36], is designated toward building extraction from aerial remote sensing images. MC-FCN (multi-constraint fully convolutional networks) consist of a bottom-up/top-down fully convolutional architecture and multi-constraints that are computed between the binary cross-entropy of prediction and the corresponding ground truth [37].

Since the ascension of deep learning methods, especially convolutional neural networks, the trend towards applying them to improve building extraction models is clearly observable [24,38,39]. Furthermore, the widespread availability of open remote sensing images, including high-resolution images, and aerial laser scanning data, have significantly contributed to the recent boost in automated building extraction algorithms. Bittner et al. [38] presented a method to fuse depth and spectral information based on a fully convolutional network (FCN) that can efficiently exploit mixed datasets of remote sensing imagery to extract building rooftops. The authors [38], however, pointed out that the FCN requirement for multiple training samples was a downside to this method and increased data processing costs and time. Maltezos et al. [24] introduced an efficient CNN-based deep learning model to extract buildings from orthoimages supported by height information obtained from point clouds from dense image matching. Results from Germany (the city of Vaihingen) and Greece (seaside resort of Perissa) showed promising potential in terms of robustness, flexibility and performance for automatic building detection. Huang et al. [33] noticed that "the commonly used feature fusion or skip-connection refine modules of FCNs often overlook the problem of feature selection and could reduce the learning efficiency of the networks". This contributed to the development of a fully convolutional neural network, namely the end-to-end trainable gated residual refinement network (GRRNet) that fused high-resolution aerial images and LiDAR point clouds for building extraction. The test conducted in four US cities demonstrated that the GRRNet has competitive building extraction performance in comparison with other approaches, with an overall accuracy of 96.20% and a mean IoU (Intersection over Union) score of 88.05% among all methods that have the encoder–decoder network architectures. Moreover, the source code of the GRRNet was made publicly available for researchers (see references) [33]. The literature analysis presented above shows many types of FCN architecture modification as well as the possibility of using various scenarios and data augmentation during network training, which in turn leads to improved accuracy and reliability of building extraction from remotely sensed data.

Developing effective methods for automatic building detection based on multisource data remains a challenge due to many factors related to the remotely sensed data (point cloud sparsity or image spatial and spectral variability) as well as the complexity of urban objects or data misalignment [40,41]. To overcome these challenges, Nguyen et al. [40] introduced the Super-Resolution-based Snake Model (SRSM) that operates on high-resolution LiDAR data that involved a balloon force model to extract buildings. The SRSM model is insensitive to image noise and details, as well as simplifying the snake model parameterization, and could be applied on a large scale. Gilani et al. [41] noticed that often, small, shaded, or partially occluded buildings are misclassified. Therefore, based on point clouds and orthoimagery, the building delineation algorithm identified the building regions and segmented them into grids. The problem of nearby trees classified as buildings was solved by synthesizing the point cloud and image data. As reported by the authors, [41] "the correctness of above 95%, demonstrating the robustness of the approach".

Recently, CNNs were integrated with regularized and structured building outline delineations. Girard et al. [42] introduced a deep learning method predicting vertices of polygons that outline the objects of interest. Zhao et al. [43] proposed the R-CNN Mask followed by regularization algorithm to create polygons from the building segmentation results. Girard et al. [44] employed a deep image segmentation model with a frame-field output which ultimately improved building extraction quality and provided structural information, facilitating more accurate polygonization. Among deep learning frameworks, PolyMapper deserves special attention, as it directly predicts a polygon representation describing geometric objects using a vector data structure [45,46]. The PolyMapper approach introduced by Li et al. [45]) performs building detection, instance segmentation and vectorization within a unified approach based on modern CNNs architectures and RNNs with convolutional long-short term memory modules. Zhao et al. [46], however, employed EffcientNet built on top of PolyMapper supported by a boundary refinement block (BRB) to "strengthen the boundary feature learning" and finally to improve the accuracy of the building's corner prediction. The CNNs shown above, integrated with the regularized and structured building outlines, demonstrate end-to-end approaches capable of delineating building boundaries to be close to reference data structure.

A vital part of image classification and segmentation is accuracy assessment [47]. Literature provides a wide range of metrics to assess the accuracy of buildings extraction from remotely sensed data. The most common are precision, completeness, overall accuracy, F1 score, Jaccard similarity and kappa indexes [37,48]. Some of them are pixel-based and others object-based; however, a few can be used in both pixel-based and object-based evaluation [20,24,40,41,43]. Moreover, RMSE and normalized median absolute deviation (NMAD) are frequently used to measure the positional accuracy [37,42,43]. Aung et al. [20] documented that pixel-based evaluation is more objective as it is based on the status of each pixel; however, the object-based quality assessment is a good measure when object shape and texture are concerned. A similar opinion was shared by [42], who also noted that the pixel-based assessment to compare polygons often requires vector data rasterization, which influences the accuracy assessment. Hence, some authors, e.g., [40,41], used both per-object and pixel-based accuracy measures.

Considering the limitations of existing methods to assess the accuracy of building extraction, the authors of [42] developed a new metric for comparison of polygons and line segments, named PoLiS. PoLiS can be used to assess the quality of extracted building footprints, provided that reference data is available. The metric considers the shortest distance from a corner of one building's boundary to any point on the other boundary, which constitutes simplicity in implementing. However, as found by Dey et al. [48], the PoLiS metric is significantly influenced by corresponding corners selection, especially for complex-shaped buildings, when the number of extracted corners differs substantially from reference data. Evaluation of buildings extracted from LiDAR data is even more complicated, as found by Dey et al. [48], due to a meandrous (zigzag) pattern without many details, hardly comparable with reference building outlines. The authors Dey et al. [48] overcame this problem by introducing a new, robust corner correspondence (RCC) metric that allowed to assess the extra- and under-lap areas of extracted and reference buildings. The RCC metric constitutes a combined measure of the positional accuracy and shape similarity and allows for a more realistic assessment of the extracted building boundaries from LiDAR data [48].

3. Study Area, Materials and Methods

3.1. Study Area

The experiment was carried out in Kobiałka, a peripheral part of the Polish capital, Warsaw (see Figure 1). Kobiałka is a housing estate in northern Warsaw, a typical residential area, dominated by detached buildings and green infrastructure. The area was selected for several reasons. Firstly, the Warsaw Municipality has been ordering high-resolution aerial orthophotos mainly for the development of the cadastre and the real estate market. Secondly, from the end of the 20th century, Kobiałka was characterized by unstoppable and in many cases uncontrolled growth of buildings due to many undeveloped areas with favorable housing conditions, e.g., proximity to a forest, good communication with the city center [49]. Thirdly, the reorganization of the cadastral department of the Warsaw municipality, which has been going on for several years, and the change of some important national regulations, including the geodetic and cartographic law and the construction law, have delayed the cadastre modernization, particularly the building register.

In particular, the study area covered approx. 800 ha, in the east from 21°02′52″.2 to 21°03′44″.7, and in the north from 52°03′44″.7 to 52°21′55″.6 (Figure 1, the red rectangle).

Figure 1. The study area: Warsaw and Kobiałka, the study region (marked by the red rectangle) and aerial orthoimage (selected part).

3.2. *Data Used*

Orthorectified RGB aerial images with a 0.1 m GSD (hereinafter referred to orthoimages), taken with the Leica DMC III camera during a photogrammetric campaign carried out in 2019 on 18 April (Figure 1), were used for buildings segmentation. iIDAR dense classified point clouds (12 points per 1 m^2) stored in LAS binary format were employed for building vectors extraction. Buildings data from the Warsaw cadastre was used for training and final evaluation of building extraction. The data comprised the building geographical location, number of storeys, source of geometry and building type (e.g., residential, outbuilding, office).

All data were downloaded from the official, publicly available geoportal of the Head Office of Geodesy and Cartography, the National Mapping Agency [33] https://mapy.geoportal.gov.pl/imap/Imgp_2.html. Data processing deploying deep learning methods was performed in the ENVI software, developed by Harris Geospatial Solutions, Inc., while geo-processing and final evaluation of remotely extracted building were done with ArcGIS software, provided by ESRI.

3.3. *Workflow and Applied*

The adopted methodology for automated building outlines extraction consisted of the three main stages and their subsequent processes. They are as follows: (1) building outlines extraction by the U-Net fully convolutional network from aerial orthoimages; (2) building vectors of rooftop surfaces, denoted as building roof outlines, extraction based on LiDAR data and (3) geoprocessing of building roof outlines and final evaluation of building outlines (see Figure 2).

The main assumption underlying this three-step methodology was as follows. FCN deep-learning algorithms for building extraction from remote sensing images, while universal in their nature, derive rough planar geometry of building outlines, which could not meet the requirements of the cadastral system in Poland. In a second step, the LiDAR densely classified data (DSM) improved the extraction of building outlines by creating vectors of the rooftop surface. Finally, the geo-processing stage transformed the roof vectors into geometrically corrected building outlines and assessed the accuracy of the buildings' locations.

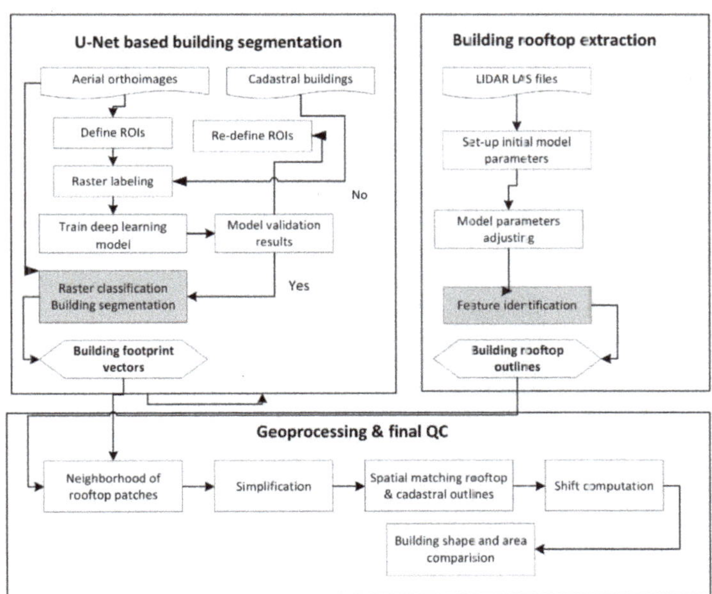

Figure 2. Schematic flow chart of the building extraction approach.

3.3.1. Building Extraction by U-Net

In remote sensing applications, convolutional networks are generally used for image classification, where the output to an image is a single class label. In this paper, we used U-Net architecture developed by Ronneberger et al. [50] and implemented in ENVI software as the ENVINet5 [51]. This U-Net architecture was modified and extended in such a way that it provides precise image segmentation with a small number of training images. ENVINet5 consists of the repeated applications of two paths, a contracting one and an expansive one. The contracting path involves the repeated application of two 3 × 3 convolutions, followed by a rectified linear unit (ReLU) and a 2 × 2 max pooling operation with stride 2 for down-sampling. The expansive path, in turn, comprises the upsampling of the feature map followed by a 2 × 2 up-convolution which reduces by half the number of feature channels, as well as a concatenation with a cropped feature map from the contracting path and a two 3 × 3 convolution followed by ReLU. Based on Ronneberger et al. [50], cropping is essential because of the loss of border pixels in every convolution. The adopted for this study ENVINet5 (see Figure 3) is characterised by 23 convolutional layers and a TensorFlow model added on the top of the network. The initialization model and the trained model in ENVI use data in HDF5 (.h5) format.

The model was trained based on building vector layers, derived from the cadastre, and 421 regions of interests (ROIs). The patch size for the training model was determined empirically based on the coincidence with label rasters (number of edge-length pixels). For our experiment, the patch size was 416 × 416 pixels. The following training parameters: number of epochs: 25; number of patches per image: 300; solid distance: 10.0; blur distance: 0. The class weight is used to highlight feature pixels at the start of training, class weights: min 1, max 2; and loss weight: 0.5.

The implemented U-net architecture loss function architecture was binary cross-entropy with the weighted map [50]:

$$E = \sum_{x \in \Omega} \omega(x) log\left(p_{l(x)}(x)\right) \quad (1)$$

where $p_{l(x)}(x)$ is the softmax loss function $l: \Omega \rightarrow \{1, \ldots, K\}$ as the true label of each pixel and $\omega : \Omega \rightarrow \mathbb{R}$ is a weight map, in order to give a higher weight to a pixel near t to the boundary point in the image. The optimizer used in this experiment was the Stochastic Gradient Descent (SGD) with lr = 0.01 and momentum = 0.99.

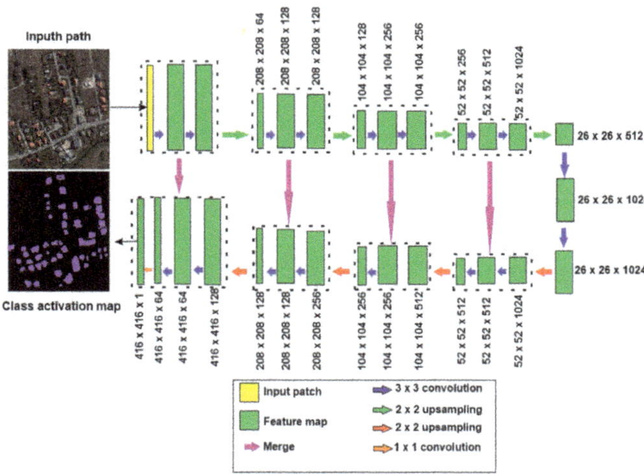

Figure 3. U-Shape Network (U-net) architecture used (based on [35]).

Five commonly known evaluation matrices were employed to assess the performance of the building outline extraction model: namely, overall accuracy, precision (correctness), recall (completeness), F1 score for pixel-based quality assessment and mean intersect over union (IoU) for per-object evaluation. The Overall Accuracy (OA) was calculated by summing the percentages of pixels that were correctly classified by the model compared to the reference labelled image (Equation (2)) [52].

$$OA = \frac{\sum_{i=0}^{k} p_{ii}}{\sum_{i=0}^{k} \sum_{j=0}^{k} p_{ij}} \tag{2}$$

where p_{ii} means the number of pixels for categories i correctly classified by the model, while p_{ij} means the number of pixels for categories i incorrectly classified into category j by the model and k is the category of building.

Precision is the fraction of true positive examples among the examples that the model classified as positive: in other words, the number of true positives divided by the number of false positives plus true positives (Equation (3)), as defined by [53,54]:

$$\text{recision} = \frac{\text{True Positives}}{\text{True Positives} + \text{False Positives}} \tag{3}$$

Recall, also known as sensitivity, is the fraction of examples classified as positive, among the total number of positive examples: in other words, the number of true positives divided by the number of true positives plus false negatives (Equation (4)):

$$\text{Recall} = \frac{\text{True Positives}}{\text{True Positives} + \text{False Negatives}} \tag{4}$$

The F1 (Equation (5)) measures the fraction of the number of true target-pixels identified in the detected target-pixels. The F1 score is the harmonic mean of the precision and recall, ranging from 0 to 1; the larger the F1, the better the prediction (Equation (5)):

$$F1 = 2 * \frac{\text{Precision} * \text{Recall}}{\text{Precision} + \text{Recall}} \tag{5}$$

The Intersect over union (IoU) metric, also known as the Jaccard similarity index, is denoted as an overlap rate of detected buildings and labelled as buildings, as it is presented in Equation (6):

$$\text{IoU} = \frac{\text{target} \cap \text{detected}}{\text{target} \cup \text{detected}} \tag{6}$$

3.3.2. Building Rooftop Extraction Using LiDAR Data

Algorithms that operate on dense LiDAR data (12 points per square meter) not only detect buildings and their approximate surface outlines but also extract flat roof surfaces, ultimately leading to the creation of building models that correctly resemble the roof structures [55]. In this study, ENVI LIDAR tools were used for building extraction and delineation of planar rooftop surfaces. The algorithm used identifies the correct position, aspect and slope of each roof plane in the work area, extracting consistent and geometrically correct 3D building models [56]. Buildings, trees and other objects (e.g., cars) were separated based on geometric criteria, such as size, height and shape characteristics. Overall, ENVI LiDAR procedures filter the data and classify each point of the cloud in a few steps (see Figure 2). Building rooftop patches were extracted, after extensive tests, by applying a threshold to the following parameter: minimum building area, Near Ground Filter Width, Buildings Points Range and Plane Surface Tolerance (PST). The values of these parameters were adopted after the analysis of the height and intensity map (Figure 4).

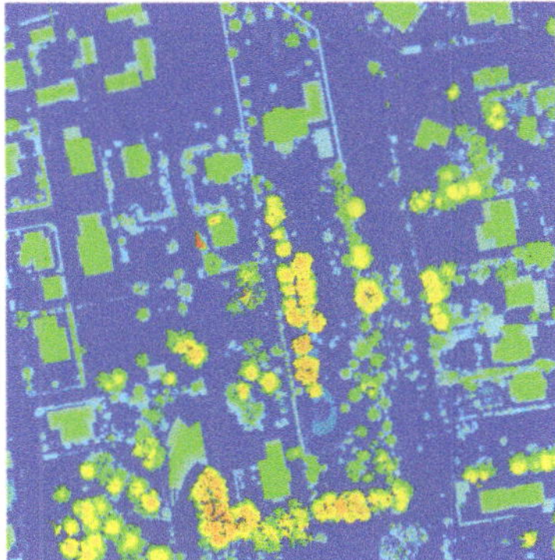

Figure 4. LiDAR data density, the selected example.

The minimum building area was set-up as 10 m^2. The Near Ground Filter Width of 5 m allowed one to classify as buildings only features located 5 m above the ground. The Buildings Points Range denotes spatial variation of a building's points and is used for a planar scan when the point density is not constant inside the analyzed area or when there are some holes in the point cloud dataset. The value of this parameter was set to 1.5 m. The Plane Surface Tolerance (PST) [51,55] recognizes curved roofs based on a series of successive planes. The algorithm used defines a new roof plane when the distance between the analyzed and previous points in the cloud reaches the declared value. Due to the importance of this parameter for distinguishing the building rooftop outlines, the classification was preceded by an analysis of the selection of the optimal value of the PST parameter. The test was performed for PSTs of 0.15 m, 0.30 and 0.5 m. As shown in Figure 5a, a low PST (i.e., 0.15 m) resulted in the incorrect classification of small construction facilities, like bowers or garden sheds. The best results were achieved for a PST value equal to 0.5 m (Figure 5b). Finally, the completeness of 95% was achieved in comparison with building cadastral data.

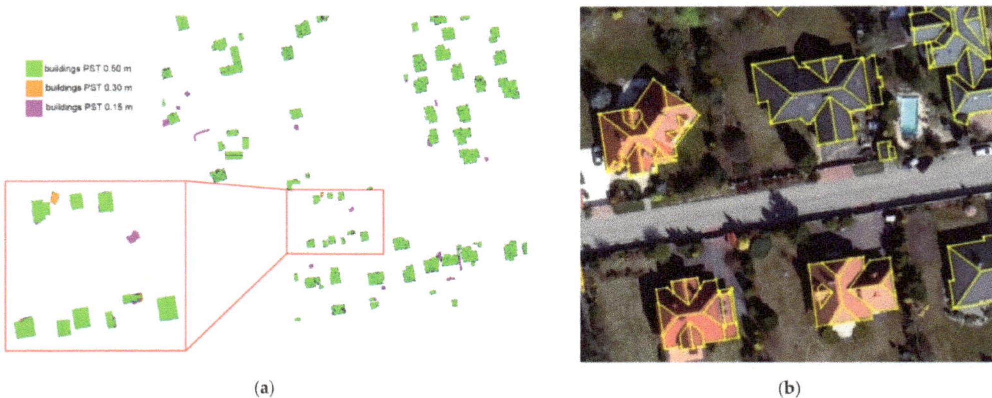

(a) (b)

Figure 5. LiDAR rooftop polygons: (**a**) adjustment of the Plane Surface Tolerance values, (**b**) selected examples of building rooftop outlines (yellow lines) superimposed on aerial orthoimagery.

3.3.3. Geoprocessing of Building Roof Outlines and Final Evaluation of Building Extraction

The results of automatic building extraction based on very high-resolution orthoimagery and LiDAR data provide the polygons for building roof outlines and planar shapes of building rooftop vectors. As mentioned in the literature [57,58], building footprints and outlines of rooftop shapes differ both in shape and location. These differences, albeit slight, make it necessary to correct the geometric building outlines in order to locate the building accurately. The geometrical adjustment employed primary determination of the neighborhood relations of rooftop patches contained by a building's outline (derived from orthoimage segmentation) and creation of a draft polygon vector layer of building outlines. For this purpose, the spatial join and dissolve functions were used. Then, the draft building rooftop outlines were simplified by identifying and removing redundant vertices, according to the Douglas–Peucker algorithm [59]. As a result, the vertices of the building outlines were reduced, and the building outline itself was simplified in accordance with the state-of-the-art of cartographic generalization [60]. The next task was to extract the edge points of buildings from the reference layer (cadastral building data), calculate the distance to the extracted building outlines and finally, assess the accuracy of remote extraction of building outlines. The following measures were used for the final accuracy assessment: the mean, standard deviation, relative standard deviation (RSD), variance-to-mean ratio (VMR). The measures were employed for the distance between the extracted buildings and buildings from cadaster, and also the differences in shape and area of the buildings.

4. Results

4.1. Buildings Segmentation

The adopted training parameters of U-Net (namely patch size: 416; number of epochs: 25; number of patches per image: 300; solid distance: 10.0; blur distance: 0; class weights: min 1, max 2; and loss weight: 0.5) resulted in a well-suited model for building extraction. As shown in Figure 6, the training validation accuracy has increased while the validation loss has decreased. Both the validation accuracy and validation loss trends indicate that the algorithm used was well optimized and can be applied for image segmentation.

Figure 7 depicts the results of consecutive processes of building extraction based on deep-learning U-Net from very high-resolution aerial imagery.

The overall accuracy is 89.5%, indicating that the U-Net model used was well suited to extracting buildings. The precision informed us that all features that were labelled as buildings in 76.5% of instances were actually buildings. Recall indicated that 80.7% of all buildings in the analyzed area were detected. For all accuracy measures, see Table 1, while Figure 7 shows some examples of correct and incorrect building extraction.

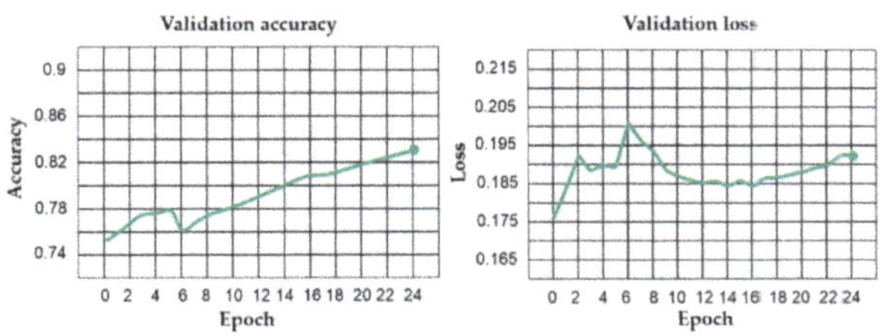

Figure 6. U-Net training validation.

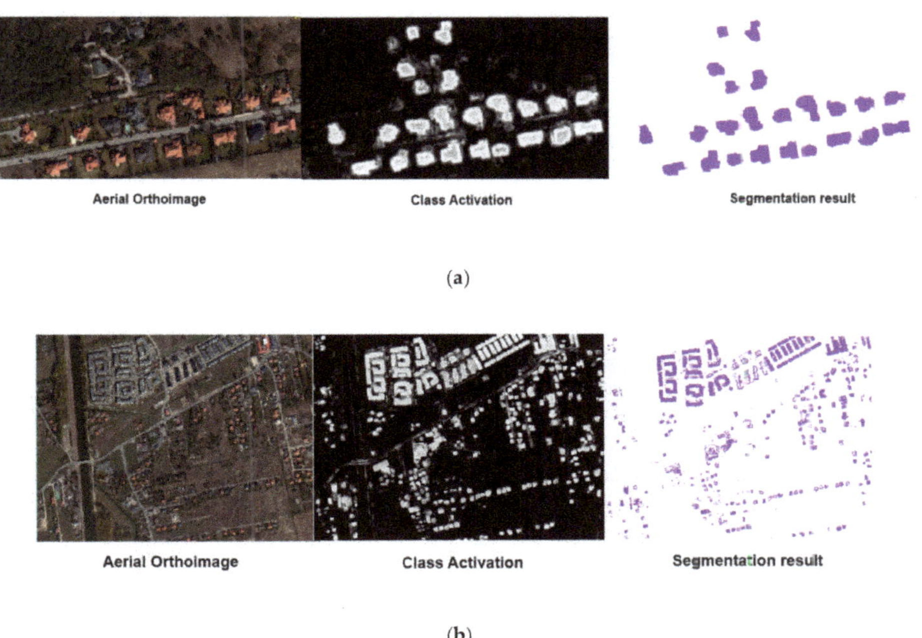

Figure 7. Buildings segmentation results: (**a**) single-family detached buildings, (**b**) multi-family building-blocks.

Table 1. Building outlines' accuracy.

Overall Accuracy (%)	Precision	Recall	F1-Score	Per-Object IoU Mean
89.5	0.765	0.807	0.785	0.748

Factors which influenced the accuracy of automated building extraction in general are as follows:

- variation in the spatial pattern of buildings and their surroundings, i.e., trees, paved roads, driveways, vehicles, porches, small garden houses or play-grounds;

- multiple colors of roofs, i.e., reddish, grayish, whitish and greenish, as well as roof installations such as satellite TV antennas, solar panels, dormers;
- building types, e.g., single-family detached or attached, semi-detached, multi-family buildings.

Figure 8a portrayals a group of multi-family houses. The building's roofs were dark grey, clearly separated from the light grayish color of the interior roads and the greenish-brown grassy surroundings as well as a few whitish cars. However, the close location of the buildings inside the estate, in rows of three, led to their extraction as one elongated building. Figure 8b shows single-family attached buildings, characterized by hipped reddish and grayish roofs, partly shadowed by trees. The ENVINet5 algorithm correctly extracted all buildings; nevertheless, the touching buildings, i.e., buildings with adjacent roofs and walls, were identified as one feature. The next example (Figure 8c) demonstrates correctly extracted single-family buildings, despite some disturbance caused by vegetation and other objects located on the land parcel. The algorithm, however, had a problem with the correct building extraction in an area characterized by a less dynamic color scheme (dark greyish) between the roofs and the parking lot and nearby streets. It was indicated as false positive, over-classified (indicated in red—Figure 8d). Another error that was observed in this area was the identification of cars as buildings.

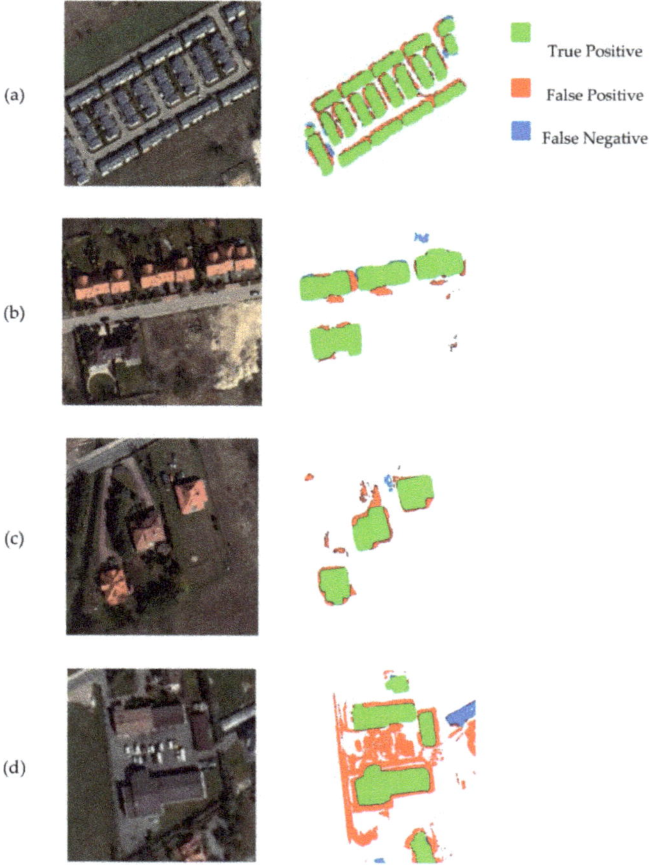

Figure 8. Results of RGB orthoimagery segmentation based on U-Net: (**a**) dense built-up area; (**b**) multiple colors of roofs; (**c**) complex roof shapes and vegetation vicinity; (**d**) complex pattern of buildings, parking lots, roads.

4.2. Building Rooftop Patches Extraction

Kobiałka, as a residential district on the outskirts of Warsaw, is characterized by a diverse building architecture, which affected the building extraction. Particularly, multi-slope roofs, the presence of dormers and the accompanying vegetation partially covering the roof edges (trees and tall shrubs) hindered the extraction algorithm (as it is seen in Figure 9) and required further geoprocessing, described in the Section 4.3.

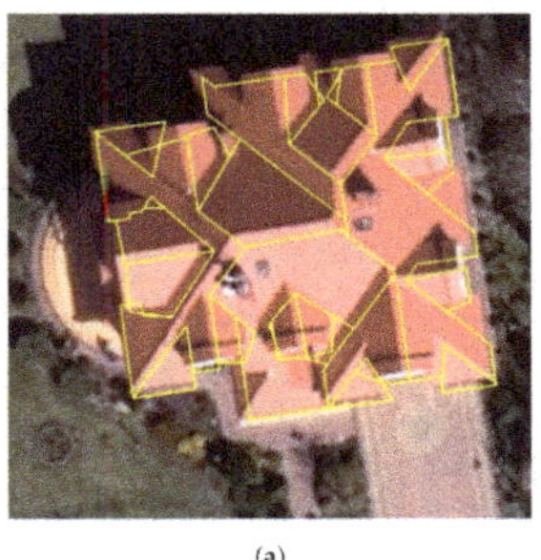

(a)

(b)

Figure 9. Buildings' rooftop polygons hampered by: (**a**) complex, multi-slope roof; (**b**) vegetation cover.

4.3. Geoprocessing of Building Rooftop Outlines

The geoprocessing of building rooftop patches was intended to provide geometrically corrected building outlines. Figure 10 shows some examples of the geoprocessing results. Building rooftop outline extracted during segmentation with the deep-learning algorithm were used to spatially match these roof patches that pointed to one building (Figure 10a) and outlines of rooftops were created (Figure 10b). These roof outlines consisted of many vertex points, which led to a distortion of the shapes of the buildings. Therefore, generalization and simplification were of utmost importance (Figure 10c). The geometrically corrected building roof outlines still did not match the building footprints stored in cadastral data, as it is seen in Figure 10d. This was due to the fact that in our approach, based on remotely sensed data, the outline of the building was considered as a building's roof footprint, while in the cadastre it is represented by the walls of the building, the outline on the ground.

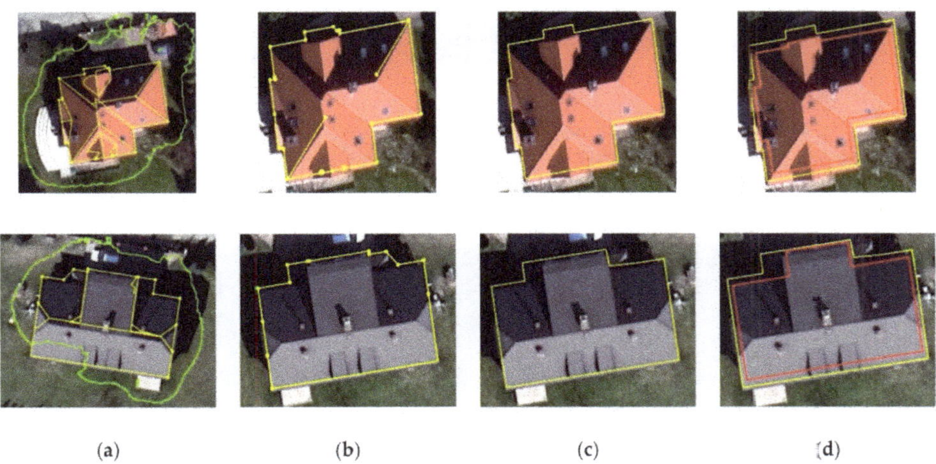

(a) (b) (c) (d)

Figure 10. Geoprocessing of buildings rooftop polygons: (**a**) building rooftop patches (yellow lines) within deep learning delineated building footprints (green line); (**b**) building the rooftop outline; (**c**) corrected and simplified rooftop outline; (**d**) shift in building outlines (yellow line) and building footprints from cadastral data.

The statistical analysis of the roof overhead showed an average shift of 1.18 m (with STD equal to 0.688) between the building outlines and the cadastral data. The dispersion index (VMR) of the shift value (near distance) amounted to 0.4, indicating a binomial distribution (under dispersion) (Figure 11a). Less than 5.6% of the edge points of the building outlines (190 out of 3406) were perceived as outliers because the near distances between corresponding edges in the compared datasets were greater than 2.44 m (see Figure 11b).

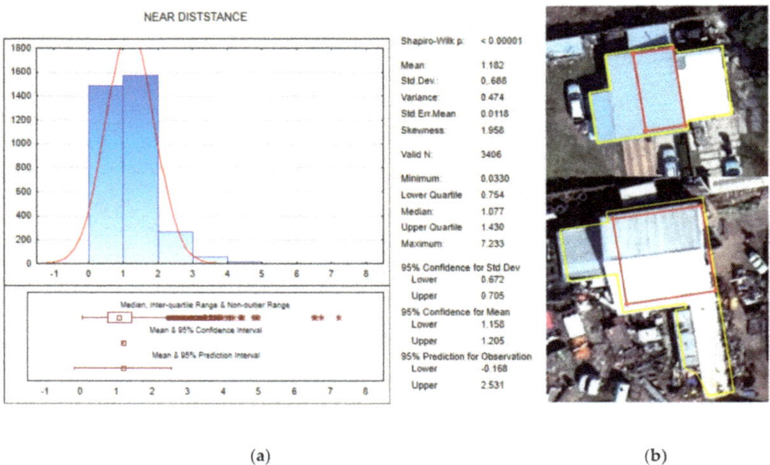

(a) (b)

Figure 11. Differences between building outlines from remote sensing and cadastral data: (**a**) descriptive statistics, (**b**) analyzed cadastral buildings contained two outlines derived from LiDAR data processing (see an example in Figure 12a).

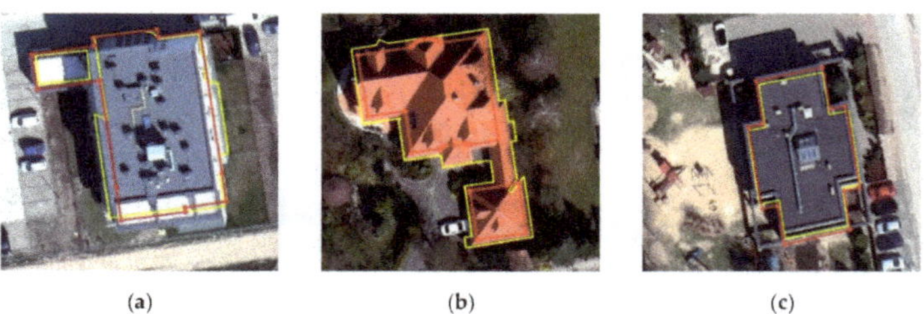

(a) (b) (c)

Figure 12. Building outlines: yellow from LiDAR, red from the cadastre: (**a**) multi-family building extracted as two building outlines; (**b**) two connected single-family houses denoted by LiDAR as one building; (**c**) outlines of a kindergarten.

For the remaining 234 buildings, the mean difference between the cadastral area and the area calculated from the LiDAR outline was 57.37 m^2, and the standard deviation was 2.59 m^2. The highest values in the differences in building area were observed for attached single-family houses, which were distinguished as one building during LiDAR data processing (as shown in Figure 12b).

The Shape index (SHI) indicates the similarity between cadastral building outlines and LiDAR outline shapes irrespective of their areas. Statistical measures of dispersion (RSD and VMR) showed a weakly dispersed binomial distribution, assuming RSD 9.45% and 14.59% and VMR 0.01 and 0.02 for the outlines of cadastral and LiDAR buildings, respectively. Statistical measures of dispersion (RSD and VMR) showed a weakly dispersed binomial distribution, assuming RSD 9.45% and 14.59% and VMR 0.01 and 0.02 for the outlines of cadastral and LiDAR buildings, respectively.

5. Discussion

Buildings are among the most important and valuable objects in cadastral systems and, due to their economic and social roles, require frequent updating of cadastral data. Moreover, the data should be complete and of high positional and thematic accuracy. Although building cadastral data is generally obtained through field surveys, in the past few decades, remote sensing techniques have increasingly replaced field surveys as being cost-effective and time-efficient (see [2,10,11,25]). Building extraction aims for the correct amount of buildings, no commission and omission, e.g., each building should be represented and only by one object. However, in practice, errors in building extraction are inevitable, and the algorithms are optimized and trained to minimize inaccuracies during image segmentation. The results of building extraction are characterized by accuracy measures that in general, indicated the efficacy of the methods used. Nevertheless, as noticed by Avbelj et al. [47], results reported in literature should not be directly compared due to the different accuracy measures, the variety of building extraction approaches, study areas types, remotely sensed and reference data used.

Maltezos et al. [24] reported that the buildings extracted by CNN were "more complete and solid" for both the Vaihingen and Perissa research areas compared to the results of the SVM (Support Vector Machine) classifiers. The overall pixel-based CNN accuracy ranged from 81% to 86%, referring to an average quality rate about 83.0%, while the accuracy for the linear SVM reached 76.3% and base function RBF SVM just 72.9%. Huang et al. [33] stated that developed GRRNet gave the best result with an overall accuracy per pixel of 96.20%. The authors also found that the mean IoU per-pixel values varied significantly depending on the test area and variants of the GRRNet model. The best results (90.59%) were achieved for the Baseline + GFL-2 (gated feature labelling-2) modification in New York and the lowest results of 67.47% for the Baseline + FF (feature fusion) modification in Arlington. The FCM model exploited by Bittner et al. [38] extracted the building footprint successfully with the IoU metric of about ~68.1% (also on pixel level), which is comparable with the accuracy obtained in our study (IoU = 64.7%). The SRSM proposed by Nguyen et al. [40] yields an average area-based quality of 62.37% and an object-based quality of 63.21% for Quebec.

Huang et al. [33] stated that developed GRRNet gave the best result with an overall accuracy of 96.20%. The authors also found that the mean IoU values varied significantly depending on the test area and variants of the GRRNet model. The best results (90.59%) were achieved for the Baseline + GFL-2 (gated feature labelling-2) modification in New York and the lowest results of 67.47% for the Baseline + FF (feature fusion) modification in Arlington. The dependence of the accuracy of building extraction on the network structure and the loss function was also noted by Shao et al. [34], with the variability of IoU/F1 per-pixel measures of 0.0582/0.0402 for network structure and 0.0121/0.0080.

Rottensteiner et al. [28] detected buildings utilizing the Dempster–Shafer method for the fusion of LiDAR data and from aerial imagery with the precision of 85% and recall (completeness) of 89%. They also noticed that the values of both measures depend on the building size (area), reaching the lowest values for small buildings of less than 40–50 m^2. Sohn and Dowman [29] received an overall accuracy of 80.5%, with a completeness of 88.3% and correctness equal to 90.1% using the Binary Space Partitioning (BSP) tree for processing fused IKONOS and LIiAR data. Kodorsa et al. [9] achieved 92% conformity of building recognition from LiDAR data using a saliency-based method. Wang et al. [21] reached a very high value of overall accuracy (94.12%) for building segmentation in the Massachusetts study region due to an innovative image processing method implementing the efficient Non-local Residual U-shape Network (ENRU-Net), composed of a U-shape encoder–decoder structure and an improved non-local block named the asymmetric pyramid non-local block (APNB). Reis et al. [39] investigated the availability of aerial orthophotos, for "cadastral works" and observed an object-based accuracy of 95% and pixel-based accuracy of 87% in comparison with cadastral data. However, the analyzed data sample included just a few buildings, so it is difficult to consider their results as representative. A profound and detailed accuracy analysis was presented in Khoshboresh-Masouleh et al. [61]. The authors evaluated building footprints' results for different types of built-up areas, e.g., shadowed, vegetation-rich, complex roofs and high-density, obtaining a mean IoU value of 76%. Nevertheless, for building footprints characterized by complex roofs, the average IoU was 74.5%. The high efficiency of the ENVI deep learning algorithm is also reported in Lai et al. [30], who noticed that the accuracy of building segmentation ranged from 80 to 90%. The regularization technique using LiDAR data and orthoimages proposed by Gilani et al. [41] obtained completeness from 83% to 93% for the area with a correctness of above 95%, which undoubtedly proved the reliability of this approach. The overall accuracy of 89.5% obtained in our study is consistent with the results reported in literature and can be considered adequate for cadastral purposes.

The land administration system, and cadaster as a part of it, are always tailored to the national possibilities and requirements. The applications of remote sensing data and deep learning techniques to update cadastral data and maps remain limited. Firstly, deep learning methods of building footprints extraction based on satellite and aerial images and LiDAR data allow one mainly to delineate the rooftop of buildings' outlines. Secondly, the obtained accuracy measures, mainly overall accuracy, completeness and RMSE, do not meet the requirements for cadastral data, and ultimately, the guarantee of the required high reliability of cadastral records is severely limited. Thirdly, it is challenging to precisely represent the shape of a building, especially for multi-walled buildings with multi-pitched roofs.

In Poland, the cadastre of land and buildings (denoted in Polish as EGiB) has a long tradition, and its transition from an analogue to a computer system, which began two decades ago, requires further improvement.

Practitioners and scientists still note many inconsistencies between cadastral data and field survey data, which means that the reliability of the building data is limited, and the position accuracy ranges from 0.7 to 1.5 m [17,62,63]. The latest regulation on technical standards for surveying measurements, in force from August 2020 [64], allow the use of modern photogrammetric techniques and remote sensing data to update cadastral data on buildings, provided that the accuracy of the location of the building footprint is not less than 0.1 m with reference to the nearest geodetic control point. As noted by Ostrowski et al. [65], such high accuracy can only be achieved when vectorizing the building contour manually on a stereomodel, what is labor-intensive, time-consuming and very expensive. Our three-step method is of great importance as it allows one to automate the building extraction and indicate areas where some discrepancies in building locations were noted, and ultimately to identify priority areas for possibility of the cadastre modernization. Particularly, the elaborated method could be applied for vector building data acquisition where it still does not operate in vector format, but as scanned analogue cadastral maps.

The entire territory of Poland is covered with high-resolution orthorectified aerial images, updated every five years, and dense LAS data with decimeter accuracy [64]. This undoubtedly creates great opportunities for the cadastre modernization based on remotely sensed data and advanced, automated feature extracting technologies, especially in areas where analogue or raster building cadastre operates. Our three-step method certainly belongs to such technologies. The applied algorithms could be used many times and on other areas, provided that for areas with a different building pattern and characteristics, the training area (ROIs) should be extended. Nevertheless, the large variation in types of built-up areas and building configurations in Poland leads to certain limitations in the fully automatic use of remote sensing data in cadastre modernization. Particularly, the problem of proper extraction of building footprints occurs for newly-built single-family houses, surrounded by greenery, with hipped roofs, dormers or solar panel installations (see Figure 9 as the example). In these areas, unfiltered noise points, as well as details on the roof surfaces, decreased the final accuracy of the roof shape reconstruction. Another challenge in many cell segmentation tasks is the separation of touching objects of the same class [21,22].

6. Conclusions

Recent developments in deep learning technology, as well as the availability of high-resolution aerial imagery and dense LAS data, offer fast and cost-effective ways of extracting buildings for cadastral purposes.

Although the automatically extracted building outlines cannot be directly uploaded the cadastral data due to differences in the outline (in the cadastre, it is the ground outline while the imagery data gives the roof outline), they could be successfully used when planning the cadastral modernization. Further research on reducing the roof outline to the outline of the building's ground floor is of utmost importance.

The experimental result showed that the proposed methodology achieved good results and was robust after adjusting the model parameters to the specifics of the analyzed area. This partially limits the possibility of transferring our approach to areas with different building characteristics and the use of other aerial images. However, this limitation could be reduced by adding ROI in such places to best portrayal the investigated area.

Summing up the discussion on the possibility of cadastre updating by remotely sensed data, it should be noted that deep learning methods for buildings extraction are a promising technology; however, in many countries, it could only be used to indicate areas where the cadastre needs updating. Final registration in the cadastral system yet requires more accurate field measurements according to national cadastral standards.

Author Contributions: Conceptualization and methodology, D.W., E.B. and O.M.; validation, D.W., O.M.; formal analysis, D.W., O.M.; investigation, D.W., O.M.; resources, O.M.; writing—original draft preparation, E.B., D.W.; writing—review and editing, D.W.; visualization, D.W., O.M. and E.B.; supervision, E.B.; project administration, O.M.; funding acquisition, O.M. All authors have read and agreed to the published version of the manuscript.

Funding: This research was funded by Military University of Technology, Faculty of Civil Engineering and Geodesy, grant number DPH: 2/DPIH/2020.

Institutional Review Board Statement: Not applicable.

Informed Consent Statement: Not applicable.

Data Availability Statement: Data used in this study, i.e.,: buildings, aerial orthoimages and LAS data, were derived from the national geoportal of Poland https://mapy.geoportal.gov.pl/imap/Imgp_2.html.

Conflicts of Interest: The authors declare no conflict of interest.

References

1. Enemark, S.; Williamson, I.; Wallace, J. Building Modern Land Administration Systems in Developed Economies. *J. Spat. Sci.* **2005**, *50*, 51–68. [CrossRef]
2. Choi, H.O. An Evolutionary Approach to Technology Innovation of Cadastre for Smart Land Management Policy. *Land* **2020**, *9*, 50. [CrossRef]
3. Williamson, L.P.; Ting, L. Land administration and cadastral trends—A framework for re-engineering. *Comput. Environ. Urban Syst.* **2001**, *25*, 339–366. [CrossRef]
4. Bennett, R.; Wallace, J.; Williamson, I. Organising land information for sustainable land administration. *Land Use Policy* **2008**, *25*, 126–138. [CrossRef]
5. Agenda 21. Action Programme—Agenda 21. 1992. Available online: https://sustainabledevelopment.un.org/content/documents/Agenda21.pdf (accessed on 25 September 2020).
6. Agenda 2030. Transforming Our World: The 2030 Agenda for Sustainable Development. Available online: https://sdgs.un.org/2030agenda (accessed on 25 September 2020).
7. UN Economic and Social Council. Progress towards the Sustainable Development Goals. Report of the Secretary-General. E/2017/66, 28 July 2016–27 July 2017. Available online: https://www.un.org/ga/search/view_doc.asp?symbol=E/2017/66&Lang=E (accessed on 25 September 2020).
8. Estoque, R.C. A Review of the Sustainability Concept and the State of SDG Monitoring Using Remote Sensing. *Remote Sens.* **2020**, *12*, 1770. [CrossRef]
9. Kodors, S.; Rausis, A.; Ratkevics, A.; Zvirgzds, J.; Teilans, A.; Ansone, J. Real Estate Monitoring System Based on Remote Sensing and Image Recognition Technologies. *Procedia Comput. Sci.* **2017**, *104*, 460–467. [CrossRef]

10. Zahir, A. Assessing Usefulness of High-Resolution Satellite Imagery (HRSI) in GIS-based Cadastral Land Information System. *J. Settl. Spat. Plan* **2012**, *3*, 111–114.
11. Janowski, A.; Renigier-Biłozor, M.; Walacik, M.; Chmielewska, A. Remote measurement of building usable floor area—Algorithms fusion. *Land Use Policy* **2021**, *100*, 104938. [CrossRef]
12. Bielecka, E.; Dukaczewski, D.; Janczar, E. Spatial Data Infrastructure in Poland–lessons learnt from so far achievements. *Geod. Cartogr.* **2018**, *67*, 3–20. [CrossRef]
13. Mika, M.; Kotlarz, P.; Jurkiewicz, M. Strategy for Cadastre development in Poland in 1989–2019. *Surv. Rev.* **2020**, *52*, 555–563. [CrossRef]
14. Mika, M. An analysis of possibilities for the establishment of a multipurpose and multidimensional cadastre in Poland. *Land Use Policy* **2018**, *77*, 446–453. [CrossRef]
15. Noszczyk, T.; Hernik, J. Understanding the cadastre in rural areas in Poland after the socio-political transformation. *J. Spat. Sci.* **2019**, *64*, 73–95. [CrossRef]
16. *Geodetic and Cartographic Law*; (Official Journal 2010 No 193, Item 1287); Official Journal of Laws: Warsaw, Poland, 2010.
17. Kocur-Bera, K.; Stachelek, M. Geo-Analysis of Compatibility Determinants for Data in the Land and Property Register (LPR). *Geosciences* **2019**, *9*, 303. [CrossRef]
18. GUGiK. *Budowa Zintegrowanego Systemu Informacji o Nieruchomościach –Faza II*; Development of the Integrated System of Real Estate Information: Warsaw, Poland, 2018. Available online: http://www.gugik.gov.pl/__data/assets/pdf_file/0009/92664/ZSIN-II.pdf (accessed on 7 January 2021).
19. EGiB Regulation. *Regulation of the Minister of Regional Development and Construction of 28 February 2019 Amending the 2001 Regulation on the Register of Land and Buildings*; Official Journal 2019, Item 397; Official Journal of Laws: Poland, Warsaw, 2019.
20. Aung, H.T.; Pha, S.H.; Takeuchi, W. Building footprint extraction in Yangon city from monocular optical satellite image using deep learning. *Geocarto Int.* **2020**. [CrossRef]
21. Wang, S.; Hou, X.; Zhao, X. Automatic building extraction from high-resolution aerial imagery via fully convolutional encoder-decoder network with non-local block. *IEEE Access* **2020**, *8*, 7313–7322. [CrossRef]
22. Liu, Y.; Zhou, J.; Qi, W.; Li, X.; Gross, L.; Shao, Q.; Zhao, Z.; Ni, L.; Fan, X.; Li, Z. ARC-Net: An Efficient Network for Building Extraction from High-Resolution Aerial Images. *IEEE Access* **2020**, *8*, 154997–155010. [CrossRef]
23. Zhang, Z.; Vosselman, G.; Gerke, M.; Persello, C.; Tuia, D.; Yang, M.Y. Detecting Building Changes between Airborne Laser Scanning and Photogrammetric Data. *Remote Sens.* **2019**, *11*, 2417. [CrossRef]
24. Maltezos, E.; Doulamis, N.; Doulamis, A.; Ioannidis, C. Deep convolutional neural networks for building extraction from orthoimages and dense image matching point clouds. *J. Appl. Remote Sens.* **2017**, *11*, 042620. [CrossRef]
25. Bennett, R.; Oosterom, P.; Lemmen, C.; Koeva, M. Remote Sensing for Land Administration. *Remote Sens.* **2020**, *12*, 2497. [CrossRef]
26. Baltsavias, E.P. Object extraction and revision by image analysis using existing geodata and knowledge: Current status and steps towards operational systems. *ISPRS J. Photogramm. Remote Sens.* **2004**, *58*, 129–151. [CrossRef]
27. Lee, D.H.; Lee, K.M.; Lee, S.U. Fusion of Lidar and imagery for reliable building extraction. *Photogramm. Eng. Remote Sens.* **2008**, *74*, 215–225. [CrossRef]
28. Rottensteiner, F.; Trinder, J.; Clode, S.; Kubik, K. Using the Dempster–Shafer method for the fusion of LIDAR data and multispectral images for building detection. *Inf. Fusion* **2005**, *6*, 283–300. [CrossRef]
29. Sohn, G.; Dowman, I. Data fusion of high-resolution satellite imagery and LIDAR data for automatic building extraction. *ISPRS J. Photogramm. Remote Sens.* **2007**, *62*, 43–63. [CrossRef]
30. Lai, X.; Yang, J.; Li, Y.; Wang, M. A Building Extraction Approach Based on the Fusion of LiDAR Point Cloud and Elevation Map Texture Features. *Remote Sens.* **2019**, *11*, 1636. [CrossRef]
31. Ma, L.; Liu, Y.; Zhang, X.; Ye, Y.; Yin, G.; Johnson, B.A. Deep learning in remote sensing applications: A meta-analysis and review. *ISPRS J. Photogramm. Remote Sens.* **2019**, *152*, 166–177. [CrossRef]
32. Zhuo, X.; Fraundorfer, F.; Kurz, F.; Reinartz, P. Optimization of OpenStreetMap Building Footprints Based on Semantic Information of Oblique UAV Images. *Remote Sens.* **2018**, *10*, 624. [CrossRef]
33. Huang, J.F.; Zhang, X.C.; Xin, Q.C.; Sun, Y.; Zhang, P.C. Automatic building extraction from high-resolution aerial images and LiDAR data using gated residual refinement network. *ISPRS J. Photogramm. Remote Sens.* **2019**, *151*, 91–105. [CrossRef]
34. Shao, Z.; Tang, P.; Wang, Z.; Saleem, N.; Yam, S.; Sommai, C. BRRNet: A Fully Convolutional Neural Network for Automatic Building Extraction from High-Resolution Remote Sensing Images. *Remote Sens.* **2020**, *12*, 1050. [CrossRef]
35. Liu, Y.; Gross, L.; Li, Z.; Li, X.; Fan, X.; Qi, W. Automatic building extraction on high-resolution remote sensing imagery using deep convolutional encoder-decoder with spatial pyramid pooling. *IEEE Access* **2019**, *7*, 128774–128786. [CrossRef]
36. Kang, W.; Xiang, Y.; Wang, F.; You, H. EU-Net: An Efficient Fully Convolutional Network for Building Extraction from Optical Remote Sensing Images. *Remote Sens.* **2019**, *11*, 2813. [CrossRef]
37. Wu, G.; Shao, X.; Guo, Z.; Chen, Q.; Yuan, W.; Shi, X.; Xu, Y.; Shibasaki, R. Automatic Building Segmentation of Aerial Imagery Using Multi-Constraint Fully Convolutional Networks. *Remote Sens.* **2018**, *10*, 407. [CrossRef]
38. Bittner, K.; Adam, F.; Cui, S.Y.; Korner, M.; Reinartz, P. Building footprint extraction from VHR remote sensing images combined with normalized DSMs using fused fully convolutional networks. *IEEE J. Sel. Top. Appl. Earth Obs. Remote Sens.* **2018**, *11*, 2615–2629. [CrossRef]

39. Reis, S.; Torun, A.T.; Bilgilioğlu, B.B. Investigation of Availability of Remote Sensed Data in Cadastral Works. In *Cadastre: Geo-Information Innovations in Land Administration*; Yomralioglu, T., McLaughlin, J., Eds.; Springer: Cham, Switzerland, 2017. [CrossRef]
40. Nguyen, T.H.; Daniel, S.; Guériot, D.; Sintès, C.; Le Caillec, J.-M. Super-Resolution-Based Snake Model—An Unsupervised Method for Large-Scale Building Extraction using Airborne LiDAR Data and Optical Image. *Remote Sens.* **2020**, *12*, 1702. [CrossRef]
41. Gilani, S.A.N.; Awrangjeb, M.; Lu, G. An Automatic Building Extraction and Regularisation Technique Using LiDAR Point Cloud Data and Orthoimage. *Remote Sens.* **2016**, *8*, 258. [CrossRef]
42. Girard, N.; Tarabalka, Y. End-to-end learning of polygons for remote sensing image classification. In Proceedings of the IGARSS 2018–2018 IEEE International Geoscience and Remote Sensing Symposium, Valencia, Spain, 22–27 July 2018; IEEE: Valencia, Spain, 2018; pp. 2083–2086.
43. Zhao, K.; Kang, J.; Jung, J.; Sohn, G. Building Extraction from Satellite Images Using Mask R-CNN with Building Boundary Regularization. In Proceedings of the 2018 IEEE/CVF Conference on Computer Vision and Pattern Recognition Workshops (CVPRW), Salt Lake City, UT, USA, 18–22 June 2018; pp. 247–251. [CrossRef]
44. Girard, N.; Smirnov, D.; Solomon, J.; Tarabalka, Y. Polygonal Building Segmentation by Frame Field Learning. *arXiv* **2020**, arXiv:2004.14875.
45. Li, Z.; Wegner, J.D.; Lucchi, A. Topological Map Extraction from Overhead Images. In Proceedings of the IEEE/CVF International Conference on Computer Vision (ICCV), Seoul, Korea, 27–28 October 2019; pp. 1715–1724. [CrossRef]
46. Zhao, W.; Ivanov, I.; Persello, C.; Stein, A. Building Outline Delineation: From Very High Resolution Remote Sensing Imagery to Polygons with an Improved End-to-End Learning Framework. *Int. Arch. Photogramm. Remote Sens. Spat. Inf. Sci.* **2020**, *43*, 731–735. [CrossRef]
47. Avbelj, J.; Muller, R.; Bamler, R. A Metric for Polygon Comparison and Building Extraction Evaluation. *IEEE Geosci. Remote Sens. Lett.* **2015**, *12*, 170–174. [CrossRef]
48. Dey, E.K.; Awrangjeb, M. A Robust Performance Evaluation Metric for Extracted Building Boundaries from Remote Sensing Data. *IEEE J. Sel. Top. Appl. Earth Obs. Remote Sens.* **2020**, *13*, 4030–4043. [CrossRef]
49. Degórska, B. Spatial growth of urbanised land within the Warsaw Metropolitan Area in the first decade of the 21st century. *Geogr. Pol.* **2012**, *85*, 77–95. [CrossRef]
50. Ronneberger, O.; Fischer, P.; Brox, T. U-Net: Convolutional Networks for Biomedical Image Segmentation. In *Medical Image Computing and Computer-Assisted Intervention—MICCAI 2015*; Navab, N., Hornegger, J., Wells, W., Frangi, A., Eds.; MICCAI 2015, Lecture Notes in Computer Science; Springer: Cham, Switzerland, 2015; Volume 9351. [CrossRef]
51. Harris. Harris Geospatial Solutions, 2020: Train Deep Learning Models. Available online: https://www.l3harrisgeospatial.com/docs/TrainDeepLearningModels.html (accessed on 2 November 2020).
52. Pan, Z.; Xu, J.; Guo, Y.; Hu, Y.; Wang, G. Deep Learning Segmentation and Classification for Urban Village Using a Worldview Satellite Image Based on U-Net. *Remote Sens.* **2020**, *12*, 1574. [CrossRef]
53. Liu, W.; Yang, M.; Xie, M.; Guo, Z.; Li, E.; Zhang, L.; Pei, T.; Wang, D. Accurate Building Extraction from Fused DSM and UAV Images Using a Chain Fully Convolutional Neural Network. *Remote Sens.* **2019**, *11*, 2912. [CrossRef]
54. DeepAI. Available online: https://deepai.org/machine-learning-glossary-and-terms/machine-learning (accessed on 30 December 2020).
55. Rottensteiner, F.; Briese, C. A New Method for Building Extraction in Urban Areas from High-Resolution LIDAR Data. *Int. Arch. Photogramm. Remote Sens. Spat. Inf. Sci.* **2002**, *34 Pt 3A*, 295–301.
56. Africani, P.; Bitelli, G.; Lambertini, A.; Minghetti, A.; Paselli, E. Integration of LIDAR data into amunicipal GIS to study solar radiation. *Int. Arch. Photogramm. Remote Sens. Spat. Inf. Sci.* **2013**, *XL-1-W1-1*, 1–6. [CrossRef]
57. Baillard, C.; Schmid, C.; Zisserman, A.; Fitzgibbon, A. Automatic line matching and 3D reconstruction of buildings from multiple views. *IAPRS* **1999**, *32*, 69–80.
58. Vosselman, G.; Dijkman, S. 3D building model reconstruction from point clouds and ground plans. In Proceedings of the ISPRS Workshop: Land Surface Mapping and Characterization Using Laser Altimetry, Annapolis, MD, USA, 22–24 October 2001; Hofton, M.A., Ed.; pp. 37–43. Available online: http://www.isprs.org/proceedings/XXXIV/3-W4/pdf/Vosselman.pdf (accessed on 29 January 2021).
59. Douglas, D.; Peucker, T. Algorithms for the reduction of the number of points required to represent a digitized line or its caricature. *Can. Cartogr.* **1973**, *10*, 112–122. [CrossRef]
60. Lupa, M.; Kozioł, K.; Leśniak, A. An Attempt to Automate the Simplification of Building Objects in Multiresolution Databases. In *Beyond Databases, Architectures and Structures*; Kozielski, S., Mrozek, D., Kasprowski, P., Małysiak-Mrozek, B., Kostrzewa, D., Eds.; BDAS 2015, Communications in Computer and Information Science; Springer: Cham, Switzerland, 2015; Volume 521. [CrossRef]
61. Khoshboresh-Masouleh, M.; Alidoost, F.; Hossein, A. Multiscale building segmentation based on deep learning for remote sensing RGB images from different sensors. *J. Appl. Remote Sens.* **2020**, *14*, 034503. [CrossRef]
62. Hanus, P.; Benduch, P.; Pęska-Siwik, A. Budynek na mapie ewidencyjnej, kontur budynku i bloki budynku. *Przegląd Geod.* **2017**, *7*, 15–20. (In Polish) [CrossRef]
63. Buśko, M. Modernization of the Register of Land and Buildings with Reference to Entering Buildings into the Real Estate Cadastre in Poland. In Proceedings of the International Conference on Environmental Engineering. Vilnius Gediminas Technical University, Vilnius, Lithuania, 27–28 April 2017. [CrossRef]

64. Ministry Regulation, 2020; Regulation of the Ministry of Development of 18 August 2020 r. On Technical Standards for the Performance of Situational and Height Measurements as Well as the Development and Transfer of the Results of These Measurements to the State Geodetic and Cartographic Resource. Official Journal 2020, item 1429. Available online: https://www.dziennikustaw.gov.pl/D2020000142901.pdf (accessed on 28 September 2020).
65. Ostrowski, W.; Pilarska, M.; Charyton, J.; Bakuła, K. Analysis of 3D building models accuracy based on the airborne laser scanning point clouds. *Int. Arch. Photogramm. Remote Sens. Spat. Inf. Sci.* **2018**, *42*, 797–804. [CrossRef]

 remote sensing

MDPI

Article

Building Change Detection Method to Support Register of Identified Changes on Buildings

Dušan Jovanović *, Milan Gavrilović, Dubravka Sladić, Aleksandra Radulović and Miro Govedarica

Faculty of Technical Sciences, University of Novi Sad, 106314 Novi Sad, Serbia; milangavrilovic@uns.ac.rs (M.G.); dudab@uns.ac.rs (D.S.); sanjica@uns.ac.rs (A.R.); miro@uns.ac.rs (M.G.)
* Correspondence: dusanbuk@uns.ac.rs; Tel.: +381-63-103-8664

Abstract: Based on a newly adopted "Rulebook on the records of identified changes on buildings in Serbia" (2020) that regulates the content, establishment, maintenance and use of records on identified changes on buildings, it is expected that the geodetic-cadastral information system will be extended with these records. The records contain data on determined changes of buildings in relation to the reference epoch of aerial or satellite imagery, namely data on buildings: (1) that are not registered in the real estate cadastre; (2) which are registered in the real estate cadastre, and have been changed in terms of the dimensions in relation to the data registered in the real estate cadastre; (3) which are registered in the real estate cadastre, but are removed on the ground. For this purpose, the LADM-based cadastral data model for Serbia is extended to include records on identified changes on buildings. In the year 2020, Republic Geodetic Authority commenced a new satellite acquisition for the purpose of restoration of official buildings registry, as part of a World Bank project for improving land administration in Serbia. Using this satellite imagery and existing cadastral data, we propose a method based on comparison of object-based and pixel-based image analysis approaches to automatically detect newly built, changed or demolished buildings and import these data into extended cadastral records. Our results, using only VHR images containing only RGB and NIR bands, showed object identification accuracy ranging from 84% to 88%, with kappa statistic from 89% to 96%. The accuracy of obtained results is satisfactory for the purpose of developing a register of changes on buildings to keep cadastral records up to date and to support activities related to legalization of illegal buildings, etc.

Keywords: image segmentation; neural network; classification; building footprint extraction; cadastre; change detection; VHR aerial images

Citation: Jovanović, D.; Gavrilović, M.; Sladić, D.; Radulović, A.; Govedarica, M. Building Change Detection Method to Support Register of Identified Changes on Buildings. *Remote Sens.* **2021**, *13*, 3150. https://doi.org/10.3390/rs13163150

Academic Editor: Mila Koeva

Received: 30 June 2021
Accepted: 29 July 2021
Published: 9 August 2021

Publisher's Note: MDPI stays neutral with regard to jurisdictional claims in published maps and institutional affiliations.

1. Introduction

With advanced technology related to collecting geospatial data in the 21st century, each organization is faced with growing volumes of spatial data. Geospatial data can originate from different satellites, airplanes or even UAV platforms. Collected data vary from large amounts of LiDAR data to the satellite and aerial images with different spatial resolutions. Very high spatial resolution (VHR) optical satellite imageries have increased their usability in applications of change detection and urban monitoring. Classification of VHR images requires a significant research task in remote sensing and image analysis; thus, it has great importance in infrastructure planning and change detection in the urban area, etc. [1]. Often, focus on these applications is on the classification of urban structures and identification, characterization and quantification of change detection on footprints of buildings or buildings' rooftops.

The manual method of collecting individual building information with attributes is very expensive and time-consuming. Automatic extraction of building information using high resolution (HR) remote sensing images is one of the widely used methods globally.

Even though building footprint extraction has received quite a bit of attention in the computer vision community, most approaches use supplemental data, such as point clouds, building height information, and multi-band imagery—all of which are too expensive to produce or unattainable for most cities worldwide [2].

In the past, many methods based on automatic or semi-automatic processes were developed for building footprint extraction. The large variations in appearance, geometry and spectral properties of buildings altogether make it a challenging task to enable automated building extraction at large scale from remote sensing satellite imagery. The spectral and geometrical properties of buildings such as rectangular shape, homogeneous surface (uniform spectra) and low-level features such as edges, lines/corners and association of shadows with the buildings are a few of the fundamental elements of buildings [3].

Within the context of increasing availability of high-resolution imagery, image segmentation is regarded as a solution to automate conversion of the raw data into tangible information, which is required in many application domains [4]. With the advancement of remote sensing technology, spatial resolution has become finer to the extent that even the smallest objects consist of a larger number of pixels, which has significantly increased variability within the class, making it difficult to classify at the pixel level and highlighting the need for segmentation and object-oriented classification of images [5]. Hossain and Chen in their review paper found 290 publications since 1999 that are related to object-based image analysis (OBIA) or image segmentation [6]. A variety of image segmentation methods have been developed. According to Schiewe [7], these methods can be grouped into three categories: edge-based, region-based, and hybrid segmentation methods. Meinel and Neubert [8] compared several segmentation methods and concluded that region merging is the most effective image segmentation method for the analysis of high-resolution remote sensing images. A characteristic region-merging method and probably the most widely used one within the OBIA domain is multi-resolution segmentation (MRS) [9], which is available in the eCognition® software package. Multiresolution segmentation (MRS) is now one of the most important algorithms in the object-oriented analysis of remote sensing imagery [10]. MRS performs region-growing segmentation, starting at the pixel level, then adjacent pixels are merged if they are homogeneous. The unitless user-set scale parameter represents the heterogeneity threshold, below which merging occurs, and above which merging stops. The higher the scale parameter, the higher the allowed within-segment heterogeneity, and thus the larger the resulting image segments. For a given scale parameter, heterogeneous regions will have smaller segments than homogeneous regions. The scale parameter is defined as the maximum standard deviation of the homogeneity criteria, which are a weighted combination of color and shape values. The user can adjust the relative weights (importance) assigned to each [11]. In terms of the segmentation software adopted in the 254 case studies, studies on segmentation using eCognition software account for 80.9% [12].

Myint et al. [13] used QuickBird satellite image with four spectral ranges of 2.4 m spatial resolution for the Phoenix area in Arizona. The target materials or classes in both types of classification were buildings, other man-made structures (roads and parking lots), land, trees/shrubs, grass, swimming pools and lakes/ponds. The authors used multi-resolution segmentation, where they used different parameters and weight factors of this method for each class. In the classification process, the authors used different parameters (scale factor, compactness and shape) in the segmentation process and created additional data that they used for each of the classes. Thus, in addition to the basic bands, the authors created the PCA layer and NDVI, and in the classification, they used membership function classifier and nearest neighbor.

Zhou et al. [14] presented a study with WorldView-2 images to provide detailed land use and land cover map at local level using OBIA method. The result shows that a total of nine LULC classes have been successfully classified with an overall classification accuracy of 79.4%. Masayu at el. [15] used Sentinel 2 images to classify buildings in Selangor, Malaysia, using suitable segmentation parameters and features. They used support vector machine (SVM) and decision tree (DT) classifiers to categorize five different classes: water, forest, green area, building, and road. Accuracy of the classified image using a different number of object features was assessed and it shown that the number of features applied will affect the classification accuracy. Additionally, the overall accuracy of SVM was 93% and for DT was 73%. Masayu at el. [16] extracted building footprint from high-resolution Worldview 3 (WV3) satellite data. They used 25 experiments with three segmentation parameters (scale, shape, and compactness), each having five varying values that directly affect the quality of segmentation. With optimal parameters used for segmentation process in eCognition, image objects were classified into five land cover classes (building, road, water, trees, and grass) by using a supervised non-parametric statistical learning technique, support vector machine (SVM) classifier.

In the last few years, algorithms for segmentation of satellite images based on convolutional neural networks have been developed. These networks learn spatial-contextual characteristics directly from the input VHR image, efficiently integrating the feature extraction step into the training classifier [17]. One of the most successful algorithms is based on the fully convolutional network (FCN) [18]. The most common way to perform semantic segmentation is to use a convolutional neural network because it achieves very good results. For image segmentation, one of the most well-known architectures used is U-Net, which has a coder-decoder type structure [19]. U-Net is a type of fully convolutional network that was originally used for medical image analysis, but later this model began to be applied to the pixel-based classification of satellite images [20].

Automatic building extraction from remote sensing data is a hot but challenging research topic for cadastre verification, modernization and updating. Deep learning algorithms are perceived as more promising in overcoming the difficulties of extracting semantic features from complex scenes and large differences in buildings' appearance.

Geospatial objects change over time and this necessitates periodic updating of the cartography that represents them. Currently, this updating is done manually, by interpreting aerial photographs, but this is an expensive and time-consuming process.

Automatic detection of illegally built or changed buildings from satellite imagery is a specific and important problem for both the research community and government agencies, which has not been sufficiently investigated since it combines the challenge of automatic remote sensing data interpretation and verification with a cadastral map. Recovery of building footprints from satellite images is a very complicated process because building areas and their surroundings are represented with various color intensities and complex features. Yanan et al. [21] in their survey, with 195 different papers related to the change detection methods based on different remote sensing images and multi-objective scenarios, found that 61% of multi-source images are multispectral. Additionally, from the point of the multi-objective scenarios, urban change detection is mostly related to the buildings and roads, where for the buildings important information includes features such as unique roof feature, and shape of parallelograms that represent buildings.

The Republic Geodetic Authority (RGA) in Serbia is a government organization responsible for professional and public administration related to state survey, real estate cadastre, utility cadastre, topographic-cartographic activities, property valuation, and geodetic-cadastral. This organization is also responsible for the legal framework related to survey and cadastre. As part of this legal framework, the rulebook on records on determined changes on buildings was adopted in 2020 [22]. This rulebook defines the content, development, maintenance and use of the register about the determined changes on buildings based on satellite imagery. For this purpose, RGA conducted the acquisition of the VHR satellite imagery data for the entire country. The register will be built for the entire territory of the Republic of Serbia and will be part of the geodetic-cadastral information system (GCIS). The register should contain data about detected changes on objects in relation to the reference era of aerial or satellite photography. These changes include data about buildings that are not recorded in the real estate cadastre, and buildings that are recorded in GCIS, but are removed from the field or changed in terms of dimensions of the footprint recorded in the database. The register contains graphical and alphanumerical parts, i.e., geospatial data about footprints of the buildings, and attribute data that describe particular buildings (such as, unique identification number, building area, type of change, etc.)

The aim of this paper is to develop a procedure that will automate the development of the register on determined changes on buildings based on satellite imagery. Since there are no additional data for Serbia other than high-resolution images, we propose the use of pixel-based and object-based classification, over VHR images from two epochs, 2016 and 2020, as the first step in detecting changes in buildings. Furthermore, one of the largest issues is the existence of more than two million illegal buildings (according to the official data [23]) and a lack of software tools that can help in detection of illegal buildings (newly built or changed). With this proposed method we want to improve the current situation and support the implementation of the new "Rulebook on the records of identified changes on buildings" in Serbia. The Rulebook on records on determined changes on buildings was used as a reference framework for the development of the proposed approach that will result in (semi)automatic extraction of changes on buildings and update of cadastral records in the cadastral database in a timely manner. Our tests with VHR images containing only RGB and NIR bands showed object identification accuracy ranging from 84% to 88%, with kappa statistic from 89% to 96%. Based on these results, the LADM-based cadastral data model for Serbia is extended to include records on identified changes on buildings and their legality assessment.

The paper is structured as follows: after the introduction in Section 1, Section 2 presents materials and methods used. This section describes datasets used for the experiments, and overall methodology used for the development of the register on determined changes on buildings based on satellite imagery, based on pixel based and object-based methods. The data model of the register is also developed and described to demonstrate what information is necessary to extract from the satellite imagery to build a register according to the rulebook. Results of the development and assessment of the model are presented in Section 3. Furthermore, the results are classified into three categories of possible changes on buildings defined in the rulebook to support the conclusion that the method is capable of solving the problem defined in the rulebook. Discussion and conclusions are presented afterwards.

2. Materials and Methods

2.1. Study Areas and Datasets

The study areas are parts of two cities, Subotica and Zrenjanin, in the province of Vojvodina, located in the northern part of Serbia. Topography in the study area is without variations and it is characterized by a low altitude from 76 to 109 m. The geographical coordinates of Subotica are 46°06′ N and 19°39′ E, and for Zrenjanin 45°23′ N and 20°23′ E (Figure 1). These two cities are typical representatives of Serbia's type of settlements, in how they consist of urban parts (with high buildings and lot of impervious surfaces such as roads, parking lots, etc.) and rural parts where there are predominantly houses with big green spaces (parks or even gardens). Urban parts of the cities are similar not only in Serbia, but also in the wider area of this region, and rooftops are usually created from concrete or similar types of materials. On the other hand, what is typical for Serbia is a different type of roof on the houses, which is built from clay tiles oriented in two or four sides, and with lots of vegetation cover. We have chosen this kind of study area because with our proposed method we want to include all possible kinds of buildings that can be found in Serbia.

An area of 0.5 km^2 was selected as a training area for developing classification rules for building rooftop detection in both cities, and test area covered about 2.5 km^2. The accuracy of building rooftops detection was estimated based on the test area divided into two categories, as previously described, comprising urban parts with high buildings and areas that predominantly contain houses. Later, the same algorithm was applied to both cities, to verify its reliability and transferability to other settlements in Serbia.

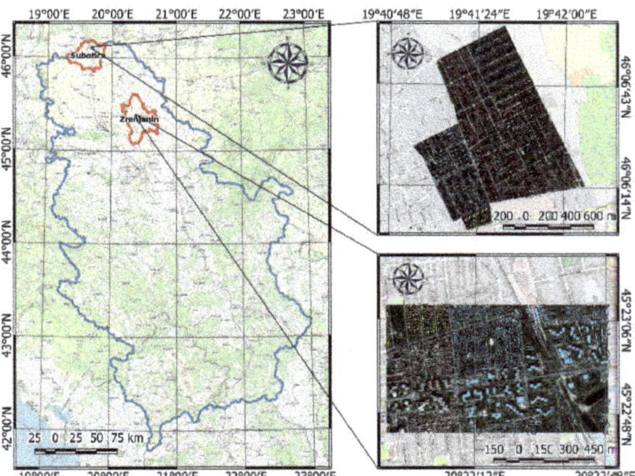

Figure 1. Area of interest.

The Worldview 2 image of the area of Zrenjanin was acquired on 12 April 2016 at an angle of 16.5°, with the cloud cover equal to 0%. The Worldview 2 image of the area of Subotica was acquired on 29 March 2020 at an angle of 14.1°, with the cloud cover equal to 1%. The WorldView-2 image includes four multispectral bands (Blue, Green, Red, and Near-Infrared-1). The data were purchased by the private company Vekom, involved as a partner in this research, and downloaded from the DigitalGlobe image archive, as a standard Ortho-Ready product projected on a plane with a UTM projection (Universal Transverse Mercator) and a WGS84 datum. The orthorectification procedure was performed by the employees of the private company and authors of this paper.

Objects' cadastral data are for the area of the city of Zrenjanin from 2016, and for the area of the city of Subotica from 2020.

2.2. Methods

The general idea and overall architecture of the proposed software solution is shown in Figure 2. According to the developed conceptual data model of the register of detected changes, a physical model and database schema are generated. The cadastral database is populated with the data obtained using a developed method for the (semi)automatic change detection, which is additionally inspected and corrected using traditional GIS editing tools.

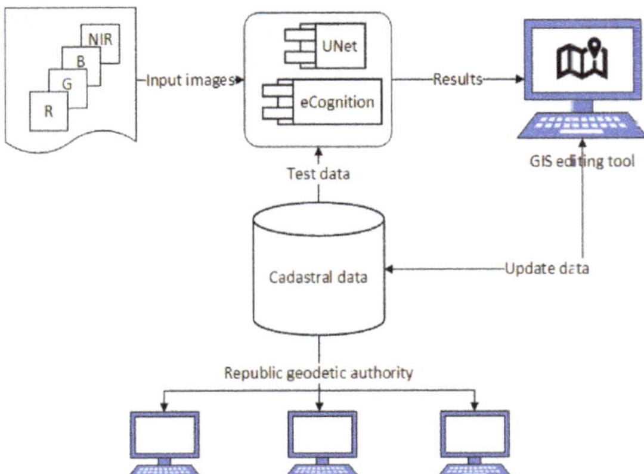

Figure 2. Overall architecture of the proposed solution.

The overview of the basic steps of the proposed methodology is shown in Figure 3. The first step after the pre-processing and validation of the proposed model is a detection of the building's footprint using satellite images with a high spatial resolution. After detection of the building's footprint and identification of changes on objects according to the rulebook and digital cadastral plan (DCP), it is necessary to check if the geometry of a detected object is valid, and to fix it if it is not. This step is manual, which means that for each detection it is necessary to validate geometry using an editing tool. After the validation, the final step is to update the cadastral database. All steps will be explained in detail.

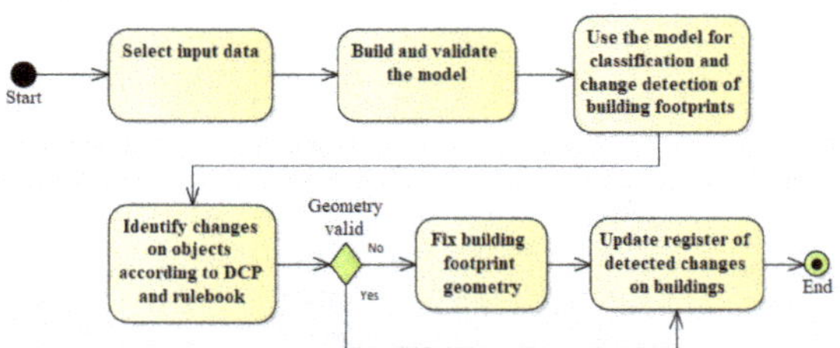

Figure 3. The overview of the proposed methodology.

The building footprint detection used in the study included two proposed methods. The first one used OBIA methods developed using eCognition, shown in Figure 4, while the pixel-based classification was carried out using U-Net.

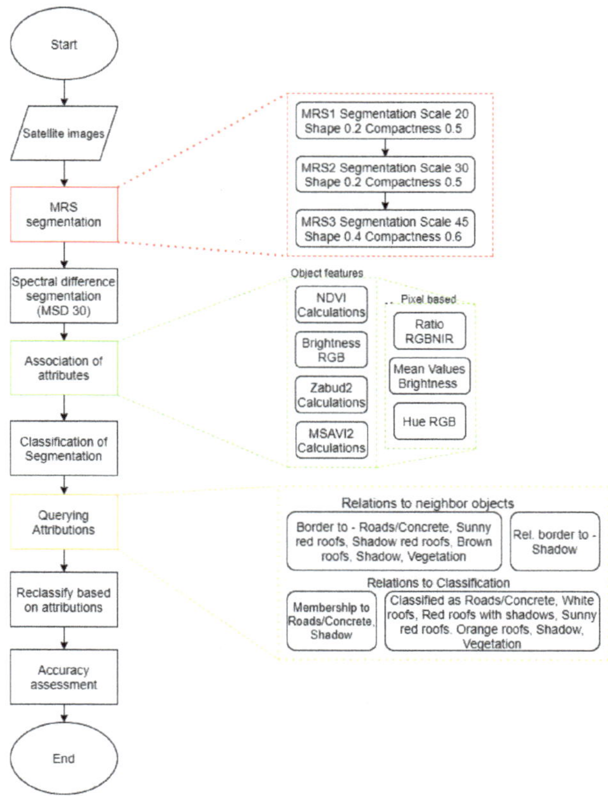

Figure 4. Extracting footprints of buildings using ecognition.

2.2.1. Object-Based Classification Method

Generally, this process consists of several phases. The first one was multi-resolution segmentation (MRS) of the satellite image. This technique was used to extract reasonable image objects that we can use in the next steps. In the segmentation stage, scale, shape, and compactness must be determined in advance (related parameters are described in detail in [4,24] and weights were 2 for NIR band and 1 for RGB bands. Generally, the parameters were determined through visual assessment as well as trial and error for all three multi-resolution segmentations. We set the scale factor as 20, 30 and 45 respectively, shape parameter was 0.2, 0.2 and 0.4, and compactness was 0.5, 0.5 and 0.6 respectively. After MRS, spectral difference segmentation was performed, which allows merging of neighbouring image objects if the difference between their layer mean intensities is below the value given by the maximum spectral difference, which in our case was 30. With this segmentation, objects produced by previous MRS segmentations were refined by merging spectrally similar image objects. Association of attributes or feature selection was carried out using several indices such as NDVI, Zabud and MSAVI2, and we calculated the ratio of RGB and NIR bands, mean brightness values and Hue of RGB bands. Specifically, for the vegetation class we used training data or values of NDVI greater than 0.43. MSAVI2 and Zabud2 was additionally used to distinguish vegetation, ground, and objects. For class shadow, we used training data or values of brightness from −1 to 230. For class sun red roof, we have used training data and size of objects greater than 15 pixels (approximately 4 m^2). In the classification part we have used all parts of the features form the previous step, and we also used relations to neighbour objects and relations to classification to reclassify all classes (shadow, sunny red roofs, shadow red roofs, brown roofs, vegetation, white roofs, orange roofs, red roofs with shadow and roads/concrete) in order to get one class that represents buildings footprints. Specifically, for example, for all sunny classes (brown, red, grey, and orange roof) the relation border to class shadow objects has a value of 0. For classes red and orange roof, values for relative border to shadow class must be below 0.09 and rectangularity fit value must be greater than 0.6. Classes red roof with shadow also must have a value of relative border to shadow greater than 0.1 and below 0.6. The most difficult part was to separate concrete roofs from concrete roads and paths. In this situation we have used relative border values with concrete (lower than 0.7), roads (lower than 0.4), vegetation (greater than 0.4) and shadow (lower than 0.2). All these values were defined after analysis of values of all clearly classified objects. Therefore, at the beginning of the classification, great attention should be given to image segmentation, and by the end of these several segmentations, we must get as clearly separated objects as possible. Reclassification and accuracy assessment was the final step.

2.2.2. Pixel-Based Method

In the last few years, convolutional neural networks have achieved superior accuracy in various areas of computer processing, such as image classification, object detection, and semantic segmentation. You et al. [25] analyzed the literature related to change detection using remote sensing images in the last five years, to summarize the current development situation and outline the possible direction of research to detect changes in an urban environment. Convolutional neural networks are a subtype of artificial neural networks, a type of deep neural network, designed to process locally dependent data coming in multiple sequences, usually images. Convolutional neural networks are widely accepted because they have proven superior to traditional methods in tasks such as image classification, object detection, and semantic segmentation [26–28].

Long, Shelhamer and Darrell were the first to develop an end-to-end model for image segmentation called fully convolutional neural network (FCN). An FCN uses a convolutional neural network to transform image pixels to pixel classes [29]. As Cheng et al. noted after covering more than 160 papers, this is still an active research topic [30]. Avoiding the use of dense layers means fewer parameters, which makes such nets faster for training. According to the structure, the most modern models of semantic segmentation can be divided into encoder-decoder and spatial association of pyramids. U-Net is a typical architecture with an encoder-decoder structure. The descending and ascending paths of the network are symmetrical, so the network has the appearance of the letter U, from which it derived its name. This architecture has shown significant improvement in several applications, especially for the detection of objects on satellite images, as evidenced by numerous papers [31–35]. U-Net has gained popularity due to its results in various semantic segmentation tasks. The main advantage of U-Net is the ability to perform precise segmentation with small training data.

The second approach for the object detection used in this paper is based on the application of convolutional neural networks, more precisely on the U-Net architecture of the neural networks. The methodology is shown in the Figure 5, and it will be explained in detail below.

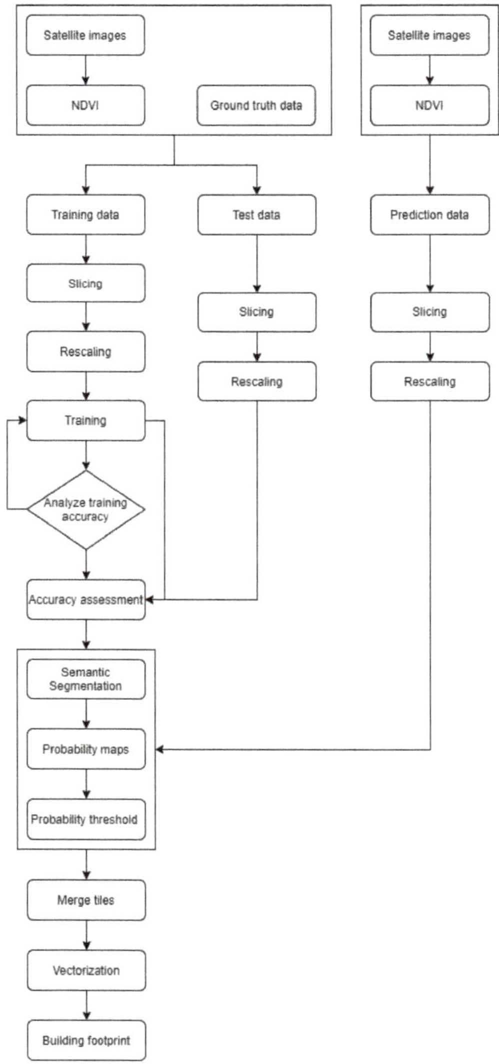

Figure 5. Extracting footprints of buildings using U-net.

The first step in this methodology is preprocessing of data, which in addition to creating vegetation indices and dividing the data into a set for training, test, and prediction, also includes slicing the raster into smaller parts of regular shape (128 × 128 pixels in our case). This step is necessary to apply the U-Net neural network architecture to our data.

As shown in Figure 6, the U-Net consists of two parts: an encoder (left) and a decoder (right). The U-Net architecture consists of an encoder that captures contextual information and a symmetric decoder that returns the spatial resolution of the initial raster. Skip connection is used to connect high-resolution maps from the encoder to the corresponding caused decoder output, allowing the network to more accurately predict the outputs based on that information.

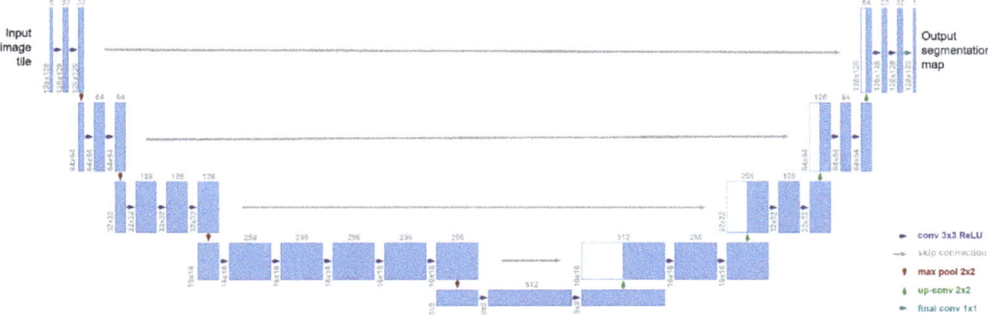

Figure 6. Proposed adapted U-net architecture.

The encoder has a typical CNN architecture (convolution, activation, maximum association). In each step of the down-sampling, we double the number of channels. This encoder architecture contains four blocks: the first two blocks consist of two convolutional layers with a 3 × 3 filter, while the third and fourth have three and five convolutional layers, respectively. The addition of new convolutional layers represents a simple modification of the basic model of the U-net architecture, which provides satisfactory accuracy. The ReLU activation function was applied to each block, as well as a maximalization operation with a 2 × 2 filter.

The decoder and encoder contain four blocks. Each decoder block consists of a re-sampling operation followed by a 2 × 2 convolution that halves the number of channels, which is combined with the corresponding encoder map. Each decoder block has two convolutional layers with a 3 × 3 filter and a ReLU activation function applied to each of them. The last layer of the grid connects each pixel with a certain class and performs a convolution operation with a 1 × 1 filter. Finally, we can say that the network architecture applied in this paper has 23 convolutional layers, and 21 ReLU activation functions.

After the data were prepared in an appropriate way, the neural network architecture was chosen and the model for classification was created, the training of that model and the assessment of its accuracy could begin. The input data set for training was divided into two parts with a ratio of 80:20. With this step, we could conduct a quantitative assessment of the model during the training of the neural network, and we could obtain information about how well the network is trained with data that did not participate in the training. To avoid excessive network training, which can lead to poor identification of objects, the early stop parameter in the training phase was used. With this parameter, the training was interrupted at the moment when the accuracy rating over the validation data starts to decrease (if better results are not obtained in the next three epochs from the moment the highest accuracy is reached).

After the completion of the training phase, testing and evaluation of the results was performed. We first applied the trained model for building identification, and then evaluated the accuracy of the classification, using the dataset for testing. After that, we applied the same model on the part of the image, for which we do not have cadastral records, and used it to perform semantic segmentation of the raster (building detection).

The goal of the semantic image segmentation is to mark each pixel of an image with an appropriate class. Image segmentation is the process of dividing a digital image into multiple segments known as image objects. Modern models for image segmentation are based on convolutional networks. Consequently, deep learning, especially the deep convolutional neural network (CNN), is a good and rewarding approach to automatic learning of object characteristics. Panboonyuen et al. [36] proposed a new method to improve the accuracy of semantic segmentation. The proposed model showed superiority over other models on all tested data. The important thing to note is that in this segmentation, instances of the same class are not separated, in other words, if we have two objects of the same category in the input image, we do not essentially distinguish them as separate objects.

After we applied the model over the input data, the result was a raster that represents probability maps with a range of (0, 1), with 0 as the lowest probability of the building's existence, and with 1 as the highest probability of the building's existence. The next step is to define the probability value, based on which buildings will be identified. The probability value used in this work is 0.5.

The last step is to merge all the classified parts of the initial raster in order to obtain a raster with two classes (object and not object) of the same dimensions as the initial satellite image.

2.2.3. Accuracy Assessment

Metrics

In addition to the visual assessment, the most important is certainly the numerical, i.e., quantitative, assessment of the accuracy of the obtained results. In classification tasks, the confusion matrix is often used to assess the accuracy and reveal information and performance of the model, where each row of the confusion matrix represents the prediction category, and each column represents the actual category to which the pixel belongs.

Overall accuracy is one indicator for evaluating the classification model. The total accuracy tells us the number of accurately classified pixels, from all reference locations. Total accuracy is usually expressed as a percentage, with an accuracy of 100% representing perfect classification where all reference locations are correctly classified [37].

Formally, accuracy has the following definition:

$$OA = \frac{Number\ of\ correct\ predictions}{Total\ number\ of\ predictions} \in [0,1] \quad (1)$$

For binary classification, accuracy can also be calculated using positives and negatives as follows: [38]

$$OA = \frac{TP + TN}{TP + TN + FP + FN} \in [0,1] \quad (2)$$

True positive (*TP*) is the number of correctly identified pixels of buildings, and true negative (*TN*) is the number of correctly identified pixels that do not belong to buildings. False positive (*FP*) pixels are those that are classified as a building in a place where it does not exist, while false negative pixels (*FN*), are those that belong to buildings and are classified in another class.

The Kappa coefficient is one of the parameters used in this paper to assess the quality of the model and it is a measure of the correspondence between the classification results and the reference data [39]:

$$k = \frac{p_0 - p_e}{1 - p_e} = 1 - \frac{1 - p_0}{1 - p_e} \quad (3)$$

where p_0 is observed accuracy and p_e is random agreement. The observed accuracy is determined by the diagonal in the confusion matrix, while the random agreement includes the members outside the diagonal. If p_0 and p_e completely agree, then the Kappa statistic is equal to 1. However, if there is no agreement between these values the Kappa statistic will be zero. Similar to most correlation statistics, Kappa can range from −1 to +1.

Although Kappa is one of the most commonly used statistics for reliability testing, it also has limitations on the level of Kappa statistics that are sufficient to accept a model. To solve this problem Landis and Koch [40] proposed the following scale:

- <0—No agreement
- 0—0.20 Slight
- 0.21—0.40 Fair
- 0.41—0.60 Moderate
- 0.61—0.80 Substantial
- 0.81–1.0—Perfect

Additionally, the values of the Precision and Recall parameters provide accuracy information. The overall performance of the model is not well described when it comes to an unbalanced data set, and the precision and recall parameters reflect the true performance of the classification [18]:

$$Precision = \frac{TP}{TP + FP} \in [0, 1] \tag{4}$$

$$Recall = \frac{TP}{TP + FN} \in [0, 1] \tag{5}$$

Precision is the measurement of accurately identified positive cases from all positive cases and will decrease if the number of false positive results is high. *Recall* is the measurement of accurately identified positive cases from all actual positive cases. This will indicate whether false negatives have a large impact on model performance. These ratings range from 0 to 1, where a higher number indicates better performance.

Loss Function

Deep learning is an iterative process, usually with many parameters. In order to make the adjustment as efficient as possible, the loss function is applied to the problem, among other things. The loss function is a cost that the optimizer will try to reduce by updating the weights, so the neural network learns and improves its performance. The loss function examines each pixel individually and compares class predictions with accurate data.

One of the most commonly used loss functions for image segmentation tasks is pixel-wise cross entropy. Cross entropy can be derived from the maximum likelihood (ML) method, which is a method for estimating model parameters [41,42]. A good estimate of the parameters is obtained by using the parameter model $p_{model}(x_i; \theta)$, where x is the input data and θ the model parameters. The ML will try to fit the function that maps the given entry as close as possible to the true function, and this is achieved by optimization via the parameter θ, with the criterion:

$$\theta_{ML} = \underset{\theta}{\text{argmax}} \prod_{i=1}^{m} p_{model}(x_i; \theta) \tag{6}$$

In the case of a binary classification problem, binary cross entropy can be used as a loss function, assuming that there are only two classes. The loss minimization in this paper was conducted using the Adam optimizer.

2.2.4. Data Model for the Register on Determined Changes on Buildings

Due to the frequent occurrence of illegal construction of buildings and due to the inconsistency of data in the field and in official registers in general, the Rulebook on established changes on buildings was adopted in July 2020 [22]. This rulebook regulates the content, establishment, maintenance and use of the records that contain data about identified changes on buildings.

The records should contain data on the determined changes in the buildings in relation to reference epoch of aerial photography. Three cases of change should be recorded:

- Buildings which are not registered in the real estate cadastre.

- Buildings which are registered in the real estate cadastre, but their base dimension has changed in relation to buildings registered in the real estate cadastre.
- Buildings which are registered in the real estate cadastre but are demolished in the field.

The buildings for which changes are determined are buildings of all types (residential, commercial and commercial buildings, cultural, sports and recreation buildings and similar buildings).

The rulebook defines the new records as an extension of real estate cadastre records. Therefore, it is necessary to define a data model for the determined changes on the buildings. A well-defined data model is the core of the cadastral information system. Thus, it must be in accordance with existing international standards as well as with national legislation. A conceptual data model for real estate cadastre in Serbia [43] was developed according to the Law on State Survey and Cadastre [44] and ISO 19152 Land Administration Domain Model [45]. A conceptual data model (UML Class diagram) for Serbian cadastre is based on four classes that represent main concepts of real estate cadastre as shown on Figure 7: parties (*RS_Party*), spatial units such as parcels, buildings and building parts (flats, business offices) (*RS_SpatialUnit*), rights and restrictions that parties can have over spatial units (*RS_RRR*) and basic administrative units that collect all the data regarding one spatial unit (*RS_BAUnit*).

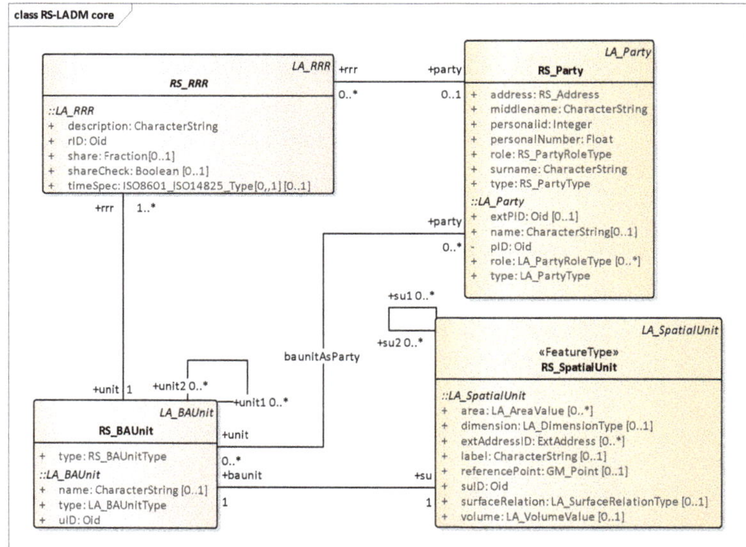

Figure 7. LADM core clasess.

This core model is further developed in detail to introduce all necessary classes and associations between them to fully represent the standardized cadastral domain for Serbia [43]. In order to extend this model to contain new records on changed buildings, a new class RS_Changed_Building was added (Figure 8).

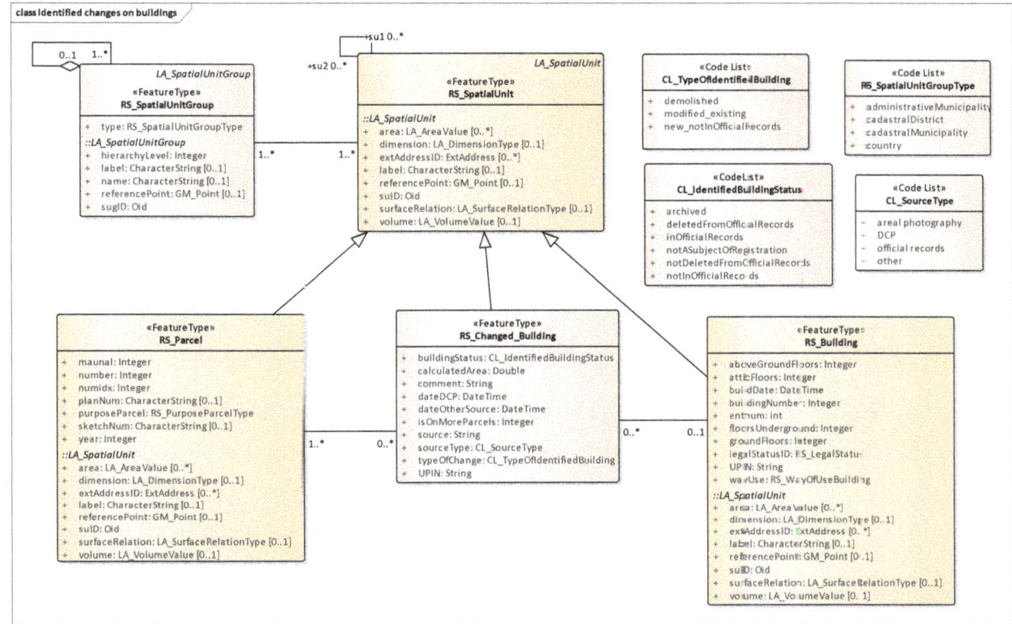

Figure 8. Proposed conceptual model for the register on determined changes on buildings.

An attribute *UPIN* represents the unique property identification number that is defined for each property in Serbia. For the new building this number is generated, and for the modified or deleted building it is the same as the *UPIN* in the real estate cadastre for a specific building. For the case of new and modified buildings, a geometry attribute will be populated, and the area of the building base will be calculated. Class *RS_Changed_Building* is derived from *RS_SpatialUnit* and contains a link to a specific spatial unit group such as cadastral municipality. When the building is built illegally it is possible that it is placed not on just one, but on multiple parcels. In order to record such data, an association between classes *RS_Parcel* and *RS_Changed_Building* is added. There is also association to the class *RS_Building* to show the connection with existing buildings in the real estate cadastre. Further, it is necessary to keep the information about the source (for example, aerial photography) that is used to compare with actual real estate cadastre data (*CL_SourceType*). Whether the identified change is the new building, demolished building or modified building, is one of the key pieces of information for these records. These data are selected from *CL_TypeOfIdentidiedBuilding* code list. Two dates, *dateDCP* and *dateOtherSource*, represent the dates of digital cadastral plan validity and other source's validity. Based on collected data, derived building status can be chosen from *CL_IdentifiedBuildingStatus*.

These records are used in the process of maintaining the real estate cadastre, in the procedure of legalization of buildings in accordance with the national laws.

3. Results

3.1. Preprocessing

In the initial phase of data collection and analysis, a large amount of data on buildings were collected; this set contains over 1500 polygons which represent the position and shape of each object. Of the total amount of data on buildings, for classification in eCognition and U-net, about 80% was taken for model training while the remaining 20% was used for model testing to assess accuracy (Figure 9).

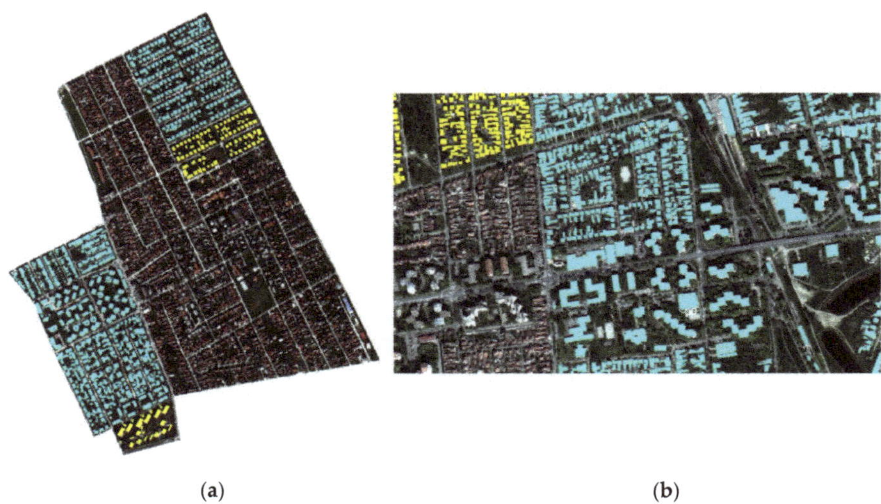

(**a**) (**b**)

Figure 9. Data used for training (blue) and model testing data (yellow): (**a**) Subotica; (**b**) Zrenjanin.

In the next phase, inaccuracies in the vector data were corrected, and then that data were converted to raster format. In this way, binary rasters were obtained where pixels with a value of 1 represent the locations where the object is located, while pixels with a value of 0 represent locations where there are no objects. The raster obtained in this way, together with satellite image, will represent the input data for the training of the U-Net neural network (Figure 10).

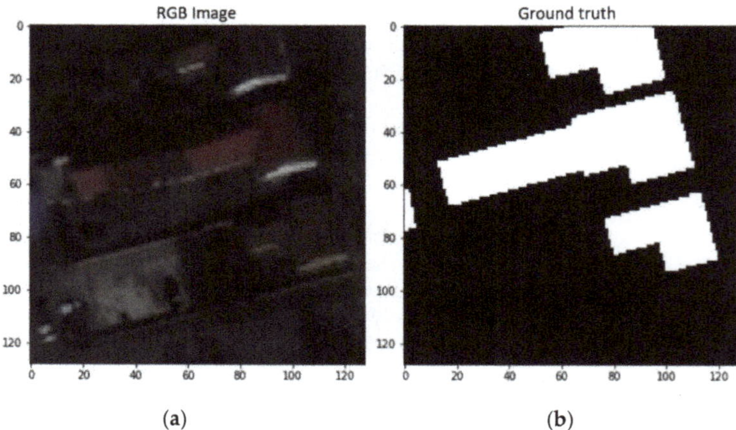

(**a**) (**b**)

Figure 10. An example of a pair of input data used for training: (**a**) Subotica; (**b**) Zrenjanin.

Since the models of deep learning for the training phase need to forward images of fixed size, it is necessary to slice satellite images and rasters with object masks into smaller parts (defined dimensions of the input data are 128 × 128 pixels).

The following table (Table 1) provides data on the number of images for training, test, and prediction for both analyzed locations, and total number of objects from cadastral data.

Table 1. Number of images used for training, test and prediction, and total number of objects.

City	Number of Images				Number of Objects
	Training	Test	Prediction	Total	
Zrenjanin	269	32	106	407	780
Subotica	321	80	604	1005	778

3.2. Training and Accuracy

The neural network training was conducted using the publicly available cloud platform Collaborators, hosted on Google Cloud [46].

Using early stop parameter, we determine that the number of epochs required for training the U-Net network with data for Zrenjanin and Subotica is 17, while the accuracy of the training model in both cases is over 90%.

In the following diagrams (Figures 11 and 12), we can see the accuracy and loss function curve, that are obtained in the process of training the neural network. From these diagrams we can see that, for both analyzed areas, slightly weaker results are obtained for data that did not participate in the training. What can still be concluded is that with the increase in the number of epochs, greater accuracy of the model will not be obtained, since after the first 10 epochs the evaluation curve of the model takes a stable appearance without major jumps, while the validation curve follows with smaller variations. Therefore, if necessary, the accuracy of the model could be increased in some other way, such as increasing the set of input data, satellite image bands, use of DSM, or change of neural network architecture (by adding new layers).

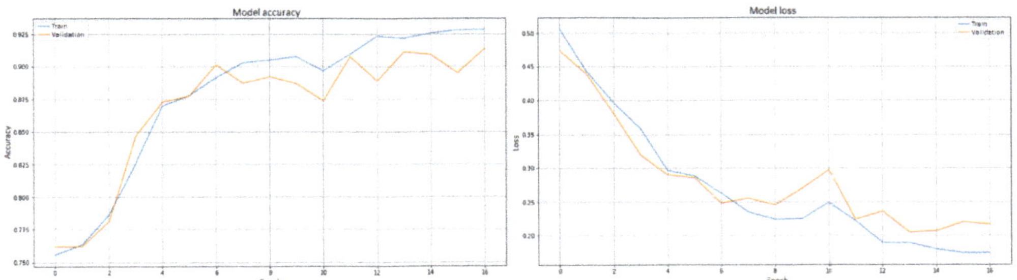

Figure 11. Subotica—accuracy and loss function curve.

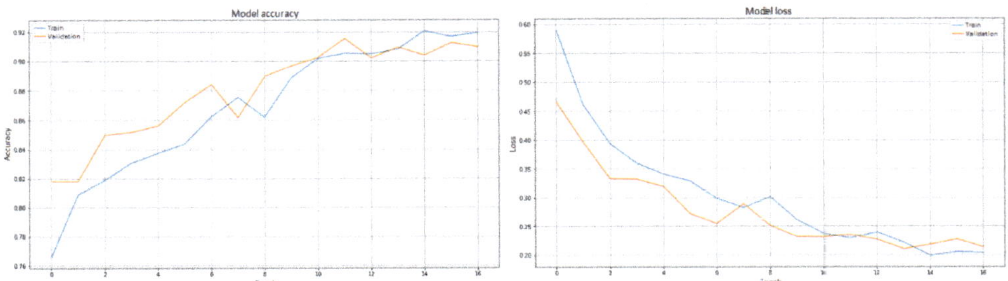

Figure 12. Zrenjanin—accuracy and loss function curve.

3.3. Building Identification Results

After the completion of the training phase, testing and evaluation of results is performed. Next, the trained model was applied to the testing data set, and accuracy assessment of the classification was performed again. The same procedure was applied on the remaining data set. The results obtained using the U-Net neural network in the test area for both locations are shown in (Table 2):

Table 2. Building identification results in test area.

City	Total Number of Objects	Number of Correctly Identified Objects	
		U-Net	eCognition
Zrenjanin	141	127	126
Subotica	120	104	111

The total number of objects in both tables (Tables 1 and 2) represents a number of all objects from the official cadastral database. As we can see, there are a number of buildings that have not been identified, but based on the results shown above, we can conclude that the accuracy of object identification on the new data set is about 90%, which follows the accuracy of the model obtained during training. Based on this, we can conclude that the model is able to identify buildings with high accuracy for areas that are outside the area used for training. In the next figure (Figure 13) we can see the result of the identification of objects in the test area of the city of Subotica.

Figure 13. Ground truth cadastral data overlap with results—Subotica.

Evaluation of the accuracy of the classification (object identification) was carried out using 2000 points for each test area, and 1000 points for each of the classes (we have two classes of buildings and not buildings). That number was chosen because with 2000 points a high density was obtained and the entire test area was evenly covered. Accuracy assessment was performed based on error matrix and Kappa statistics. The results of the accuracy assessment are shown in the following table (Table 3).

Table 3. Accuracy assessment.

City	Accuracy		Kappa Statistic	
	U-Net	eCognition	U-Net	eCognition
Zrenjanin	86.08	86.02	89	89
Subotica	83.99	88.04	96.84	96

During the analysis of the results, the raster data obtained after the applied classification were compared with the reference data, and a new raster was obtained in which the pixels were classified into three categories: TP—true positive, FP—false positive and FN—false negative. A visual overview for one of the three test areas is given in the following figure (Figure 14). Green polygons represent correctly identified objects (TP), blue polygons represent an object that was not recognized the with model but does exist (FN), and red parts are recognized as objects but do not exist (FP).

Figure 14. Example of object identification in Subotica: (**a**) U-Net; (**b**) eCognition.

In the next figure (Figure 15) RGB view of the satellite image, ground truth data, prediction results and the difference between this data and ground truth data are presented. This is an example of how the resulting changes can be identified.

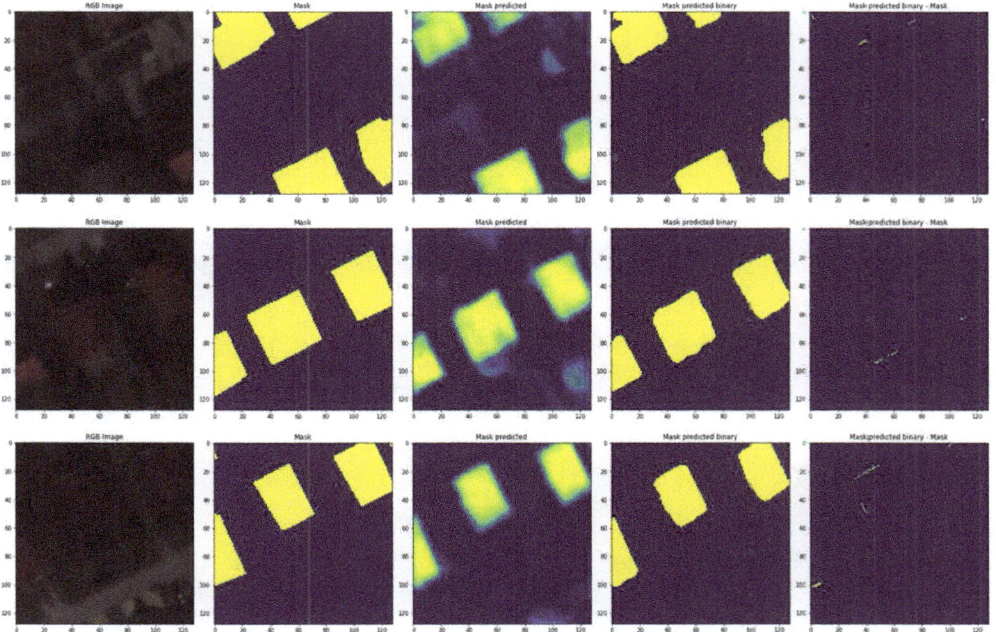

Figure 15. Test data—Subotica.

Very often with this type of analysis, a problem in identifying roofs can be caused by the size of the building, which varies greatly, so the building can be defined with a few pixels or can cover most of the image. Additionally, another problem consists of old buildings that have dark roofs and are visually difficult to distinguish from other buildings and structures. The orientation of the object can also play an important role in identifying objects, because the part of the roof that is oriented towards the sun (in the shade) has a weak reflection, while the opposite side is light and shows a high reflection in all lanes; thus, obtained high contrast can make it impossible to identify the part of the roof that is in the shade (this is typical for classification results from eCognition). However, in the following figure (Figure 16), which displays the result of identifying objects without ground truth data, we can see that the proposed U-Net neural network architecture copes very well with these problems and easily overcomes them.

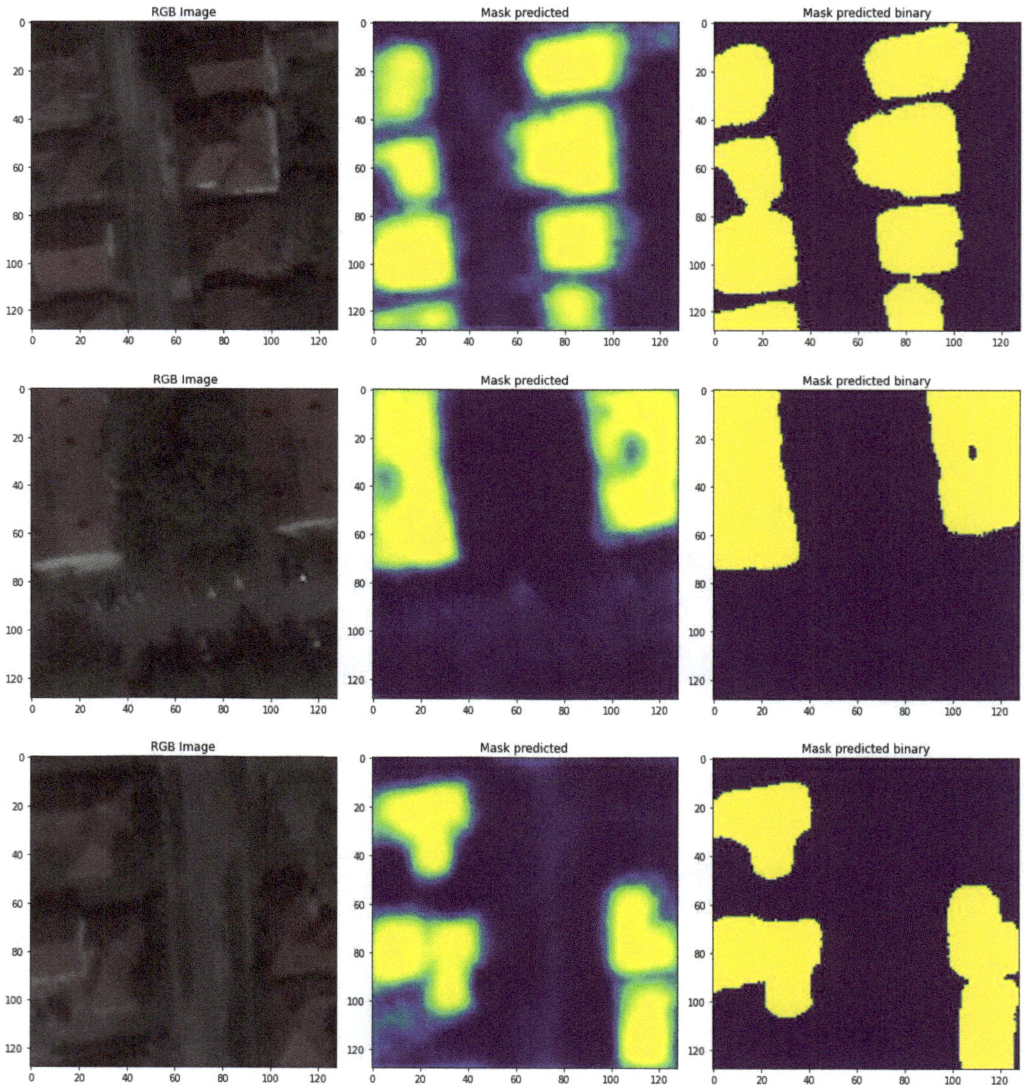

Figure 16. Examples of object identification without ground truth data—Zrenjanin.

Additionally, the proposed architecture of the U-Net neural network is capable of solving problems in the input data, when some parts of the image are incorrectly marked as parts under the object. Such errors most often occur in densely built-up parts of the city when several buildings are located next to each other. In these cases, it may happen in the input data, that all these near objects are marked as one, although this is not the case. In the following figure (Figure 17), we can see the building blocks and the error in the input data, which was partially corrected after classification. As can be seen, in the end, individual polygons are obtained, which represent separate objects and not one common polygon for all buildings, as was the case in the input data.

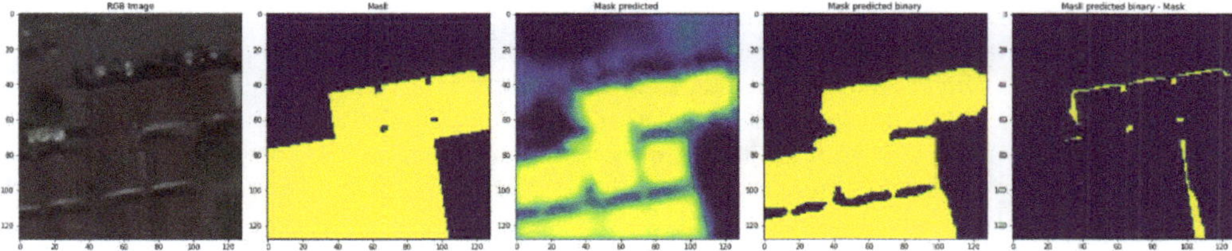

Figure 17. Corrected errors in the input data.

We can conclude that the proposed architecture of the U-Net neural network can easily cope with errors in the input data, which is the main advantage of this type of object identification compared to other models of machine learning that are much more sensitive to input data errors.

Errors in this type of building identification can also be caused by red cars that have a reflection similar to the reflection of roofs (Figure 18).

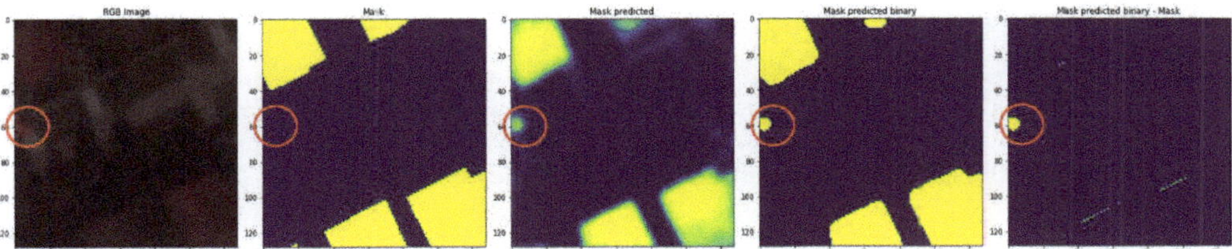

Figure 18. A car identified as a building.

Moreover, due to the existence of residential buildings in the area of analysis that have gray roofs, errors are visible in concrete surfaces (such as playgrounds, football fields, etc.). Examples of these errors are given in the figure below (Figure 19).

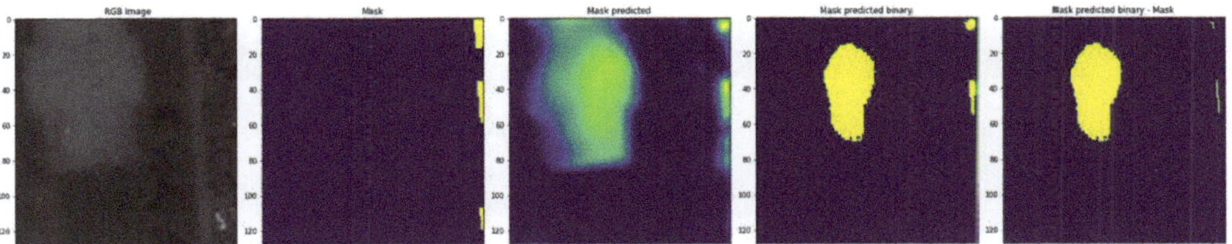

Figure 19. Concrete terrain identified as a building.

3.4. Identification of Objects According to the Rulebook

The following results show how identified objects can be classified according to the Rulebook and demonstrate the applicability of the method to address the requirements given in this official document.

3.4.1. Objects That Exist in Cadastral Records but Are Not Visible on the Orthophoto

The lack of up to date data in the cadastre is a big problem for the further development of the cadastre itself and the successful collection of taxes. In addition to not registering newly built facilities, another problem is outdated records of facilities, i.e., we can find information about buildings that have been demolished and do not exist anymore but are still registered in the cadastre. To solve this problem, two different methods of satellite image classification have been applied. The following figures indicate the errors in the input data and show the classification results.

Figure 20 shows that both methods correctly identified the non-existence of objects that are still registered in the real estate cadastre. Here we can also see the correct identification of newly built objects that are not registered in the cadastre, so we conclude that U-Net and eCognition give quite similar results.

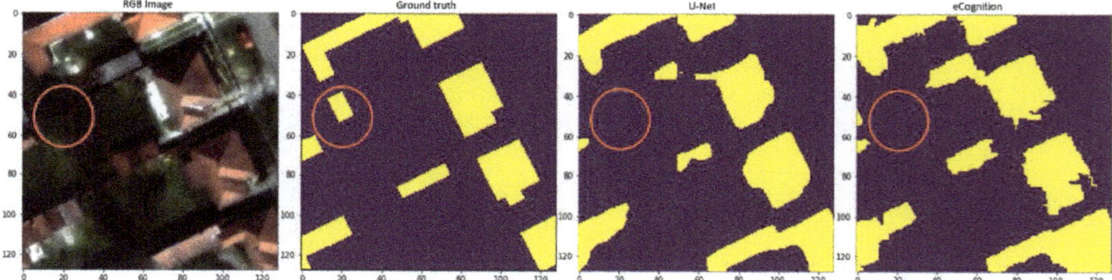

Figure 20. Non-existence of objects that are still registered in the real estate cadastre—Subotica.

Figure 21 show the success of both methods in overcoming this problem, with very similar results obtained using U-Net and eCognition.

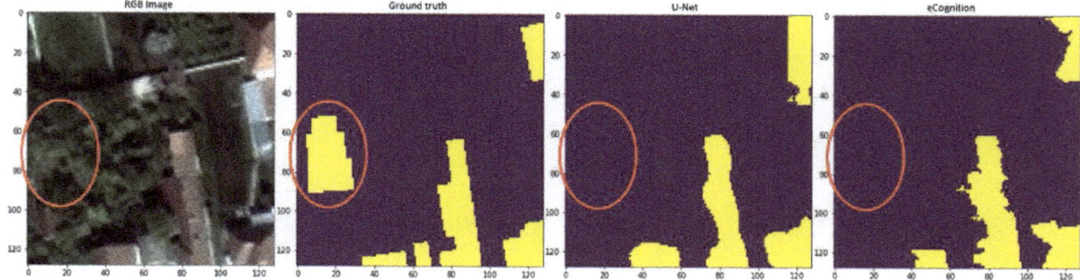

Figure 21. Non-existence of objects that are still registered in the real estate cadastre—Zrenjanin.

3.4.2. Objects That do Not Exist in Cadastre but Are Visible on the Orthophoto

One of the reasons for the non-existence of some objects in the input data set (cadastral data), which can be seen on the orthophoto, is the result of illegally constructed buildings that need to be identified, which with the estimate of two million illegally constructed buildings [47] represent a major issue in Serbia. As already mentioned, the biggest advantage of the U-Net neural network architecture is that it can handle errors in the input data set used for training without major problems. Therefore, if some buildings were not given in the training set (cadastral data), after training and running the model over the same data all objects that were omitted from the input data set will be successfully identified, which can be seen in the following figures (Figures 22 and 23) circled in red.

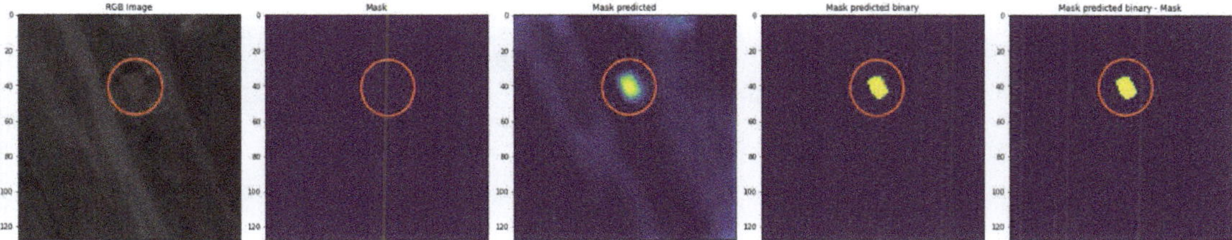

Figure 22. Correctly identified objects that were not in the training set—Zrenjanin.

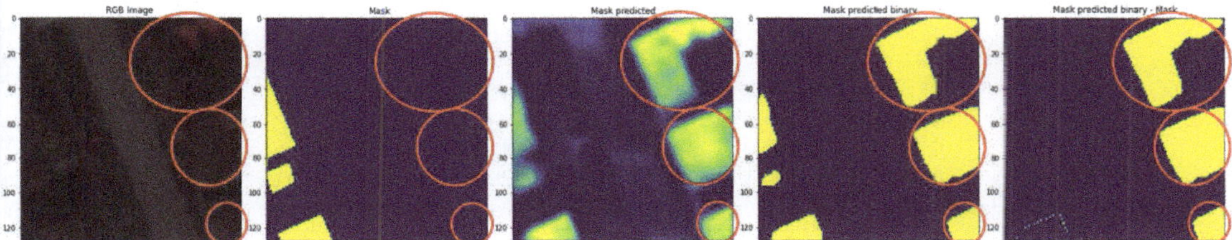

Figure 23. Correctly identified objects that were not in the training set—Subotica.

Due to the comparison of this type of classification (U-Net neural network) with the object-oriented classification from eCognition, the following images show the same part of the satellite image with the results in the identification of illegal objects obtained in both ways. In the following example, we notice that two smaller objects are successfully recognized by both methods, but we can also see that in addition to these two objects, there is a certain number of pixels that are incorrectly classified. Depending on the classification method and the area of analysis, the number of these misclassified pixels varies, so for this area we see that the best results were achieved using the U-Net neural network (Figure 24).

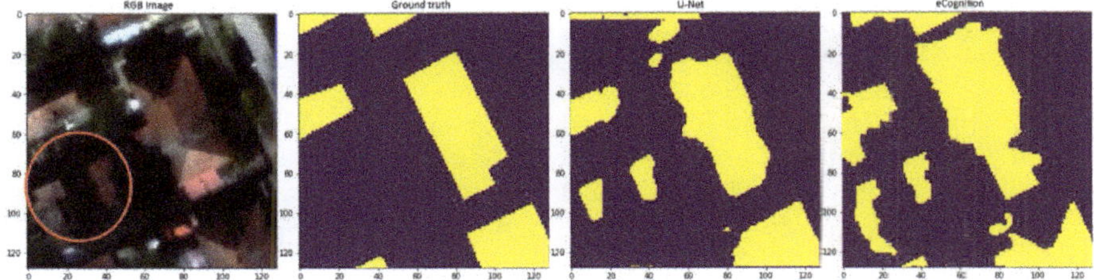

Figure 24. New identified objects.

The following examples (Figure 25) also show a comparison of the two classification models, where we can see that both methods successfully recognized the upgrade of one object. In this example as well as in the previous one, we see that the best results (with the least number of misclassified pixels) are achieved using the proposed architecture of the U-Net neural network.

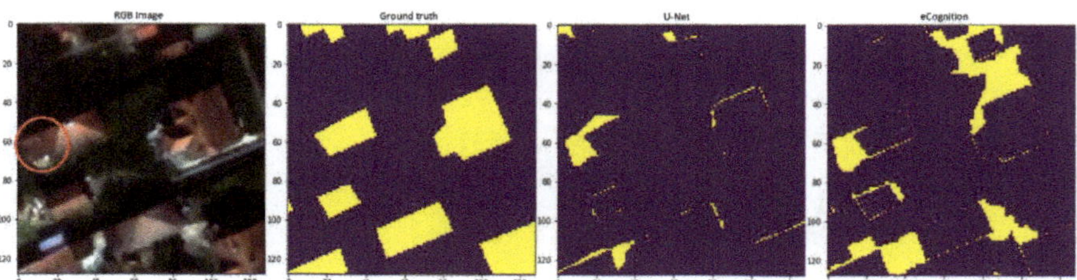

Figure 25. Comparison of the two classification models—misclassified pixels.

3.4.3. Objects Exist in Cadastre and in Orthophoto, but with Different Surfaces

A problem that can occur when comparing the data obtained from the cadastre and the data obtained during the identification of buildings is the identification of a slightly wider zone around the building and the assignment to the same class of buildings. The reason for the appearance of this "buffer" is often the acquisition angle, which can make it difficult to separate the roof from the side walls of the object. U-Net neural network architecture can solve this problem, by reducing the probability limit. The following figure (Figure 26) gives an example of this problem for all classification methods. This problem, which is especially pronounced in tall buildings, leads to incorrect identification of objects, i.e., an object with a larger area than the actual area of the object in cadastral records; moreover, spatial position of the identified object will be shifted in relation to the building (the position of the object will differ from the cadastre).

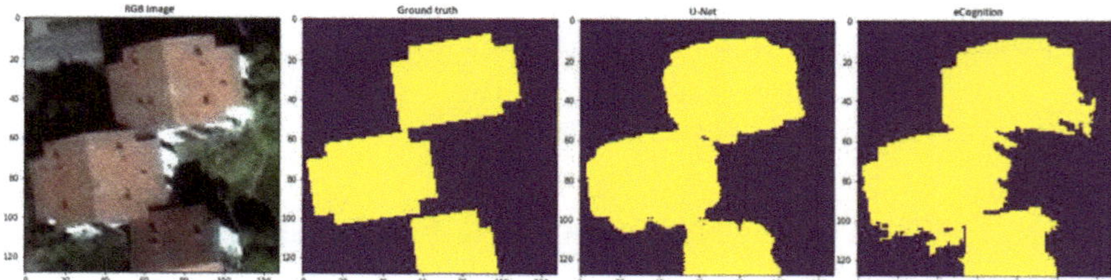

Figure 26. Examples of identifying tall objects.

On the other side, proposed methods for classification of objects show very good results in identification of objects that exist in the cadastre and in orthophoto, but with different surfaces. The following figure (Figure 27) shows the difference between the objects identified by both analyzed methods and the objects in the cadastre. We can now see much more clearly that both classification methods found a newly constructed building near an existing one.

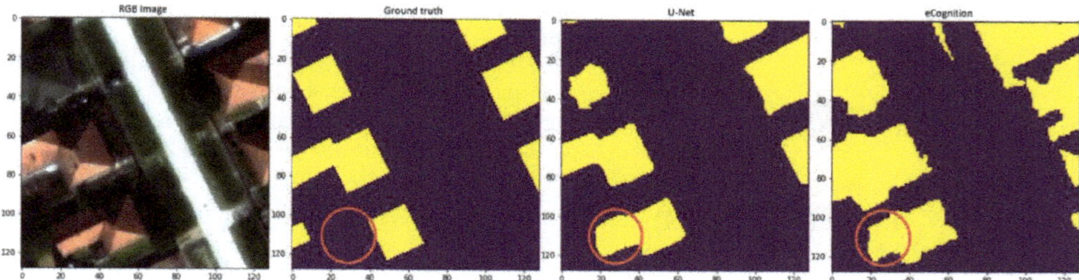

Figure 27. Difference of identified buildings and cadastral data.

3.5. Verification of the Results in the Register on Determined Changes on Buildings

Based on the obtained data, a register on determined changes on buildings can be established. Such a register should be based on the data model defined in Section 2.2.4. The main class (that represents a database table after conversion) that will be populated with detected data is RS_Changed_Building. An algorithm for this process will be based on spatial operators in order to create associations with parcels and buildings that are stored in the official records and also to create additional values and codes that will be stored in attributes of the RS_Changed_Building database table. The next three figures present three cases that can arise from the obtained data. These cases are important to recognize according to the Rulebook.

The first case is a situation when the detected data show that the building is demolished. A building 4602/1/3 is stored in the official records. The proposed method showed that there is no building on that location anymore. Since the date of satellite acquisition is later than the validity date in the official records, it can be concluded that the building has been demolished in the meantime. An instance diagram and building in the official records that represent such situation in Subotica are presented on Figure 28.

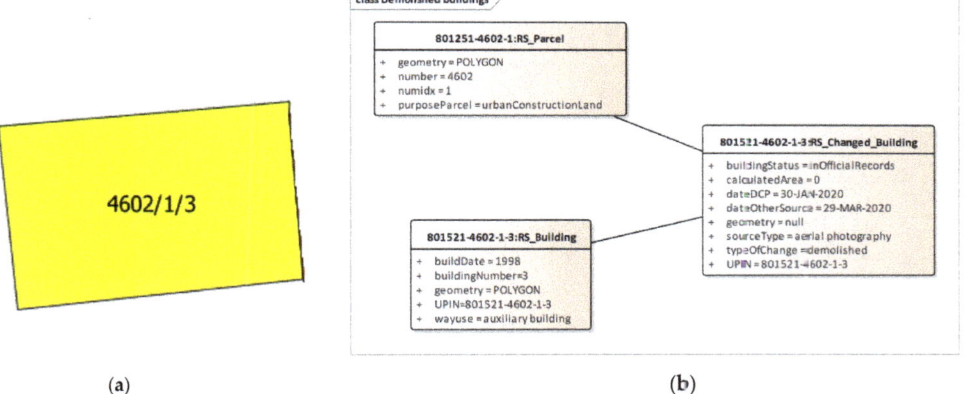

(a) (b)

Figure 28. (a) Building in the official records. (b) An instance diagram of the detected demolished building.

The second case is a situation when the detected data show the existence of new buildings that do not exist in the official records. An example on Figure 29 shows that two buildings have been built on agricultural parcels in Subotica in the period between the date of satellite acquisition and the validity date in the official records. Additionally, the new buildings are located on two parcels which means that these buildings are built without proper building permit, and that further processes of decision-making should be conducted by the surveying and mapping authority.

The third case is a situation when the detected data show that the building was modified since its footprint differs from the one in the official records. Figure 30 shows an example in Subotica where the existing building was expanded, which was also carried out without proper permits and requires an appropriate response from the surveying and mapping authority.

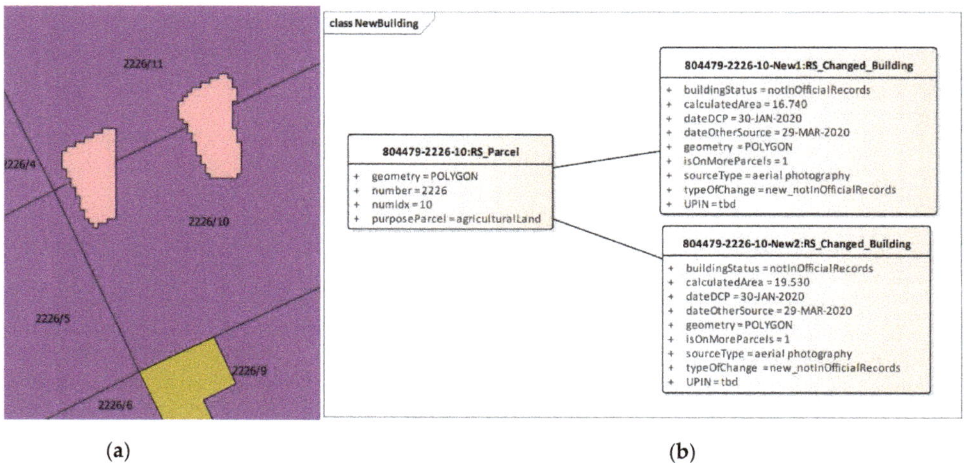

Figure 29. (**a**) Purple—parcels in the official records, yellow—buildings in the official records, pink—detected buildings. (**b**) An instance diagram of detected new buildings.

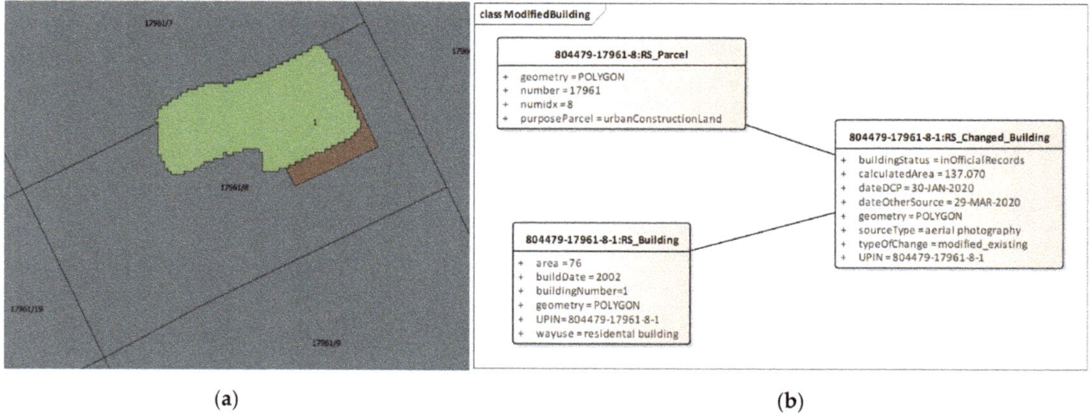

Figure 30. (**a**) Blue—parcels in the official records, red—buildings in the official records, green—detected buildings. (**b**) An instance diagram of detected change on the building.

4. Discussion

Organized and well-structured cadastral maps are a prerequisite for better services in land administration. Following best practices with using remote sensing techniques instead of a filed survey, Republic Geodetic Authority of the Republic of Serbia has acquired very high resolution satellite images for the years 2015/2016, and 2020. New very high resolution satellite images were acquired within the project "Improvement of Land Administration in Serbia" that is being implemented with the support of the World Bank [47]. These images will provide numerous benefits for both citizens and the economy, through the provision of up to date information on real properties.

Very high resolution satellite images from the years 2015/2016 and 2020 were of great importance for the implementation of infrastructure projects, spatial planning, and projects of national importance in the fields of agriculture, water management, forestry, environmental protection, mining and energy, risk management, and establishment of spatial information systems at the national and local levels.

By using orthophotos, made on the basis of satellite images that were collected in 2020 and 2015/2016, it will be possible to determine changes on real properties in order to update official registers and records on real properties of the responsible state institutions. The results of building extraction presented in this paper can be compared to other results reported in the literature, but also not directly due to different study areas, data that are used, variety of buildings and finally approaches that are used and the purpose of the study.

Lucian et al. [48] evaluated the impact of the spatial extent on the geometric accuracy of the objects delineated through multiresolution image segmentation. The experiments revealed that the geometric accuracy improved by 8–19% in quality rate when multiresolution segmentation was performed in smaller extents, as compared to the segmentation of whole images. Mariana and Lucian [49] also compared supervised and unsupervised segmentation approaches in OBIA by using them to classify buildings from three test areas in Salzburg, Austria, using QuickBird and WorldView-2 imagery. All three of the methods evaluated achieved similar classification accuracies, with overall accuracies between 82.3% and 86.4% and Kappa coefficients between 0.64 and 0.72. They also concluded that segmentation has an impact on classification with very different image objects, but accuracies were very similar. This result suggests that, as long as under-segmentation remains at acceptable levels, imperfections in segmentation can be ignored so that a high level of classification accuracy can still be achieved. Lei et al. [12], using information available in 173 scientific publications, among other things have found that high spatial resolution remote-sensing imagery remains the most frequently used data source for supervised object-based land-cover image classification, and the dominant image resolutions are 0–2 m. Divyesh in his thesis [50] used two datasets from East Asia and Munich, Germany (Ikonos and WV2) and DSM. He used information similarity measures for change detection of buildings, using VHR satellite images and DSM (incorporating height information from DSMs to assess changes in both horizontal as well as vertical direction). The effectiveness of the presented approach was evaluated through pixel-based assessment, as well as object-based assessment, with overall accuracy ranging from 86.307% to 93.75 % for object-based and 96.99 to 99.1291 for pixel-based quality assessment.

Khosravi et al. [51] evaluated and compared four building detection algorithms: two pixel-based and two object-based algorithms, using a diverse set of high-resolution satellite imagery. The results indicated that the performance and the reliability of object-based algorithms were better than pixel-based algorithms. Kriti et al. [52] compared several deep learning techniques with different architecture in automatic building footprint extraction. The evaluation over the test datasets with different networks showed accuracy from 85.2% to 91.5% in global results, and in urban areas of slums, isolated and dense built-up areas, accuracy went from 60% to 96.75% due to low spectral and textural variance among the buildings. Additionally, distances between buildings are less than 2 pixels which are difficult to delineate even through visual interpretation, resulting in creation of relatively poor training data for slum areas, with model accuracy of 72.5%. On the other hand, Wang et al. [31] reached a very high value of overall accuracy (94.12%) for building segmentation with their proposed innovative image processing method implementing the efficient Non-local Residual U-shape Network (ENRU-Net).

The main issue in Serbia is a lack of (or limited access to) up to date VHR images and other remote sensing data such as LiDAR and other derived products, not just for practical application, but also for research. Furthermore, one of the largest issues is the existence of more than two million illegal buildings [23] (newly built or changed), which results in an inaccurate cadastral database and a lack of tools that can help in detection of those illegally built or changed buildings in a (semi)automatic manner. With our proposed method we want to improve the current situation and support the implementation of a newly adopted "Rulebook on the records of identified changes on buildings" in Serbia, which requires the development of a special register of such buildings. This method will significantly speed up the entire process of detecting such buildings, entering data in the register, and, consequently, it will lead to an up to date cadastral database. The methodology proposed in this paper for automatic building extraction is simple, fast, efficient and achieves accuracy from 84% to 88% (Table 3). It does not require additional information, such as digital surface models (DSM), and gives good results even when we have only a satellite image with RGB and NIR bands. The proposed methodology can be further used for various applications, not only in Serbia, but also in all developing countries, which have a problem of lack of funds and access to additional spatial data that can help. In addition, it can be used in the process of the identification of illegally constructed or changed buildings, for which it is applied in this paper, and to assess damage by identifying damaged and undamaged buildings. Interpretation of the obtained results shows that buildings with very light and dark roofs have been successfully identified. Additionally, interpretation of obtained results shows that proposed models can be used in a typical Serbian type of settlements, in an urban part, but also in the rural part of settlements where there are dominantly houses with different types of roofs and big green spaces. Similarly, some non-building structures such as cars are classified in the class of buildings because of their similar reflection value and structural properties.

Future work should consider the development of an appropriate software solution with a fully automated proposed methodology for storing and maintaining acquired data and using U-Net as a tool for change detection, since it is simple for implementation and achieves results comparable to the object-based method. The database for the solution should be organized according to the extended LADM country profile defined in the Methods and Material section. Available data not just for research, but also for practical application are limited, so future work will include improvements of proposed methods for building detection in terms of accuracy based on those available data.

5. Conclusions

The paper proposes a building change detection method to support the development of the register of identified changes on buildings defined in the official Rulebook of the Government of Serbia. Our results, using only VHR images containing only RGB and NIR bands, showed object identification accuracy ranging from 84% to 88%, with kappa statistic from 89% to 96%.

The proposed method is simple, efficient and achieves sufficient accuracy in both building detection and legality assessment, without the need of additional information such as LiDAR data or a digital terrain model. The proposed method can greatly increase the speed of the development of such a register compared to the manual procedure, considering a very large number of objects that need to be identified. The results are classified into three categories of possible changes on buildings defined in the rulebook to support the conclusion that the method is capable of solving the set of requirements specified in this Rulebook. Further improvements of the method will consider achieving higher accuracy based on the available data. Furthermore, the development of an appropriate software solution for storing and maintaining acquired data is anticipated.

Author Contributions: Conceptualization, D.J.; Data curation, M.G. (Milan Gavrilović); Investigation, M.G. (Milan Gavrilović) and A.R.; Methodology, D.J., M.G. (Milan Gavrilović), D.S. and A.R.; Supervision, D.J. and M.G. (Miro Govedarica); Writing—original draft, D.J., D.S. and A.R.; Writing—review and editing, M.G. (Milan Gavrilović), D.S., A.R. and M.G. (Miro Govedarica). All authors have read and agreed to the published version of the manuscript.

Funding: This research received no external funding.

Acknowledgments: Results presented in this paper are part of the research conducted within the Grant No. 37017, Ministry of Education and Science of the Republic of Serbia.

Conflicts of Interest: The authors declare no conflict of interest.

References

1. Yan, L.; Fan, B.; Liu, H.; Huo, C.; Xiang, S.; Pan, C. Triplet Adversarial Domain Adaptation for Pixel-Level Classification of VHR Remote Sensing Images. *IEEE Trans. Geosci. Remote Sens.* **2020**, *58*, 3558–3573. [CrossRef]

2. Li, W.; He, C.; Fang, J.; Zheng, J.; Fu, H.; Yu, L. Semantic Segmentation-Based Building Footprint Extraction Using Very High-Resolution Satellite Images and Multi-Source GIS Data. *Remote Sens.* **2019**, *11*, 403. [CrossRef]
3. Inglada, J. Automatic recognition of man-made objects in high resolution optical remote sensing images by SVM classification of geometric image features. *ISPRS J. Photogramm. Remote Sens.* **2007**, *62*, 236–248. [CrossRef]
4. Blaschke, T. Object based image analysis for remote sensing. *ISPRS J. Photogramm. Remote Sens.* **2010**, *65*, 2–16. [CrossRef]
5. Cheng, G.; Han, J.; Lu, X. Remote Sensing Image Scene Classification: Benchmark and State of the Art. *Proc. IEEE Inst. Electr. Electron Eng.* **2017**, *105*, 1865–1883. [CrossRef]
6. Hossain, M.D.; Chen, D. Segmentation for Object-Based Image Analysis (OBIA): A review of algorithms and challenges from remote sensing perspective. *ISPRS J. Photogramm. Remote Sens.* **2019**, *150*, 115–134. [CrossRef]
7. Schiewe, J. Segmentation of high-resolution remotely sensed data-concepts, applications and problems. *Int. Arch. Photogramm. Remote Sens. Spat. Inf. Sci.* **2002**, *4*, 380–385.
8. Meinel, G.; Neubert, M. A Comparison of segmentation programs for high resolution remote sensing data. *Int. Arch. Photogramm. Remote Sens. Spat. Inf. Sci.* **2004**, *35*, 1097–1102.
9. Baatz, M.; Schäpe, A. Multiresolution Segmentation—An Optimization Approach for High Quality Multi-Scale Image Segmentation. In *Angewandte Geographische Informations-Verarbeitung*; Strobl, J., Blaschke, T., Griesebner, G., Eds.; Wichmanm Verlag: Karlsruhe, Germany, 2000; Volume 38, pp. 12–23.
10. Cheng, G.; Han, J. A survey on object detection in optical remote sensing images. *ISPRS J. Photogramm. Remote. Sens.* **2016**, *117*, 11–28. [CrossRef]
11. Trimble. *Trimble eCognition Developer for Windows Operating System*; Trimble Germany GmbH: Munich, Germany, 2017; ISBN 2008000834. Available online: https://bit.ly/3ldP6bc (accessed on 20 April 2020).
12. Ma, L.; Li, M.; Ma, X.; Cheng, L.; Du, P.; Liu, Y. A review of supervised object-based land-cover image classification. *ISPRS J. Photogramm. Remote Sens.* **2017**, *130*, 277–293. [CrossRef]
13. Myint, S.W.; Gober, P.; Brazel, A.; Grossman-Clarke, S.; Weng, Q. Per-pixel vs. object-based classification of urban land cover extraction using high spatial resolution imagery. *Remote Sens. Environ.* **2011**, *115*, 1145–1161. [CrossRef]
14. Zhou, X.; Jancso, T.A.; Chen, C.H.; Verone, M.W. Urban land cover mapping based on object oriented classification using WorldView 2 satellite remote sensing images. In Proceedings of the International Scientific Conference on Sustainable Development & Ecological Footprint, Sopron, Hungary, 26–27 March 2012; p. 10. Available online: https://bit.ly/3j23Ozs (accessed on 1 June 2021).
15. Norman, M.; Shahar, H.M.; Mohamad, Z.; Rahim, A.; Mohd, F.A.; Shafri, H.Z.M. Urban building detection using object-based image analysis (OBIA) and machine learning (ML) algorithms. *IOP Conf. Ser. Earth Environ. Sci.* **2021**, *620*, 012010. [CrossRef]
16. Norman, M.; Shafri, H.Z.M.; Idrees, M.O.; Mansor, S.; Yusuf, B. Spatio-statistical optimization of image segmentation process for building footprint extraction using very high-resolution WorldView 3 satellite data. *IEEE Trans. Geosci. Remote Sens.* **2019**, *35*, 1124–1147. [CrossRef]
17. Bergado, J.R.; Persello, C.; Stein, A. Recurrent Multiresolution Convolutional Net-works for VHR Image Classification. *IEEE Trans. Geosci. Remote Sens.* **2018**, *56*, 6361–6374. [CrossRef]
18. Wu, G.; Shao, X.; Guo, Z.; Chen, Q ; Yuan, W.; Shi, X.; Xu, Y.; Shibasaki, R. Automatic Building Segmentation of Aerial Imagery Using Multi-Constraint Fully Convolutional Networks. *Remote Sens.* **2018**, *10*, 407. [CrossRef]
19. Kokeza, Z.; Vujasinović, M.; Govedarica, M.; Milojević, B.; JakovljeviĆ, G. Automatic building footprint extraction from UAV images using neural networks. *Geod. Vestn.* **2020**, *64*, 4. [CrossRef]
20. Ronneberger, O.; Fischer, P.; Brox, T. U-Net: Convolutional Networks for Biomedical Image Segmentation. *ArXiv* **2015**, *9351*, 234–241. [CrossRef]
21. Xiao, P.; Yuan, M.; Zhang, X.; Feng, X.; Guo, Y. Cosegmentation for Object-Based Building Change Detection from High-Resolution Remotely Sensed Images. *IEEE Trans. Geosci. Remote Sens.* **2017**, *55*, 1587–1603. [CrossRef]
22. Official Gazette of the Republic of Serbia No 102/20. The Rulebook on Established Changes on Buildings. Available online: https://bit.ly/3j6pxpF (accessed on 10 April 2021).
23. Database of Illegally Constructed Buildings. Ministry of Building, Transportation and Infrastructure of Serbia, Belgrade. 2017. Available online: https://bit.ly/3rI00XP (accessed on 1 April 2021).
24. Devereux, B.J.; Amable, G.S.; Posada, C.C. An efficient image segmentation algorithm for landscape analysis. *Int. J. Appl. Earth Obs. Geoinf.* **2004**, *6*, 47–61. [CrossRef]
25. You, Y.; Cao, J.; Zhou, W. A Survey of Change Detection Methods Based on Remote Sensing Images for Multi-Source and Multi-Objective Scenarios. *Remote Sens.* **2020**, *12*, 2460. [CrossRef]
26. Huang, J.; Zhang, X.; Xin, Q.; Sun, Y.; Zhang, P. Automatic building extraction from high-resolution aerial images and LiDAR data using gated residual refinement network. *ISPRS J. Photogramm. Remote Sens.* **2019**, *151*, 91–105. [CrossRef]
27. Shao, Z.; Tang, P.; Wang, Z.; Saleem, N.; Yam, S.; Sommai, C. BRRNet: A Fully Convolutional Neural Network for Automatic Building Extraction from High-Resolution Remote Sensing Images. *Remote Sens.* **2020**, *12*, 1050. [CrossRef]
28. Wierzbicki, D.; Matuk, O.; Bielecka, E. Polish Cadastre Modernization with Remotely Extracted Buildings from High-Resolution Aerial Orthoimagery and Airborne LiDAR. *Remote Sens.* **2021**, *13*, 611. [CrossRef]
29. Long, J.; Shelhamer, E.; Darrell, T. Fully convolutional networks for semantic segmentation. In Proceedings of the 2015 IEEE Conference on Computer Vision and Pattern Recognition (CVPR), Boston, MA, USA, 7–12 June 2015.

30. Cheng, G.; Xie, X.; Han, J.; Guo, L.; Xia, G.-S. Remote Sensing Image Scene Classification Meets Deep Learning: Challenges, Methods, Benchmarks, and Opportunities. *IEEE J. Sel. Top. Appl. Earth Obs. Remote Sens.* **2020**, *13*, 3735–3756. [CrossRef]
31. Wang, S.; Hou, X.; Zhao, X. Automatic building extraction from high-resolution aerial imagery via fully convolutional encoder-decoder network with non-local block. *IEEE Access* **2020**, *8*, 7313–7322. [CrossRef]
32. Wu, C.; Zhang, F.; Xia, J.; Xu, Y.; Li, G.; Xie, J.; Du, Z.; Liu, R. Building Damage Detection Using U-Net with Attention Mechanism from Pre and Post-Disaster Remote Sensing Datasets. *Remote Sens.* **2021**, *13*, 905. [CrossRef]
33. Jin, Y.; Xu, W.; Zhang, C.; Luo, X.; Jia, H. Boundary-Aware Refined Network for Automatic Building Extraction in Very High-Resolution Urban Aerial Images. *Remote Sens.* **2021**, *13*, 692. [CrossRef]
34. Ivanovsky, L.; Khryashchev, V.; Pavlov, V.; Ostrovskaya, A. Building Detection on Aerial Images Using U-NET Neural Networks. In Proceedings of the 24th Conference of Open Innovations Association (FRUCT), Moscow, Russia, 8–12 April 2019; pp. 116–122. Available online: https://bit.ly/2UWpQf2 (accessed on 1 June 2021).
35. De Jong, K.L.; Sergeevna Bosman, A. Unsupervised Change Detection in Satellite Images Using Convolutional Neural Networks. In Proceedings of the International Joint Conference on Neural Networks (IJCNN), Budapest, Hungary, 14–19 July 2019; pp. 1–8. Available online: https://bit.ly/3fbOuii (accessed on 15 June 2021). [CrossRef]
36. Panboonyuen, T.; Jitkajornwanich, K.; Lawawirojwong, S.; Srestasathiern, P.; Vateekul, P. Semantic Labeling in Remote Sensing Corpora Using Feature Fusion-Based En-hanced Global Convolutional Network with High-Resolution Representations and Depthwise Atrous Convolution. *Remote Sens.* **2020**, *12*, 1233. [CrossRef]
37. Ulmas, P.; Liiv, I. Segmentation of Satellite Imagery using U-Net Models for Land Cover Classification. *Comput. Sci.* **2020**, *3*, 1–2. [CrossRef]
38. Kulkarni, A.; Chong, D.; Batarseh, F.A. Foundations of data imbalance and solutions for a data democracy. *ScienceDirect* **2020**, *6*, 83–106. [CrossRef]
39. Julius, S.; Wright, C.C. The Kappa Statistic in Reliability Studies: Use, Interpretation, and Sample Size Requirements. *Phys. Ther.* **2005**, *85*, 257–268. [CrossRef]
40. Landis, J.R.; Koch, G.G. The measurement of observer agreement for categorical data. *Int. J. Biom.* **1977**, *3*, 159–174. [CrossRef]
41. Goodfellow, I.; Bengio, Y.; Courville, A. *Deep Learning*; MIT Press: Cambridge, MA, USA, 2016.
42. Fritz, K. Instance Segmentation of Buildings in Satellite Images. Master's Thesis, Linköping University, Linköping, Sweden, January 2020. Available online: https://bit.ly/3fbdHcV (accessed on 15 April 2021).
43. Radulović, A.; Sladić, D.; Govedarica, M. Towards 3D Cadastre in Serbia: Development of Serbian Cadastral Domain Model. *ISPRS Int. J. Geo-Inf.* **2017**, *6*, 312. [CrossRef]
44. Official Gazette of the Republic of Serbia. The Law on State Survey and Cadastre. 2009. Available online: https://bit.ly/3xjMXxe (accessed on 30 May 2021).
45. ISO 19152:2012 Geographic Information—Land Administration Domain Model (LADM). Available online: https://bit.ly/3fcfM8r (accessed on 24 May 2021).
46. Cloud Computing Services. Available online: https://cloud.google.com/ (accessed on 15 April 2021).
47. Serbia-Real Estate Management Project. The World Bank. Available online: https://bit.ly/3j7EUOZ (accessed on 15 April 2021).
48. Lucian, D.; Mariana, B.; George, P.; Peter, B. Sensitivity of multiresolution segmentation to spatial extent. *Int. J. Appl. Earth Obs. Geoinf.* **2019**, *81*, 146–153. [CrossRef]
49. Mariana, B.; Lucian, D. Comparing supervised and unsupervised multiresolution segmentation approaches for extracting buildings from very high resolution imagery. *ISPRS J. Photogramm. Remote Sens.* **2014**, *96*, 65–75. [CrossRef]
50. Divyesh, V. Change Detection of Buildings Using Satellite Images and DSMs. Master's Thesis, Technische Universität München, München, Germany, 2011.
51. Khosravi, I.; Momeni, M.; Rahnemoonfar, M. Performance Evaluation of Object-based and Pixel-based Building Detection Algorithms from Very High Spatial Resolution Imagery. *Photogramm. Eng. Remote Sens.* **2014**, *6*, 519–528. [CrossRef]
52. Kriti, R.; Pankaj, B.; Shashikant, A.S. Automatic Building Footprint extraction from Very High-Resolution Imagery using Deep Learning Techniques. *Geocarto Int.* **2020**, *5*, 1–4. [CrossRef]